THE VARIATIONAL PRINCIPLES OF MECHANICS

FOURTH EDITION

by

CORNELIUS LANCZOS

DOVER PUBLICATIONS, INC.
NEW YORK

Manufactured in the United States of America
Dover Publications, Inc., 31 East 2nd Street, Mineola, N.Y. 11501

Library of Congress Cataloging-in-Publication Data

Lanczos, Cornelius, 1893–
   The variational principles of mechanics.

   Reprint. Originally published: Toronto : University of Toronto Press, c1970. (Mathematical expositions ; no. 4).
   Bibliography: p.
   Includes index.
   1. Mechanics, Analytic. 2. Variational principles. I. Title.
II. Series: Mathematical expositions ; no. 4.
QA805.L278   1986   531'.01'51   85-29168
ISBN 0-486-65067-7 (pbk.)

Was du ererbt von deinen Vätern hast,
Erwirb es, um es zu besitzen.

GOETHE

PREFACE

FOR a number of years the author has conducted a two-semester lecture course on the variational principles of mechanics at the Graduate School of Purdue University. Again and again he experienced the extraordinary elation of mind which accompanies a preoccupation with the basic principles and methods of analytical mechanics. There is hardly any other branch of the mathematical sciences in which abstract mathematical speculation and concrete physical evidence go so beautifully together and complement each other so perfectly. It is no accident that the principles of mechanics had the greatest fascination for many of the outstanding figures of mathematics and physics. Nor is it an accident that the European universities of earlier days included a course in theoretical mechanics in the study plan of every prospective mathematician and physicist. Analytical mechanics is much more than an efficient tool for the solution of dynamical problems that we encounter in physics and engineering. However great may be the importance of the gyroscope as a practical instrument of navigation or engineering, it is not needed as an excuse to demonstrate the importance of theoretical mechanics. The very existence of the general principles of mechanics is their justification.

The present treatise on the variational principles of mechanics should not be regarded as competing with the standard textbooks on advanced mechanics. Without questioning the excellent quality of these primarily technical and formalistic treatments, the author feels that there is room for monographs which exhibit the fundamental skeletons of the exact sciences in an elementary and philosophically oriented fashion.

Many of the scientific treatises of today are formulated in a half-mystical language, as though to impress the reader with the uncomfortable feeling that he is in the permanent presence of a superman. The present book is conceived in a humble spirit and is written for humble people. The author knows from past experience that one outstanding weakness of our present system of college education is the custom of classing certain fundamental and apparently simple concepts as "elementary," and of relegating them to an age-level at which the student's mind is not mature enough to grasp their true meaning. The fruits of this error can be observed daily. The student who is thoroughly acquainted with the smallest details of an atom-smashing apparatus often has entirely confused ideas concerning the difference between mass and weight, or between heavy mass and inertial mass. In mechanics, which is a fundamental science, the confusion is particularly conspicuous.

To a philosophically trained mind, the difference between actual and virtual displacement appears entirely obvious and needs no further comment. But a student of today is anything but philosophically minded. To him the difference is not only not obvious, but he cannot grasp the meaning of a "virtual displacement" without experimenting with the concept for a long time by applying it to a variety of familiar situations. Hence, the author could think of no better introduction to the application of the calculus of variations than letting the student deduce a number of familiar results of vectorial mechanics from the principle of virtual work. As a by-product of these exercises, the student notices how previously unconnected and more or less axiomatically stated properties of forces and moments all follow from one single, all-embracing principle. His interest is now aroused and he would like to go beyond the realm of statics. Here d'Alembert's principle comes to his aid, showing him how the same method of virtual displacements can serve to obtain the equations of motion of an arbitrarily complicated dynamical problem.

The author is well aware that he could have shortened his exposition considerably, had he started directly with the Lagrangian equations of motion and then proceeded to Hamilton's theory. This procedure would have been justified had the purpose of this book been primarily to familiarize the student with a certain formalism and technique in writing down the differential equations which govern a given dynamical problem, together with certain "recipes" which he might apply in order to solve them. But this is exactly what the author did *not* want to do. There is a tremendous treasure of philosophical meaning behind the great theories of Euler and Lagrange, and of Hamilton and Jacobi, which is completely smothered in a purely formalistic treatment, although it cannot fail to be a source of the greatest intellectual enjoyment to every mathematically-minded person. To give the student a chance to discover for himself the hidden beauty of these theories was one of the foremost intentions of the author. For this purpose he had to lead the reader through the entire historical development, starting from the very beginning, and felt compelled to include problems which familiarize the student with the new concepts. These problems, of a simple character, were chosen in order to exhibit the general principles involved.

As regards content, an important topic not included in the present book is the perturbation theory of the dynamical equations. Also, an originally planned chapter on Relativistic Mechanics had to be omitted, owing to limitations of space. Yet even the material as it stands can well form the subject matter of a two-semester graduate course of three hours weekly, and it may suffice for a student of mathematics or physics or engineering who does not intend to specialize in mechanics, but wants to get a thorough grasp of the underlying principles.

The author apologizes for not giving many specific references. This material has grown with him for many years and consequently he is often unable to tell where he got his information. He is primarily indebted to Whittaker's *Analytical Dynamics*, Nordheim's article in the *Handbuch der Physik*, and Prange's article in the *Encyklopaedie der mathematischen Wissenschaften*— all of which are mentioned in the Bibliography and recommended for advanced reading.

The author is deeply indebted to Professor J. L. Synge for the painstaking care with which he reviewed the manuscript and pointed out many weak points in the author's presentation. In some instances a divergence of viewpoints was apparent, but in others an improved formulation could be given, thus enhancing substantially the readability of the book. The author is also indebted to Professor G. de B. Robinson and Professor A. F. C. Stevenson who revised the entire manuscript, Mrs. Ida Rhodes of the Mathematical Tables Project, New York City, and Mr. G. F. D. Duff of the University of Toronto, who helped read the proof and prepare the Index. The generous co-operation of all these friends and colleagues has been a heart-warming experience.

The variational principles of mechanics are firmly rooted in the soil of that great century of Liberalism which starts with Descartes and ends with the French Revolution and which has witnessed the lives of Leibniz, Spinoza, Goethe, and Johann Sebastian Bach. It is the only period of cosmic thinking in the entire history of Europe since the time of the Greeks. If the author has succeeded in conveying an inkling of that cosmic spirit, his efforts will be amply rewarded.

CORNELIUS LANCZOS

Los Angeles
May, 1949

PREFACE TO THE SECOND EDITION

APART from Appendix I—which gives a simpler derivation of the transition from the Lagrangian to the Hamiltonian form of mechanics—this edition differs from the previous one by the addition of a new chapter on Relativistic Mechanics (numerous minor inaccuracies of the first printing were eliminated earlier on the occasion of the Fourth Printing*).

The author is grateful to the Directors of the University of Toronto Press for giving him the opportunity to incorporate in his book this material, which deals with one of the most amazing discoveries of the speculative genius of mankind. Although in condensed form, yet all the relevant concepts, principles, and results of Einstein's revolutionary theories have been adequately treated, as far as the kinematics and dynamics of a particle are involved. The general theory of the Lorentz transformations is developed on the basis of Hamilton's quaternions, which are so eminently suited to this task that one could hardly find any other mathematical tool of equal simplicity and conciseness. The field equations of General Relativity were naturally outside the scope of the book, but the dynamical aspects of Einstein's gravitational theory, including the three crucial tests of general relativity, have been fully treated, since they lie entirely within the framework of the Lagrangian and Hamiltonian form of dynamics.

It is the author's hope that the addition of this chapter will enhance the usefulness of the book and add new friends to the already large family of readers who found the study of this introduction to one of the great chapters of science rewarding and a source of inspiration.

Dublin, Ireland   C.L.
July, 1962

*The author wishes to thank particularly Dr. Leonard S. Abrams, Woodland Hills, Calif., for providing him with a very complete list of errors and misprints.

PREFACE TO THE THIRD EDITION

The present edition differs from the previous one by the addition of a section on Noether's invariant variational problems (cf. Appendix II, p. 401). The original paper of Noether is not easy reading. In fact, however, the discussion can be subordinated to the well-known theory of "ignorable variables," and this is the method followed in the present exposition. In this approach the parameters of Noether's transformation appear as added action variables of the variational problem, for which the Euler-Lagrange equations can be found.

C.L.

Dublin, Ireland
October, 1965

PREFACE TO THE FOURTH EDITION

MORE than twenty years have passed since the author's book was first published and he is grateful to a benevolent fate which has permitted him to see the appearance of the fourth edition. On this occasion, with the kind cooperation of the University of Toronto Press, he has been able to realize his old dream of adding a last chapter on Continuum Mechanics (i.e. the physics of the 19th century), although this material is, strictly speaking, beyond the pale of the "Variational Principles of Mechanics" in the traditional sense.

The curriculum of the physics student of today is so much loaded with elementary particle physics and all the other branches of Quantum Theory that subjects like Hydrodynamics, Elasticity, or even the Electromagnetic Theory of Faraday and Maxwell, are often only tangentially treated. If the student wants a more substantial glimpse of these subjects, he is forced to take refuge in the specialized literature with its many hundreds of pages, for which he may lack time and inclination. To consider these great chapters of theoretical physics from the viewpoint of the variational principle, dealing only with basic and historically interesting matters, seemed thus a worthy enterprise.

A remark may be permitted concerning the last part of this new chapter, which deals in detail with several applications of Emmy Noether's far-reaching idea, which shows how all conservation laws are the consequence of certain invariance properties of the basic Lagrangian. The author's approach of conceiving Noether's principle as a special application of the properties of kinosthenic (in particle mechanics usually called "ignorable") variables greatly clarifies the issue and gives the simplest way of deriving all results from a unified viewpoint. This part of the chapter is thus an independent contribution to a frequently occurring problem of contemporary physics, which many physicists may find particularly useful.

Dublin, Ireland   C.L.
June, 1970

CONTENTS

INTRODUCTION

I. THE BASIC CONCEPTS OF ANALYTICAL MECHANICS

II. THE CALCULUS OF VARIATIONS

IX. RELATIVISTIC MECHANICS

X. HISTORICAL SURVEY

XI. MECHANICS OF THE CONTINUA

INTRODUCTION

1. **The variational approach to mechanics.** Ever since Newton laid the solid foundation of dynamics by formulating the laws of motion, the science of mechanics developed along two main lines. One branch, which we shall call "vectorial mechanics,"[1] starts directly from Newton's laws of motion. It aims at recognizing all the forces acting on any given particle, its motion being uniquely determined by the known forces acting on it at every instant. The analysis and synthesis of forces and moments is thus the basic concern of vectorial mechanics.

While in Newton's mechanics the action of a force is measured by the momentum produced by that force, the great philosopher and universalist Leibniz, a contemporary of Newton, advocated another quantity, the *vis viva* (living force), as the proper gauge for the dynamical action of a force. This *vis viva* of Leibniz coincides—apart from the unessential factor 2—with the quantity we call today "kinetic energy." Thus Leibniz replaced the "momentum" of Newton by the "kinetic energy." At the same time he replaced the "force" of Newton by the "work of the force." This "work of the force" was later replaced by a still more basic quantity, the "work function." Leibniz is thus the originator of that second branch of mechanics, usually called "analytical mechanics,"[2] which bases the entire study of equilibrium and motion on two fundamental scalar quantities, the "kinetic energy" and the "work function," the latter frequently replaceable by the "potential energy."

Since motion is by its very nature a *directed* phenomenon, it seems puzzling that two scalar quantities should be sufficient to determine the motion. The energy theorem, which states that the sum of the kinetic and potential energies remains unchanged

---

[1]This use of the term does not necessarily imply that vectorial methods are used.

[2]See note on terminology at end of chap. I, section 1.

during the motion, yields only *one* equation, while the motion of a single particle in space requires *three* equations; in the case of mechanical systems composed of two or more particles the discrepancy becomes even greater. And yet it is a fact that these two fundamental scalars contain the complete dynamics of even the most complicated material system, provided they are used as the basis of a *principle* rather than of an *equation*.

2. The procedure of Euler and Lagrange.

In order to see how this occurs, let us think of a particle which is at a point P_1 at a time t_1. Let us assume that we know its velocity at that time. Let us further assume that we know that the particle will be at a point P_2 after a given time has elapsed. Although we do not know the path taken by the particle, it is possible to establish that path completely by mathematical experimentation, provided that the kinetic and potential energies of the particle are given for any possible velocity and any possible position.

Euler and Lagrange, the first discoverers of the exact principle of least action, proceed as follows. Let us connect the two points P_1 and P_2 by *any* tentative path. In all probability this path, which can be chosen as an arbitrary continuous curve, will *not* coincide with the actual path that nature has chosen for the motion. However, we can gradually *correct* our tentative solution and eventually arrive at a curve which can be designated as the *actual* path of motion.

For this purpose we let the particle move along the tentative path in accordance with the energy principle. The sum of the kinetic and potential energies is kept constant and always equal to that value E which the actual motion has revealed at time t_1. This restriction assigns a definite velocity to any point of our path and thus determines the motion. We can choose our path freely, but once this is done the conservation of energy determines the motion uniquely.

In particular, we can calculate the time at which the particle will pass an arbitrarily given point of our fictitious path and hence the time-integral of the *vis viva* i.e., of double the kinetic energy, extended over the entire motion from P_1 to P_2. This time integral is called "action." It has a definite value for our tentative path and likewise for any other tentative path, these paths being always drawn between the same two end-points P_1, P_2 and always traversed with the same given energy constant E.

The value of this "action" will vary from path to path. For some paths it will come out larger, for others smaller. Mathematically we can imagine that *all* possible paths have been tried. There must exist one definite path (at least if P_1 and P_2 are not too far apart) for which the action assumes a minimum value. The principle of least action asserts that *this particular path is the one chosen by nature as the actual path of motion.*

We have explained the operation of the principle for one single particle. It can be generalized, however, to any number of particles and any arbitrarily complicated mechanical system.

3. Hamilton's procedure. We encounter problems of mechanics for which the work function is a function not only of the position of the particle but also of the time. For such systems the law of the conservation of energy does not hold, and the principle of Euler and Lagrange is not applicable, but that of Hamilton is.

In Hamilton's procedure we again start with the given initial point P_1 and the given end-point P_2. But now we do not restrict the trial motion in any way. Not only can the path be chosen arbitrarily—save for natural continuity conditions—but also the motion in time is at our disposal. All that we require now is that our tentative motion shall start at the observed time t_1 of the actual motion and end at the observed time t_2. (This condition is not satisfied in the procedure of Euler-Lagrange, because there the energy theorem restricts the motion, and the time taken to go from P_1 to P_2 in the tentative motion will generally differ from the time taken in the actual motion.)

The characteristic quantity that we now use as the measure of action—there is unfortunately no standard name adopted for this quantity—is the time-integral of the *difference between the kinetic and potential energies*. The Hamiltonian formulation of the principle of least action asserts that *the actual motion realized*

*in nature is that particular motion for which this action assumes its
smallest value.*

One can show that in the case of "conservative" systems, i.e.
systems which satisfy the law of the conservation of energy, the
principle of Euler-Lagrange is a consequence of Hamilton's
principle, but the latter principle remains valid even for non-
conservative systems.

4. The calculus of variations. The mathematical problem
of minimizing an integral is dealt with in a special branch of the
calculus, called "calculus of variations." The mathematical
theory shows that our final results can be established without
taking into account the infinity of tentatively possible paths.
We can restrict our mathematical experiment to such paths as
are *infinitely near* to the actual path. A tentative path which
differs from the actual path in an arbitrary but still *infinitesimal*
degree, is called a "variation" of the actual path, and the calculus
of variations investigates the changes in the value of an integral
caused by such infinitesimal variations of the path.

**5. Comparison between the vectorial and the variational
treatments of mechanics.** The vectorial and the variational
theories of mechanics are two different mathematical descriptions
of the same realm of natural phenomena. Newton's theory bases
everything on two fundamental vectors: "momentum" and
"force"; the variational theory, founded by Euler and Lagrange,
bases everything on two scalar quantities: "kinetic energy" and
"work function." Apart from mathematical expediency, the
question as to the equivalence of these two theories can be raised.
In the case of *free* particles, i.e. particles whose motion is not
restricted by given "constraints," the two forms of description
lead to equivalent results. But for systems with constraints the
analytical treatment is simpler and more economical. The given
constraints are taken into account in a natural way by letting the
system move along all the tentative paths in harmony with them.
The vectorial treatment has to take account of the forces which
maintain the constraints and has to make definite hypotheses

concerning them. Newton's third law of motion, "action equals reaction," does not embrace all cases. It suffices only for the dynamics of rigid bodies.

On the other hand, Newton's approach does not restrict the nature of a force, while the variational approach assumes that the acting forces are derivable from a scalar quantity, the "work function." Forces of a frictional nature, which have no work function, are outside the realm of variational principles, while the Newtonian scheme has no difficulty in including them.

Such forces originate from inter-molecular phenomena which are neglected in the macroscopic description of motion. If the macroscopic parameters of a mechanical system are completed by the addition of microscopic parameters, forces not derivable from a work function would in all probability not occur.

6. Mathematical evaluation of the variational principles. Many elementary problems of physics and engineering are solvable by vectorial mechanics and do not require the application of variational methods. But in all more complicated problems the superiority of the variational treatment becomes conspicuous. This superiority is due to the complete freedom we have in choosing the appropriate coordinates for our problem. The problems which are well suited to the vectorial treatment are essentially those which can be handled with a rectangular frame of reference, since the decomposition of vectors in curvilinear coordinates is a cumbersome procedure if not guided by the advanced principles of tensor calculus. Although the fundamental importance of invariants and covariants for all phenomena of nature has been discovered only recently and so was not known in the time of Euler and Lagrange, the variational approach to mechanics happened to anticipate this development by satisfying the principle of invariance automatically. We are allowed sovereign freedom in choosing our coordinates, since our processes and resulting equations remain valid for an arbitrary choice of coordinates. The mathematical and philosophical value of the variational method is firmly anchored in this freedom of choice and the corresponding freedom of arbitrary coordinate transformations. It greatly facilitates the formulation of the differential equations of motion, and likewise their solution. If we hit on a certain type of coordinates, called "cyclic" or "ignorable," a partial integration of the basic differential equations is at once accomplished. If *all* our coordinates are ignorable, our problem is completely solved. Hence, we can formulate the entire problem of solving the differential equations of motion as a problem of coordinate transformation. Instead of trying to integrate the differential equations of motion directly, we try to produce more and more ignorable coordinates. In the Euler-Lagrangian form of mechanics it is more or less accidental if we hit on the right coordinates, because we have no systematic way of producing ignorable coordinates. But the later developments of the theory by Hamilton and Jacobi broadened the original procedures immensely by introducing the "canonical equations," with their much wider transformation properties. Here we are able to produce a *complete* set of ignorable coordinates by solving one single partial differential equation.

Although the actual solution of this differential equation is possible only for a restricted class of problems, it so happens that many important problems of theoretical physics belong precisely to this class. And thus the most advanced form of analytical mechanics turns out to be not only esthetically and logically most satisfactory, but at the same time very practical by providing a tool for the solution of many dynamical problems which are not accessible to elementary methods.[1]

7. Philosophical evaluation of the variational approach to mechanics. Although it is tacitly agreed nowadays that scientific treatises should avoid philosophical discussions, in the case of the variational principles of mechanics an exception to the rule may be tolerated, partly because these principles are rooted in a century which was philosophically oriented to a very high degree, and partly because the variational method has often been the focus of philosophical controversies and misinterpretations.

---

[1]The present book does not discuss other integration methods which are not based on the transformation theory. Concerning such methods the reader is referred to the advanced text-books mentioned in the Bibliography.

Indeed, the idea of enlarging reality by including "tentative" possibilities and then selecting one of these by the condition that it minimizes a certain quantity, seems to bring a *purpose* to the flow of natural events. This is in contradiction to the usual *causal* description of things. Yet we must not be surprised that for the more universal approach which was current in the 17th and 18th centuries, the two forms of thinking did not necessarily appear contradictory. The keynote of that entire period was the seemingly pre-established harmony between "reason" and "world." The great discoveries of higher mathematics and their immediate application to nature imbued the philosophers of those days with an unbounded confidence in the fundamentally intellectual structure of the world. Thus "deus intellectualis" was the basic theme of the philosophy of Leibniz, no less than that of Spinoza. At the same time Leibniz had strong teleological tendencies, and his activities had no small influence on the development of variational methods.[1] But this is not surprising if we observe how the purely esthetic and logical interest in maximum-minimum problems gave one of the strongest impulses to the development of infinitesimal calculus, and how Fermat's derivation of the laws of geometrical optics on the basis of his "principle of quickest arrival" could not fail to impress the philosophically-oriented scientists of those days. That the dilettante misuse of these tendencies by Maupertuis and others for theological purposes has put the entire trend into disrepute, is not the fault of the great philosophers.

The sober, practical, matter-of-fact nineteenth century—which carries over into our day—suspected all speculative and interpretative tendencies as "metaphysical" and limited its programme to the pure description of natural events. In this philosophy mathematics plays the role of a shorthand method, a conveniently economical language for expressing involved relations. Hence, it is not surprising, but quite consistent with the "positivistic" spirit of the nineteenth century, to meet with the following appraisal of analytical mechanics by one of the leading

[1]See the attractive historical study of A. Kneser, *Das Prinzip der kleinsten Wirkung von Leibniz bis zur Gegenwart* (Leipzig: Teubner, 1928).

figures of that trend, E. Mach, in "The Science of Mechanics" (*Open Court*, 1893, p. 480): "No fundamental light can be expected from this branch of mechanics. On the contrary, the discovery of matters of principle must be substantially completed before we can think of framing analytical mechanics the sole aim of which is a perfect *practical* mastery of problems. Whosoever mistakes this situation will never comprehend Lagrange's great performance, which here too is essentially of an *economical* character." (Italics in the original.) According to this philosophy the variational principles of mechanics are not more than alternative mathematical formulations of the fundamental laws of Newton, without any primary importance.

However, philosophical trends float back and forth and the last word is never spoken. In our own day we have witnessed at least *one* fundamental discovery of unprecedented magnitude, namely Einstein's Theory of General Relativity, which was obtained by mathematical and philosophical speculation of the highest order. Here was a discovery made by a kind of reasoning that a positivist cannot fail to call "metaphysical," and yet it provided an insight into the heart of things that mere experimentation and sober registration of facts could never have revealed. The Theory of General Relativity brought once again to the fore the spirit of the great cosmic theorists of Greece and the eighteenth century.

In the light of the discoveries of relativity, the variational foundation of mechanics deserves more than purely formalistic appraisal. Far from being nothing but an alternative formulation of the Newtonian laws of motion, the following points suggest the supremacy of the variational method:

1. The Principle of Relativity requires that the laws of nature shall be formulated in an "invariant" fashion, i.e. independently of any special frame of reference. The methods of the calculus of variations automatically satisfy this principle, because the minimum of a scalar quantity does not depend on the coordinates in which that quantity is measured. While the Newtonian equations of motion did not satisfy the principle of relativity, the principle of least action remained valid, with the

only modification that the basic action quantity had to be brought into harmony with the requirement of invariance.

2. The Theory of General Relativity has shown that matter cannot be separated from field and is in fact an outgrowth of the field. Hence, the basic equations of physics must be formulated as partial rather than ordinary differential equations. While Newton's particle picture can hardly be brought into harmony with the field concept, the variational methods are not restricted to the mechanics of particles but can be extended to the mechanics of continua.

3. The Principle of General Relativity is automatically satisfied if the fundamental "action" of the variational principle is chosen as an invariant under any coordinate transformation. Since the differential geometry of Riemann furnishes us such invariants, we have no difficulty in setting up the required field equations. Apart from this, our present knowledge of mathematics does not give us any clue to the formulation of a co-variant, and at the same time consistent, system of field equations. Hence, in the light of relativity the application of the calculus of variations to the laws of nature assumes more than accidental significance.

THE VARIATIONAL
PRINCIPLES OF MECHANICS

CHAPTER I

THE BASIC CONCEPTS OF ANALYTICAL MECHANICS

1. The principal viewpoints of analytical mechanics. The analytical form of mechanics, as introduced by Euler and Lagrange, differs considerably in its method and viewpoint from vectorial mechanics. The fundamental law of mechanics as stated by Newton: "mass times acceleration equals moving force" holds in the first instance for a single particle only. It was deduced from the motion of a particle in the field of gravity on the earth and was then applied to the motion of planets under the action of the sun. In both problems the moving body could be idealized as a "mass point" or a "particle," i.e. a single point to which a mass is attached, and thus the dynamical problem presented itself in this form: "A particle can move freely in space and is acted upon by a given force. Describe the motion at any time." The law of Newton gave the differential equation of motion, and the dynamical problem was reduced to the integration of that equation.

If the particle is not free but associated with other particles, as for example in a solid body, or a fluid, the Newtonian equation is still applicable if the proper precautions are observed. One has to isolate the particle from all other particles and determine the force which is exerted on it by the surrounding particles. Each particle is an independent unit which follows the law of motion of a free particle.

This force-analysis sometimes becomes cumbersome. The unknown nature of the interaction forces makes it necessary to introduce additional postulates, and Newton thought that the principle "action equals reaction," stated as his third law of motion, would take care of all dynamical problems. This, however, is not the case, and even for the dynamics of a rigid body the additional hypothesis that the inner forces of the body are of the nature of central forces had to be made. In more complicated situations the Newtonian approach fails to give a unique answer to the problem.

The analytical approach to the problem of motion is quite different. The particle is no longer an isolated unit but part of a "system." A "mechanical system" signifies an assembly of particles which interact with each other. The single particle has no significance; it is the system as a whole which counts. For example, in the planetary problem one may be interested in the motion of one particular planet. Yet the problem is unsolvable in this restricted form. The force acting on that planet has its source principally in the sun, but to a smaller extent also in the other planets, and cannot be given without knowing the motion of the other members of the system as well. And thus it is reasonable to consider the dynamical problem of the *entire* system, without breaking it into parts.

But even more decisive is the advantage of a unified treatment of force-analysis. In the vectorial treatment each point requires special attention and the force acting has to be determined independently for each particle. In the analytical treatment it is enough to know one single function, depending on the positions of the moving particles; this "work function" contains implicitly all the forces acting on the particles of the system. They can be obtained from that function by mere differentiation.

Another fundamental difference between the two methods concerns the matter of "auxiliary conditions." It frequently happens that certain kinematical conditions exist between the particles of a moving system which can be stated a *priori*. For example, the particles of a solid body may move as if the body were "rigid," which means that the distance between any two points cannot change. Such kinematical conditions do not actually exist on a *priori* grounds. They are maintained by strong forces. It is of great advantage, however, that the analytical treatment does not require the knowledge of these forces, but can take the given kinematical conditions for granted. We can develop the dynamical equations of a rigid body without knowing what forces produce the rigidity of the body. Similarly we need not know in detail what forces act between the particles of a fluid. It is enough to know the empirical fact that a fluid opposes by very strong forces any change in its volume, while the forces which oppose a change in shape of the fluid without changing the volume are slight. Hence, we can discard the unknown inner forces of a fluid and replace them by the kinematical conditions that during the motion of a fluid the volume of any portion must be preserved. If one considers how much simpler such an *a priori* kinematical condition is than a detailed knowledge of the forces which are required to maintain that condition, the great superiority of the analytical treatment over the vectorial treatment becomes apparent.

However, more fundamental than all the previous features is the *unifying principle* in which the analytical approach culminates. The equations of motion of a complicated mechanical system form a large number—even an infinite number—of separate differential equations. The variational principles of analytical mechanics discover the unifying basis from which all these equations follow. There is a *principle* behind all these equations which expresses the meaning of the entire set. Given one fundamental quantity, "action," the principle that this action be stationary leads to the entire set of differential equations. Moreover, the statement of this principle is independent of any special system of coordinates. Hence, the analytical equations of motion are also invariant with respect to any coordinate transformations.

Note on terminology. The word "analytical" in the expression "analytical mechanics" has nothing to do with the philosophical process of analyzing, but comes from the mathematical term "analysis," referring to the application of the principles of infinitesimal calculus to problems of pure and applied mathematics. While the French and German literature reserves the term "analytical mechanics" for the abstract mathematical treatment of mechanical problems by the methods of Euler, Lagrange, and Hamilton, the English and particularly the American literature frequently calls even very elementary applications of the calculus to problems of simple vectorial mechanics by the same name. The term "mechanics" includes "statics" and "dynamics," the first dealing with the *equilibrium* of particles and systems of particles, the second with their *motion*. (A separate application of mechanics deals with the "mechanics of continua"—which includes fluid mechanics and elasticity—based on partial, rather than ordinary, differential equations. These problems are not included in the present book.)

Summary. These, then, are the four principal viewpoints in which vectorial and analytical mechanics differ:

1.　Vectorial mechanics isolates the particle and considers it as an individual; analytical mechanics considers the system as a whole.

2.　Vectorial mechanics constructs a separate acting force for each moving particle; analytical mechanics considers one single function: the work function (or potential energy). This one function contains all the necessary information concerning forces.

3.　If strong forces maintain a definite relation between the coordinates of a system, and that relation is empirically given, the vectorial treatment has to consider the forces necessary to maintain it. The analytical treatment takes the given relation for granted, without requiring knowledge of the forces which maintain it.

4.　In the analytical method, the entire set of equations of motion can be developed from one unified principle which implicitly includes all these equations. This principle takes the form of minimizing a certain quantity, the "action." Since a minimum principle is independent of any special reference system, the equations of analytical mechanics hold for any set of coordinates. This permits one to adjust the coordinates employed to the specific nature of each problem.

2.　Generalized coordinates. In the elementary vectorial treatment of mechanics the abstract concept of a "coordinate" does not enter the picture. The method is essentially geometrical in character.

Vector methods are eminently useful in problems of statics. However, when it comes to problems of motion, the number of such problems which can be solved by pure vector methods is relatively small. For the solution of more involved problems, the geometrical methods of vectorial mechanics cease to be adequate and have to give way to a more abstract analytical treatment. In this new analytical foundation of mechanics the coordinate concept in its most general aspect occupies a central position.

Analytical mechanics is a completely mathematical science. Everything is done by calculations in the abstract realm of quantities. The physical world is translated into mathematical relations. This translation occurs with the help of coordinates. The coordinates establish a one-to-one correspondence between the points of physical space and numbers. After establishing this correspondence, we can operate with the coordinates as algebraic quantities and forget about their physical meaning. The end result of our calculations is then finally translated back into the world of physical realities.

During the century from Fermat and Descartes to Euler and Lagrange tremendous developments in the methods of higher mathematics took place. One of the most important of these was the generalization of the original coordinate idea of Descartes. If the purpose of coordinates is to establish a one-to-one correspondence between the points of space and numbers, the setting up of three perpendicular axes and the determination of length, width, and height relative to those axes is but *one* way of establishing that correspondence. Other methods can serve equally well. For example, the polar coordinates r, θ, ϕ may take the place of the rectangular coordinates x, y, z. It is one of the characteristic features of the analytical treatment of mechanics that we do not specify the nature of the coordinates which translate a given physical situation into an abstract mathematical situation.

Let us first consider a mechanical system which is composed of N free particles, "free" in the sense that they are not restricted by any kinematical conditions. The rectangular coordinates of these particles:

$$x_i, y_i, z_i, (i = 1, 2, \ldots, N), \tag{12.1}$$

characterize the position of the mechanical system, and the problem of motion is obviously solved if x_i, y_i, z_i are given as functions of the time t.

The same problem is likewise solved, however, if the x_i, y_i, z_i are expressed in terms of some other quantities

$$q_1, q_2, \ldots, q_{3N}, \tag{12.2}$$

and then these quantities q_k are determined as functions of the time t.

This indirect procedure of solving the problem of motion provides great analytical advantages and is in fact the decisive factor in solving dynamical problems. Mathematically, we call it a "coordinate transformation." It is a generalization of the transition from rectangular coordinates x, y, z of a single point in space to polar coordinates r, θ, ϕ. The relations

$$\begin{aligned} x &= r \sin \theta \cos \phi, \\ y &= r \sin \theta \sin \phi, \\ z &= r \cos \theta, \end{aligned} \tag{12.3}$$

are generalized so that the old variables are expressed as arbitrary functions of the new variables. The number of variables is not 3 but $3N$, since the position of the mechanical system requires $3N$ coordinates for its characterization. And thus the general form of such a coordinate transformation appears as follows:

$$\begin{aligned} x_1 &= f_1 (q_1, \ldots, q_{3N}), \\ &\cdots \cdots \cdots \cdots \cdots \\ &\cdots \cdots \cdots \cdots \cdots \\ z_N &= f_{3N} (q_1, \ldots, q_{3N}). \end{aligned} \tag{12.4}$$

We can prescribe these functions, $f_1, \ldots, f_{3N}$, in any way we wish and thus shift the original problem of determining the x_i, y_i, z_i as functions of t to the new problem of determining the $q_1, \ldots, q_{3N}$ as functions of t. With proper skill in choosing the right coordinates, we may solve the new problem more easily than the original one. The flexibility of the reference system makes it possible to choose coordinates which are particularly suitable for the given problem. For example, in the planetary problem, i.e. a particle revolving around a fixed attracting centre, polar coordinates are better suited to the problem of motion than rectangular ones.

The advantage of generalized coordinates is even more obvious if mechanical systems with given kinematical conditions are considered. Such conditions find their mathematical expression in certain functional relations between the coordinates. For example, two atoms may form a molecule, the distance between the two atoms being determined by strong forces which are in equilibrium at that distance. Dynamically this system can be considered as composed of two mass points with coordinates x_1, y_1, z_1 and x_2, y_2, z_2 which are kept at a constant distance a from one another. This implies the condition

$$(x_1 - x_2)^2 + (y_1 - y_2)^2 + (z_1 - z_2)^2 = a^2. \tag{12.5}$$

Because of this condition, the 6 coordinates $x_1, \ldots, z_2$ cannot be prescribed arbitrarily. It suffices to give 5 coordinates; the sixth coordinate is then determined by the auxiliary condition (12.5). However, it is obviously inappropriate to designate one of the rectangular coordinates as a dependent variable when the relation (12.5) is symmetrical in all coordinates. It is more natural to prescribe the three rectangular coordinates of the centre of mass of the system and add two angles which characterize the direction of the axis of the diatomic molecule. The 6 rectangular coordinates $x_1, \ldots, z_2$ are expressible in terms of these 5 parameters.

As another example, consider the case of a rigid body, which can be composed of any number of particles. But whatever the number of particles may be, it is sufficient to give the three coordinates of the centre of mass and three angles which define the position of the body relative to the centre of mass. These 6 parameters determine the position of the rigid body completely. The coordinates of each of the component particles can be expressed as functions of these 6 parameters.

In general, if a mechanical system consists of N particles and there are m independent kinematical conditions imposed, it will be possible to characterize the configuration of the mechanical system uniquely by

$$n = 3N - m \qquad (12.6)$$

independent parameters

$$q_1, q_2, \ldots, q_n, \qquad (12.7)$$

in such a way that the rectangular coordinates of all the particles are expressible as functions of the variables (12.7):

$$x_1 = f_1(q_1, \ldots, q_n),$$
$$\cdots \cdots \cdots \cdots \qquad (12.8)$$
$$\cdots \cdots \cdots \cdots$$
$$z_N = f_{3N}(q_1, \ldots, q_n).$$

The number n is a characteristic constant of the given mechanical system which cannot be altered. Less than n parameters are not enough to determine the position of the system. More than n parameters are not required and could not be assigned without satisfying certain conditions. We express the fact that n parameters are necessary and sufficient for a unique characterization of the configuration of the system by saying that it has "n degrees of freedom." Moreover, we call the n parameters $q_1, q_2, \ldots, q_n$ the "generalized coordinates" of the system. The number N of particles which compose the mechanical system is immaterial for the analytical treatment, as are also the coordinates of these particles. It is the generalized coordinates $q_1, q_2, \ldots, q_n$ and certain basic functions of them which are of importance. A rigid body may be composed of an infinity of mass points, yet for the mechanical treatment it is a system of not more than 6 independent coordinates.

Examples:

One degree of freedom: A piston moving up and down. A rigid body rotating about a fixed axis.

Two degrees of freedom: A particle moving on a given surface.

Three degrees of freedom: A particle moving in space. A rigid body rotating about a fixed point (top).

Four degrees of freedom: Two components of a double star revolving in the same plane.

Five degrees of freedom: Two particles kept at a constant distance from each other.

Six degrees of freedom: Two planets revolving about a fixed sun. A rigid body moving freely in space.

The generalized coordinates $q_1, q_2, \ldots, q_n$ may or may not have a geometrical significance. It is necessary, however, that the functions (12.8) shall be finite, single valued, continuous and differentiable, and that the Jacobian of at least *one* combination of n functions shall be different from zero. These conditions may be violated at certain singular points, which have to be excluded from consideration. For example, the transformation (12.3) from rectangular to polar coordinates satisfies the general regularity conditions, but special care is required at the values $r = 0$ and $\theta = 0$, for which the Jacobian of the transformation vanishes.

In addition to these restrictions "in the small," certain conditions "in the large" have to be observed. It is necessary that a proper continuous range of the variables $q_1, q_2, \ldots, q_n$ shall permit a sufficiently wide range of the original rectangular coordinates, without restricting them more than the given kinematical conditions require. For example, the transformation (12.3) guarantees the complete, infinite range of the variables x, y, z if r varies between 0 and ∞, θ between 0 and π, and ϕ between 0 and 2π. However, such conditions in the large seldom burden our considerations because, even if we cannot cover the entire range of motion, we may still be able to describe a characteristic portion of it.

It is not always advisable to eliminate all the kinematical conditions of a problem by introducing suitable generalized coordinates. We sometimes prefer to eliminate only *some* of the kinematical conditions, and to leave the others as additional restricting conditions. Hence the general problem of mechanical systems with kinematical conditions presents itself in the following form. We have the equations (12.8) and in addition m equations of the form:

$$\phi_i(q_1, q_2, \ldots, q_n) = 0, \ (i = 1, \ldots, m). \tag{12.9}$$

The number of degrees of freedom is here

$$n' = n - m.$$

Summary. Analytical problems of motion require a generalization of the original coordinate concept of Descartes. Any set of parameters which can characterize the position of a mechanical system may be chosen as a suitable set of coordinates. They are called the "generalized coordinates" of the mechanical system.

3. The configuration space. One of the most imaginative concepts of the abstract analytical treatment of mechanical problems is the concept of the configuration space. If we associate with the three numbers x, y, z a point in a three-dimensional space, there is no reason why we should not do the same with the n numbers q_1, q_2, $\ldots$, q_n, considering these as the rectangular coordinates of a "point" P in an n-dimensional space. Similarly, if we can associate with the equations

$$\begin{aligned}
x &= f(t), \\
y &= g(t), \\
z &= h(t),
\end{aligned} \tag{13.1}$$

the idea of a "curve" and the motion of a point along that curve, there is no reason why we should not do the same with the corresponding equations

$$\begin{aligned}
q_1 &= q_1(t), \\
&\cdots \cdots \\
&\cdots \cdots \\
q_n &= q_n(t).
\end{aligned} \tag{13.2}$$

These equations represent the solution of a dynamical problem. In the associated geometrical picture, we have a point P of an n-dimensional space which moves along a given curve of that space.

This geometrical picture is a great aid to our thinking. No matter how numerous the particles constituting a given mechanical system may be, or how complicated the relations existing between them, the entire mechanical system is pictured as a single point of a many-dimensional space, called the "configuration space." For example, the position of a rigid body—with all the infinity of mass points which form it—is symbolized as a single point of a 6-dimensional space. This 6-dimensional space has, to be sure, nothing to do with the physical reality of the rigid body. It is merely *correlated* to the rigid body in the sense of a one-to-one correspondence. The various positions of the rigid body are "mapped" as various points of the 6-dimensional space, and conversely, the various points of a 6-dimensional space can be physically interpreted as various positions of a rigid body. For the sake of brevity we shall refer to the point which symbolizes the position of the mechanical system in the configuration space as the "C-point," while the curve traced out by that point during the motion will be referred to as the "C-curve."

The picture that we have formed here of the configuration space needs further refinement. We based our discussions on the analytical geometry of a Euclidean space of n dimensions. Accordingly we have represented the n generalized coordinates of the mechanical system as rectangular coordinates of that space. A much more appropriate picture of the geometrical structure of the configuration space can be obtained if we change from analytical geometry to differential geometry, as we shall do in section 5 of this chapter. However, even our preliminary picture can serve some useful purposes. We demonstrate this by the following example.

The position of a free particle in space may be characterized by the polar coordinates (r, θ, ϕ). If these values are taken as *rectangular* coordinates of a point, we get a space whose geometrical properties are obviously greatly distorted compared with the actual space. Straight lines become curves, angles and distances are changed. Yet certain important properties of space remained unaltered by this mapping process. A point remains a point, the neighborhood of a point remains the neighborhood of a point, a curve remains a curve, adjacent curves remain adjacent curves. Continuous and differentiable curves remain continuous and differentiable curves. Now for the processes of the calculus of variations such "topological" properties of space are the really important things, while the "metrical" properties of space, such as distances, angles, areas, etc., are irrelevant. For this reason even the simplified picture of a configuration space without a corresponding geometry is still an excellent aid in visualizing some abstract analytical processes.

Summary. The pictorial language of n-dimensional geometry makes it possible to extend the mechanics of a single mass-point to arbitrarily complicated mechanical systems. Such a system may be replaced by a single point for the study of its motion. But the space which carries this point is no longer the ordinary physical space. It is an abstract space with as many dimensions as the nature of the problem requires.

4.   Mapping of the space on itself.   Since the significance of the n generalized coordinates

$$q_1, q_2, \ldots, q_n \tag{14.1}$$

is not specified beyond the requirement that it shall allow a complete characterization of the system, we may choose another set of quantities

$$\bar{q}_1, \bar{q}_2, \ldots, \bar{q}_n \tag{14.2}$$

as generalized coordinates. There must, then, exist a functional relationship between the two sets of coordinates expressible in the form:

$$\bar{q}_1 = f_1(q_1, \ldots, q_n),$$
$$\cdots \cdots \cdots$$
$$\cdots \cdots \cdots \tag{14.3}$$
$$\bar{q}_n = f_n(q_1, \ldots, q_n).$$

The functions $f_1, f_2, \ldots, f_n$ must satisfy the ordinary regularity conditions. They must be finite, single valued, continuous and differentiable functions of the q_k, with a Jacobian Δ which is different from zero.

Differentiation of the equations (14.3) gives

$$
\begin{aligned}
d\bar{q}_1 &= \frac{\partial f_1}{\partial q_1} dq_1 + \ldots + \frac{\partial f_1}{\partial q_n} dq_n, \\
&\; \cdots\cdots\cdots \\
&\; \cdots\cdots\cdots \\
d\bar{q}_n &= \frac{\partial f_n}{\partial q_1} dq_1 + \ldots + \frac{\partial f_n}{\partial q_n} dq_n.
\end{aligned}
\tag{14.4}
$$

These equations show that no matter what functional relations exist between the two sets of coordinates, their *differentials* are always *linearly* dependent.

We can connect a definite geometrical picture with this "transformation of coordinates." If the q_i are plotted as rectangular coordinates of an n-dimensional space, the same can be done with the $\bar{q}_i$. We then think of the points of the q-space, and the points

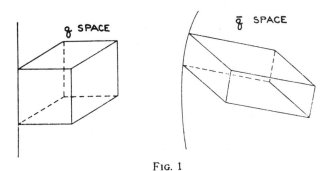

FIG. 1

of the $\bar{q}$-space. To a definite point P of the q-space corresponds a definite point $\bar{P}$ of the $\bar{q}$-space. For this reason a transformation of the form (14.3) is called a "point transformation." In a certain region the points of the q-space are in a one-to-one correspondence with the points of the q̄-space. We have a "mapping" of the n-dimensional space on itself. This mapping satisfies more than the general requirements of one-to-one correspondence. Continuity is preserved. The neighborhood of the point P is mapped on the neighborhood of $\bar{P}$ and vice versa. But even more can be said. A straight line of the q-space is no longer a straight line of the q̄-space. However, the relations get more and more regular as we consider smaller and smaller regions. In an *infinitesimal region* around P straight lines *are* mapped on straight lines and parallel lines remain parallel lines, although lengths and angles are not preserved. A small parallelepiped in the neighborhood of P is mapped on a small parallelepiped in the neighborhood of $\bar{P}$. The functional determinant Δ is just the determinant of the linear equations (14.4). Geometrically this determinant represents the ratio of the volume $\bar{\tau}$ of the new parallelepiped to the volume τ of the original one. The non-vanishing of Δ takes care of the condition that the entire neighborhood of P shall be mapped on the entire neighborhood of $\bar{P}$, which is indeed necessary if the one-to-one correspondence between the points of the two spaces shall be reversible. If Δ vanishes, an n-dimensional region around P is mapped into a region of smaller dimensionality around $\bar{P}$, with the consequence that some points around $\bar{P}$ have no image in the q-space while others have an infinity of images.

A physical realization of such a mapping of the space on itself can be found in the *motion of a fluid*. If the fluid particles are marked and a snapshot is taken at a certain instant, and then again at a later instant, the corresponding positions of the fluid particles represent a mapping of the space on itself. If we cut out a little fluid parallelepiped, the motion of the fluid will distort the angles and the lengths of the figure but it still remains a parallelepiped. Incidentally, if the fluid is incompressible, the volume of the parallelepiped will remain unchanged during the motion. The analytical picture of the motion of such a fluid is a coordinate transformation with a functional determinant which has everywhere the value 1.

Summary. In view of the arbitrariness of co-ordinates, one set of generalized coordinates can be changed to another set. This "transformation of coordinates" can be conceived geometrically as a mapping of an n-dimensional space on itself. The mapping does not preserve angles or distances. Straight lines are bent into curves. Yet in infinitesimal regions, in the neighborhood of a point, all mappings "straighten out": Straight lines are mapped on straight lines, parallel lines on parallel lines, and the ratio of volumes is preserved.

5. Kinetic energy and Riemannian geometry.[1] The use of arbitrary generalized coordinates in describing the motion of a mechanical system is one of the principal features of analytical mechanics. The equations of analytical mechanics are of such a structure that they can be stated in a form which is independent of the coordinates used. This property of the general equations of motion links up analytical mechanics with one of the major developments of nineteenth century mathematics: the theory of invariants and covariants. This theory came to maturity in our own day when Einstein's theory of relativity showed how the laws of nature are tied up with the problem of invariance. The principal programme of relativity is precisely that of *formulating the laws of nature independently of any special coordinate system.* The mathematical solution of this problem has shown that the matter is intimately linked with the Riemannian foundation of geometry. According to Einstein's general rela-

---

[1] The first discoverer of the intimate relation between dynamics and the geometry of curved spaces is Jacobi (1845) (cf. "Jacobi's principle," v). More recent investigations are due to Lipka, Berwald and Frank, Eisenhart, and others. The most exhaustive investigation of the subject, based on the systematic use of tensor calculus, is due to J. L. Synge, "On the geometry of dynamics" (*Philos. Transactions* (A) 226 (1927), p. 31).

tivity, the true geometry of nature is not of the simple Euclidean, but of the more general Riemannian, character; this geometry links space and time together as one unified four-dimensional manifold.

Descartes' discovery that geometry may be treated analytically was a landmark in the history of that subject. However, Descartes' geometry assumed the Euclidean structure of space. In order to establish a rectangular coordinate system, we must assume a geometry in which the congruence postulates and the parallel postulate hold.

A second landmark is the geometry of Riemann, which grew out of the ingenious investigations of Gauss concerning the intrinsic geometry of curved surfaces. Riemann's geometry is based on one single differential quantity called the "line element" $\overline{ds}$.[1] The significance of this quantity is *the distance between two neighboring points of space, expressed in terms of the coordinates and their differentials.* Let us consider for example the infinitesimal distance $\overline{ds}$ between two points P and P^1 with the coordinates x, y, z and $x + dx,\ y + dy, z + dz$. According to the Pythagorean theorem we have

$$\overline{ds}^2 = dx^2 + dy^2 + dz^2. \qquad (15.1)$$

This expression is a consequence of the Euclidean postulates and the definition of the coordinates x, y, z.

But let us assume now that we know nothing about any postulates, but are willing to accept (15.1) without proof as the expression for the line element. If we know in addition that the variables x, y, z vary between $-\infty$ and $+\infty$, we can then deduce all the laws of Euclidean geometry, including the interpretation of x, y, z as rectangular coordinates. Similarly, if the square of the line element happens to be given in the form

[1]Here we adopt the custom of denoting non-integrable differentials, which cannot be considered as the infinitesimal change of something, by putting a dash above the symbol. Thus dq_k or dt means the "d of q_k," or the "d of t," while $\overline{ds}$ for example, has to be conceived as an *infinitesimal expression* and not as the "d of s." Indeed, if $\overline{ds}$ were integrable, it would be impossible to find the shortest distance between two points because the length of any curve between those two points—depending on the initial and end-positions only—would be the same.

$$\overline{ds^2} = dx^2 + x^2 dy^2 + x^2 \sin^2 y \, dz^2, \tag{15.2}$$

where x varies[1] between 0 and ∞, y between 0 and π, z between 0 and 2π, the geometry established by this line element is once more Euclidean, but now the three variables x, y, z have the significance of *polar coordinates*, previously denoted by r, θ, ϕ. All this is contained in the single differential expression (15.2), augmented by the boundary conditions.

Generally, if the three coordinates of space are denoted by x_1, x_2, x_3 (instead of x, y, z), and they represent some arbitrary "curvilinear" coordinates—curvilinear because the coordinate lines of the reference system are no longer straight lines as in Descartes' scheme, but arbitrary curves—then the square of the line element appears in the following general form:

$$\begin{aligned}
\overline{ds^2} = &(g_{11}dx_1 + g_{12}dx_2 + g_{13}dx_3)dx_1 \\
+ &(g_{21}dx_1 + g_{22}dx_2 + g_{23}dx_3)dx_2 \\
+ &(g_{31}dx_1 + g_{32}dx_2 + g_{33}dx_3)dx_3.
\end{aligned} \tag{15.3}$$

An expression of this nature is called a "quadratic differential form" in the variables x_1, x_2, x_3. The "coefficients" of this form, $g_{11}, \ldots, g_{33}$ are not constants but functions of the three variables x_1, x_2, x_3. The sum (15.3) may be written in the more concise form:

$$\overline{ds^2} = \sum_{i,\,k=1}^{3} g_{ik}dx_i dx_k. \tag{15.4}$$

Since the terms $g_{ik}dx_i dx_k$ and $g_{ki}dx_i \, dx_k$ can be combined into one single term, the number of independent terms is not 9 but 6. We can cope with this reduction of terms by imposing the following condition on the coefficients:

$$g_{ki} = g_{ik} \tag{15.5}$$

In the "absolute calculus," which develops systematically the covariants and invariants of Riemannian geometry, the quantities g_{ik} form the components of a "tensor." The quantity $\overline{ds^2}$ has an absolute significance because the distance between two points does not change, no matter what coordinates are employed. It is an "absolute" or "invariant" quantity which is

[1]We have to add the boundary condition that the two points $(x,y,0)$ and $(x,y,2\pi)$ shall coincide for all x and y.

independent of any special reference system. A tensor is defined by the components of an invariant differential form. For example, an invariant differential form of the first order

$$\overline{dw} = F_1 dx_1 + F_2 dx_2 + \ldots + F_n dx_n, \qquad (15.6)$$

defines a vector. The quantities $F_1, F_2, \ldots, F_n$ are the "components" of the vector; they change with the reference system employed and are therefore "covariant" quantities. The whole differential form, however, is an invariant. Here is an abstract, purely analytical definition of a vector which dispenses entirely with the customary picture of an "arrow." A differential form of the second order defines a tensor of the second order, and so on.

The special tensor g_{ik} which lies at the foundation of geometry, is called the "metrical tensor." It allows the development of a complete geometry not only in three, but in any number of dimensions. The geometry of an n-dimensional space is determined by postulating the line element given by

$$\overline{ds^2} = \sum_{i,\,k=1}^{n} g_{ik} dx_i dx_k, \qquad (15.7)$$

with the additional condition

$$g_{ik} = g_{ki}, \qquad (15.8)$$

which makes the metrical tensor a "symmetric tensor." Einstein and Minkowski have shown that the geometry of nature includes space and time and thus forms a four-dimensional world, the four variables being x, y, z, t.

The quantities g_{ik} are generally not constants but functions of the variables $x_1, x_2, \ldots, x_n$. They are constants only if rectangular, or more generally "rectilinear," coordinates (with oblique instead of rectangular axes) are employed. For curvilinear coordinates the g_{ik} change from point to point. They are comparable to the components of a vector, except that they depend on *two* indices i and k and thus form a manifold of quantities which can be arranged in a two-dimensional instead of a one-dimensional scheme.

The fact that geometry can be established analytically and independently of any special reference system is only *one* of the merits of Riemannian geometry. Even more fundamental is the discovery by Riemann that the definition (15.7) of the line element gives not only a new, but a much more *general*, basis for building geometry than the older basis of Euclidean postulates. The g_{ik} have to belong to a *certain class of functions* in order to yield the Euclidean type of geometry. If the g_{ik} are not thus restricted, a new type of geometry emerges, characterized by two fundamental properties:

1. The properties of space change from point to point, but only in a *continuous* fashion.

2. Although Euclidean geometry does not hold in large regions, it holds in *infinitesimal* regions.

Riemann has shown how we may obtain by differentiation a characteristic quantity, the "curvature tensor," which tests the nature of the geometry. If all the components of the curvature tensor vanish, the geometry is Euclidean, otherwise not.

When the "special relativity" of Einstein and Minkowski combined time with space and showed that the geometry of nature comprises four rather than three dimensions, that geometry was still of the Euclidean type. But the "general relativity" of Einstein demonstrated that the line element of constant coefficients has to be replaced by a Riemannian line element with ten g_{ik} which are functions of the four coordinates x, y, z, t. The mysterious force of universal attraction was interpreted as a purely geometrical phenomenon—a consequence of the Riemannian structure of the space-time world.

The abstractions of Riemannian geometry are utilized not only in the theory of relativity, but also in analytical mechanics. The concepts of Riemannian geometry and the methods of the absolute calculus provide natural tools for dealing with the coordinate transformations encountered in the analytical treatment of dynamical problems.

Geometry enters the realm of mechanics in connection with the inertial quality of mass. This inertial quality is expressed on the left-hand side of the equation of Newton in the form "mass times acceleration," or "rate of change of momentum." The principles of analytical mechanics have shown that the really fundamental quantity which characterizes the inertia of mass is not the momentum but the kinetic energy. This kinetic energy is a scalar quantity and defined as $\frac{1}{2}mv^2$ for a single particle, and as

$$T = \tfrac{1}{2} \sum_{i=1}^{n} m_i v_i^2, \tag{15.9}$$

for a system of particles; v is here the velocity of the particle, defined by

$$v^2 = \left(\overline{\frac{ds}{dt}}\right)^2 = \frac{dx^2 + dy^2 + dz^2}{dt^2} \tag{15.10}$$

Thus the kinetic energy of a particle involves explicitly the line element $\overline{ds}$ and hence depends on the geometry of space.

Now let us define the line element of a $3N$-dimensional space by the equation

$$\overline{ds}^2 = 2T dt^2 = \sum_{i=1}^{N} m_i v_i^2 dt^2 \tag{15.11}$$

$$= \sum_{i=1}^{N} m_i (dx_i^2 + dy_i^2 + dz_i^2).$$

If we do that, then the total kinetic energy of the mechanical system may be written in the form

$$T = \tfrac{1}{2} m \left(\overline{\frac{ds}{dt}}\right)^2, \tag{15.12}$$

with

$$m = 1. \tag{15.13}$$

The meaning of this result is that *the kinetic energy of the whole system may be replaced by the kinetic energy of one single particle of mass* 1. This imaginary particle is a point of that $3N$-dimensional configuration space which symbolizes the position of the mechanical system. In this space one point is sufficient to represent the mechanical system, and hence we can carry over the mechanics of a free particle to any mechanical system if we place that particle in a space of the proper number of dimensions and proper geometry.

The form (15.11) of the line element shows that the $3N$-dimensional configuration space of N free particles has a Euclidean structure, and that the quantities

$$\sqrt{m}x_i, \ \sqrt{m}y_i, \ \sqrt{m}z_i \tag{15.14}$$

have to be considered as rectangular coordinates of that space.

If the rectangular coordinates (15.14) are changed to arbitrary curvilinear coordinates according to the transformation equations (12.4), the geometry remains Euclidean, although the line element is now given by the m re general Riemannian form (15.7) (with $n = 3N$).

But let us now consider a mechanical system with *given kinematical conditions* between the coordinates. We can handle such a system in two different ways. We may consider once more the previous configuration space of $3N$ dimensions but restrict the free movability of the C-point by the given kinematical conditions which take the form

$$f_1(x_1, \ldots, z_N) = 0,$$
$$\ldots\ldots\ldots\ldots\ldots \quad (15.15)$$
$$f_m(x_1, \ldots, z_N) = 0.$$

Geometrically each one of these restricting equations signifies a curved hyper-surface of the $3N$-dimensional space. The intersection of these hyper-surfaces determines a subspace of $3N - m = n$ dimensions, in which the C-point is forced to stay. This subspace is no longer a flat Euclidean but a curved Riemannian space.

Another way of attacking the same problem is to express from the very beginning the rectangular coordinates of the particles in terms of n parameters $q_1, \ldots, q_n$, as we have done in (12.8). These parameters are now the curvilinear coordinates of an n-dimensional space whose line element can be obtained by taking the differentials of each side of the equations (12.8) and substituting them in the expression (15.11). This results in a line element of the form

$$\overline{ds}^2 = \sum_{i, k = 1}^{n} a_{ik} dq_i dq_k. \quad (15.16)$$

The a_{ik} are here given functions of the q_i. The line element is now truly Riemannian, not only because the q_i are curvilinear coordinates but because the geometry of the configuration space does not preserve the Euclidean structure of the original $3N$-dimensional space, except in infinitesimal regions. For example, the motion of a perfectly balanced top (i.e. a top pivoted about its centre of gravity) can be conceived as the motion of the C-point in a certain curved Riemannian space of 3 dimensions, tracing out a shortest line of that space. Similarly the motion of a diatomic molecule can be conceived as the problem of tracing out a shortest line in a certain five-dimensional Riemannian space, and so on. *The mechanical problem is translated into a problem of differential geometry.*

Notice that in the first method of attack the n-dimensional configuration space is considered as imbedded in a Euclidean space of $3N$ dimensions. In the second method of attack the configuration space stands by itself, without being considered as imbedded in a higher-dimensional space.

Summary. The consideration of arbitrary coordinate systems and the invariance of the mechanical equations relative to arbitrary coordinate transformations links analytical mechanics intimately with the concepts and methods of Riemannian geometry. The motion of an arbitrary mechanical system can be studied as the motion of a free particle in a certain n-dimensional space of definite Riemannian structure. The kinetic energy of the system provides a definite Riemannian line element for the configuration space.

6. Holonomic and non-holonomic mechanical systems. Given kinematical conditions do not always show up as equations between the coordinates of the particles. Occasionally, conditions of a more general nature occur which can only be stated in infinitesimal form. A characteristic example is the rolling of a ball on a table. The ball, moving freely in space, has six degrees of freedom. Since the ball rests on the surface, the height of the centre is a given constant, which reduces the number of degrees of freedom to five. We may characterize the position of the ball by the two rectangular coordinates x and y of the point of contact, plus three angles α, β, γ which fix the position of the ball relative to its centre. If the ball can slide along the surface, it can make use of all of its five degrees of freedom. However, if it is confined to rolling, the point of contact has to be momentarily at rest. This requires that the instantaneous axis of rotation has to go through the point of contact. If the instantaneous axis is confined to some line which has to stay within the surface of the table, we have "pure rolling," otherwise "rolling and pivoting." Pure rolling cuts down the degrees of freedom to two. If the path of the point of contact is determined by giving x and y as functions of the time t, the condition of rolling determines the position of the ball at any time uniquely. This would suggest that perhaps the angles α, β, γ can be given as functions of x and y. This, however, is not possible. The *differentials* of α, β, γ are expressible in terms of the *differentials* of x and y, but these differential relations are not integrable. They cannot be changed into finite relations between the coordinates. Indeed, if the ball starts from a certain position and rolls along two different paths so that the final position of the piont of contact is the same in each case, the two final positions of the ball are rotated relative to each other. If the angles α, β, γ could be given as functions of x and y, the final positions of the ball would be the same in both cases.

Such kinematical conditions, which can be given only as relations between the differentials of the coordinates and not as finite relations between the coordinates themselves, were called by Hertz "non-holonomic" conditions, in contrast to the ordinary "holonomic" conditions. A kinematical condition of the form

$$f(q_1, \ldots, q_n) = 0, \qquad (16.1)$$

is holonomic. From this it follows by differentiation that

$$\frac{\partial f}{\partial q_1} dq_1 + \ldots + \frac{\partial f}{\partial q_n} dq_n = 0. \qquad (16.2)$$

However, if we start with a differential relation of the form

$$A_1 dq_1 + \ldots + A_n dq_n = 0, \qquad (16.3)$$

(with coefficients A_k which are given functions of the q_k) this relation is convertible into the form (16.1) only if certain integrability conditions are satisfied. The only exception is the case $n = 2$, because a differential relation between two variables is always integrable.

It is always possible to decide by mere differentiation and elimination whether set a given of differential conditions is holonomic or not. We show the procedure for the simplest case of one condition between three variables; the generalization to more variables and more than one equation follows automatically. At first we rewrite the given condition in the form:

$$dq_3 = B_1 \, dq_1 + B_2 \, dq_2. \tag{16.4}$$

If there is a finite relation between the variables q_1, q_2, q_3 on account of the given condition, we must have

$$\frac{\partial B_1}{\partial q_2} = \frac{\partial B_2}{\partial q_1} \tag{16.5}$$

This relation has to be reinterpreted in view of the fact that the coefficients B_1 and B_2 may depend implicitly on q_3. We thus replace (16.5) by

$$\frac{\partial B_1}{\partial q_2} + \frac{\partial B_1}{\partial q_3}\frac{\partial q_3}{\partial q_2} = \frac{\partial B_2}{\partial q_1} + \frac{\partial B_2}{\partial q_3}\frac{\partial q_3}{\partial q_1}\, ; \tag{16.6}$$

the quantities $\partial q_3/\partial q_1$ and $\partial q_3/\partial q_2$ are given by (16.4), and we get

$$\frac{\partial B_1}{\partial q_2} + \frac{\partial B_1}{\partial q_3} B_2 = \frac{\partial B_2}{\partial q_1} + \frac{\partial B_2}{\partial q_3} B_1. \tag{16.7}$$

This relation has to be satisfied for all values of q_1 and q_2, which means that it must be an identity. If that is the case, the relation (16.4) is integrable, otherwise not. However, it may happen that q_3 does not drop out from the resulting equation and can be expressed in terms of q_1 and q_2. With this solution for q_3 we then go back to (16.4) and test whether the partial derivatives of q_3 with respect to q_1 and q_2 agree with B_1 and B_2 or not. If they do, we have proved that the given differential relation is holonomic and we have replaced it by a finite relation between the coordinates. In the case of more than two independent variables all the integrability conditions

$$\frac{\partial B_i}{\partial q_k} = \frac{\partial B_k}{\partial q_i}, \tag{16.8}$$

have to be tested in a similar manner. If a whole system of independent differential relations is given, we first solve them for a suitable set of dependent differentials and then proceed in a similar way with testing the integrability conditions.

Problem. Investigate the integrability of the following differential relation:

$$x \, dz + (y^2 - x^2 - z) \, dx + (z - y^2 - xy) \, dy = 0. \tag{16.9}$$

(This condition is holonomic and can be replaced by the finite relation

$$z = x^2 - xy + y^2.) \tag{16.10}$$

Holonomic kinematical conditions can be attacked in *two* ways. If there are m equations between n variables, we can eliminate m of these and reduce the problem to $n - m$ independent variables. Or we can operate with a surplus number of variables and retain the given relations as auxiliary conditions.

Non-holonomic conditions necessitate the *second* form of treatment. A reduction in variables is here not possible and we have to operate with more variables than the degrees of freedom of the system demand. The configuration space is here imbedded in a higher-dimensional space but without forming a definite subspace of it, because the kinematical conditions prescribe certain pencils of directions, but these directions do not envelop any surface.

From the viewpoint of the variational principles of mechanics, holonomic and non-holonomic conditions display a noticeably different behavior. Although the equations of motion can be extended to the case of non-holonomic auxiliary conditions, yet these equations cease to be derivable from the principle that the variation of a certain quantity is equated to zero. (Chap. II, section 13.)

Summary. Kinematical conditions do not always take the form of definite relations between the coordinates. If they do, we call them "holonomic conditions." But occasionally the condition can be stated only as a relation between the differentials of the coordinates with no chance of changing it into a finite relation. Such kinematical conditions are called "nonholonomic conditions." The rolling of a solid body on a surface is an example.

7. Work function and generalized force. The two sides of the Newtonian equation of motion correspond to two fundamentally different aspects of mechanical problems. On the left side we have the inertial quality of mass. This is absorbed by the kinetic energy in the analytical treatment. The right side of the equation, the "moving force," describes the dynamical behavior of an external field in its action on the particle. Although we are inclined to believe that force is something primitive and irreducible, the analytical treatment of mechanics shows that it is not the force but the *work* done by the force which is of primary importance, while the force is a secondary quantity derived from the work.

Let us assume that each particle of mass m_i and rectangular coordinates x_i, y_i, z_i is acted on by a force G_i with components

$$X_i, \ Y_i, \ Z_i. \tag{17.1}$$

These forces come either from an external field or from the mutual interaction of the particles, but they are not to include the forces which maintain the given kinematical conditions. The analytical treatment of mechanics does not require a knowledge of the latter forces.

Let us change the coordinates of each one of the particles by arbitrary infinitesimal amounts dx_i, dy_i, dz_i. The total work of all the impressed forces is

$$\overline{dw} = \sum_{i=1}^{N} (X_i dx_i + Y_i dy_i + Z_i dz_i). \tag{17.2}$$

If now the rectangular coordinates are expressed as functions of the generalized coordinates $q_1, \ldots, q_n$ according to equations (12.8), the differentials $dx_i \ dy_i$, dz_i can be expressed in terms of the dq_i, and the infinitesimal work $\overline{dw}$ comes out as a linear differential form of the variables q_i:

$$\overline{dw} = F_1 dq_1 + F_2 dq_2 + \ldots + F_n dq_n. \tag{17.3}$$

This differential form is of prime importance for the analytical treatment of force. The original force-components are of no importance, but only the coefficients

$$F_1, \ F_2, \ldots, \ F_n \tag{17.4}$$

of the differential form (17.3). They take the place of the components (17.1) of the single forces which act on each particle. The quantities (17.4) determine completely the dynamical action of all the forces. These quantities form the components of a vector, but it is a vector of the n-dimensional configuration space and not one of ordinary space. We call this vector the "generalized force," and the quantities F_i the "components of the generalized force."

Just as the motion of a mechanical system can be symbolized as the motion of a single particle in an n-dimensional Riemannian space—the inertia of the entire system being included in the kinetic energy of that symbolic particle—the dynamical action of all the forces can likewise be represented by a single vector acting on that particle. This vector has n components, in agreement with the n dimensions of the configuration space. They are analytically defined as the coefficients of an invariant differential form of the first order which gives the total work of all the impressed forces for an arbitrary infinitesimal change of the position of the system.

Now the forces acting on a mechanical system fall automatically into two categories. It is possible that all we can say about the work $\overline{dw}$ is that it is a differential form of the first order. But it is also possible that $\overline{dw}$ turns out to be the true differential of a certain function. The latter case is the usual one in analytical problems.

Let us assume that the infinitesimal work $\overline{dw}$ is the true differential of a certain function so that the dash above dw may be omitted. That function is called the "work function" and is frequently denoted by the letter U. Hence we put

$$\overline{dw} = dU, \tag{17.5}$$

where

$$U = U(q_1, q_2, \ldots, q_n). \tag{17.6}$$

We thus have

$$\Sigma F_i dq_i = \Sigma \frac{\partial U}{\partial q_i} dq_i, \tag{17.7}$$

which gives

$$F_i = \frac{\partial U}{\partial q_i}. \tag{17.8}$$

The present practice is to use the *negative* work function and call this quantity V:

$$V = -U. \tag{17.9}$$

The advantage of this change of sign is that in view of the law of the conservation of energy V can be interpreted as the "potential energy" of the system. The equation (17.8) can now be rewritten in the form

$$F_i = -\frac{\partial V}{\partial q_i}, \tag{17.10}$$

where V is a function of the position coordinates:

$$V = V(q_1, q_2, \ldots, q_n). \tag{17.11}$$

Forces of this category are remarkable for two reasons. Firstly, they satisfy the law of the conservation of energy; for this reason such forces are called "conservative forces." Secondly, although the generalized force has n components, they can be calculated from *a single scalar function*, U. The variational treatment of mechanics takes account of the latter property only, while the requirement that the conservation of energy shall be satisfied, is irrelevant.

The definition of a work function on the basis of the equation (17.6) is too restricted. We have forces in nature which are derivable from a *time-dependent* work function $U = U(q_1, q_2, \ldots, q_n, t)$. Then the equation (17.5) still holds, with the understanding that in forming the differential dU the time t is considered as a constant. The equations (17.7) and (17.8) remain valid, but the conservation of energy is lost. The generalized force possesses a work function *without* being conservative. An electrically charged particle revolving in a cyclotron returns to the same point with increased kinetic energy, so that energy is not conserved. This is not because the work function does not exist, but because the work function is time-dependent. On the other hand, a generalized force may have no work function and still satisfy the conservation of energy, as for example the force which maintains rolling.

It seems desirable to have a distinctive name for forces which are derivable from a scalar quantity, irrespective of whether they are conservative or not. This book adopts the name "monogenic" (which means "single-generated") as a distinguishing name for this category of forces, while forces which are not derivable from a scalar function — such as friction — can be called "polygenic."

The work function associated with a monogenic force is in the most general case a function of the coordinates *and the velocities*:

$$U = U(q_1, \ldots, q_n; \dot{q}_1, \ldots, \dot{q}_n; t). \tag{17.12}$$

For example, the electro-magnetic force of Lorentz, which acts on a charged particle in the presence of an electric and a magnetic field, is derivable from a work function of this kind. Such forces are still susceptible to the variational treatment. As the differential equation of Euler-Lagrange shows (see later, chap. II, section 10), the connection between force and the work function is now given by the equation

$$F_i = \frac{\partial U}{\partial q_i} - \frac{d}{dt} \frac{\partial U}{\partial \dot{q}_i}, \tag{17.13}$$

which is a generalization of the simpler relation (17.8).

Summary. The quantity of prime importance for the analytical treatment of mechanics is not force, but the *work* done by the impressed forces for arbitrary infinitesimal displacements. Particularly important for the variational treatment are forces which are derivable from a single scalar quantity, the work function U. Such forces may be called "monogenic." If the work function is time-independent, we obtain a class of forces called "conservative" because they satisfy the conservation of energy. In the frequent case where the work function is not only time- but also velocity-independent, the negative of the work function can be interpreted as the potential energy, and the force becomes the negative gradient of the potential energy. Forces which are not derivable from a work function can still be characterized by the work done in a small displacement, but they do not yield to the general minimizing procedures of analytical mechanics.

8. Scleronomic and rheonomic systems. The law of the conservation of energy. In all our previous considerations we have not taken into account the most characteristic variable of all dynamical problems: the time t. The procedures of analytical mechanics are decisively influenced by the explicit presence or absence of t in the basic scalars of mechanics. All quantities in mechanics are functions of the time; the question is, however, whether the time appears *explicitly* in the kinetic energy or the work function.

In our earlier discussion (16) we have assumed that a holonomical kinematical condition takes the form of a given relation between the coordinates of the mechanical system. It might happen that this relation changes continuously with the time t, so that the equation which expresses the kinematical condition takes the form

$$f(q_1, \ldots, q_n, t) = 0. \tag{18.1}$$

This is, for example, the situation if a mass point moves on a surface which itself is moving according to a given law. The equation of that surface takes the form

$$f(x, y, z, t) = 0. \tag{18.2}$$

Or consider a pendulum whose length is being constantly changed by pulling the thread in a definite manner. This again amounts to an auxiliary condition which contains the time explicitly.

Boltzmann used the terms "rheonomic" and "scleronomic" as distinguishing names for kinematical conditions which do or do not involve the time t. If rheonomic conditions are present among the given kinematical conditions, the elimination of these conditions by a proper choice of curvilinear coordinates will have the consequence that the equations (12.8) will contain t explicitly:

$$x_1 = f_1(q_1, \ldots, q_n, t),$$
$$\cdots \cdots \cdots \cdots$$
$$\cdots \cdots \cdots \cdots \tag{18.3}$$
$$z_N = f_{3N}(q_1, \ldots, q_n, t).$$

A similar situation arises even without time-dependent kinematical conditions, if the coordinates chosen belong to a reference system *which is in motion*.

Rheonomic systems are within the scope of the analytical approach, but we lose certain characteristic consequences which hold only for scleronomic systems. In the first place, differentiation of the equations (18.3) with respect to t yields:

$$\dot{x}_1 = \frac{\partial f_1}{\partial q_1}\,\dot{q}_1 + \ldots + \frac{\partial f_1}{\partial q_n}\,\dot{q}_n + \frac{\partial f_1}{\partial t},$$

$$\cdots \cdots \cdots \cdots \cdots \cdots \cdots$$
$$\cdots \cdots \cdots \cdots \cdots \cdots \cdots \qquad (18.4)$$

$$\dot{z}_N = \frac{\partial f_{3N}}{\partial q_1}\,\dot{q}_1 + \ldots + \frac{\partial f_{3N}}{\partial q_n}\,\dot{q}_n + \frac{\partial f_{3N}}{\partial t}.$$

If these expressions are substituted in the definition (15.9) of the kinetic energy, it will not come out as a purely quadratic form of the generalized velocities $\dot{q}_i$; we obtain additional new terms which are linear in the velocities, and others which are independent of the velocities. The Riemannian geometry of the configuration space ceases to play the role that it has played before.

Again, the time may enter explicitly in the work function U. Whether the coefficients of the kinetic energy or the work function contain the time explicitly or not, the analytical consequences are the same and the system becomes rheonomic in both cases. The essential difference between a scleronomic and a rheonomic system is that in a scleronomic system—as we shall prove later—a fundamental quantity, interpreted as the "total energy" of the system, remains constant during the motion. The total energy is equal to the sum of the kinetic and the potential energy, provided that the potential energy V of the mechanical system is defined as follows:

$$V = \Sigma\,\frac{\partial U}{\partial \dot{q}_i}\,\dot{q}_i - U. \qquad (18.5)$$

In the frequent case where the work function is independent of the velocities, the potential energy becomes simply the negative work function, and we can consider the kinetic energy T and the potential energy V as the two basic scalars of mechanics, dispensing with the work function U. In the general case (18.5), however, we have to retain the work function U, but if U is time-dependent, a reasonable definition of the "potential energy" as a second form of energy which compensates for the loss or gain of kinetic energy is no longer possible.

The really fundamental quantity of analytical mechanics is not the potential energy but the work function, although the physicist and the engineer are more familiar with the former concept. In all cases where we mention the potential energy, it is tacitly assumed that the work function has the special form (17.6), together with the connection $V = -U$.

Rheonomic systems do not satisfy any conservation law but scleronomic systems do. For this reason scleronomic systems are frequently referred to as "conservative systems."

Summary. It is possible that the two basic quantities of mechanics, the kinetic energy and the work function, contain the time t explicitly. This will happen if some of the given kinematical conditions are time-dependent. It will also happen if the work function is an explicit function of the time as well as of the position coordinates (and perhaps of velocities). If both the kinetic energy and the work function are scleronomic, i.e. time-independent, the equations of motion yield a fundamental theorem called the law of the conservation of energy. If either kinetic energy or work function or both are rheonomic, i.e. time-dependent, such a conservation law cannot be found.

CHAPTER II

THE CALCULUS OF VARIATIONS

1. The general nature of extremum problems. Extremum problems have aroused the interest and curiosity of all ages. Our walk in a straight line is the instinctive solution of an extremum problem: we want to reach the end point of our destination with as little detour as possible. The proverbial "path of least resistance" is another acknowledgment of our instinctive desire for minimum solutions. The same desire expresses itself in the public interest attached to record achievements, to do something which "cannot be beaten."

Mathematically we speak of an "extremum problem" whenever the largest or smallest possible value of a quantity is involved. For example we may want to find the highest peak of a mountain or the lowest point of a valley, or the shortest path between two points, or the largest volume of a container cut out of a piece of metal, or the smallest possible waste in a problem of lighting or heating, and many other problems. For the solution of such problems a special branch of mathematics, called the "calculus of variations," has been developed.

From the formal viewpoint the problem of minimizing a definite integral is considered as the proper domain of the calculus of variations, while the problem of minimizing a function belongs to ordinary calculus. Historically, the two problems arose simultaneously and a clear-cut distinction was not made till the time of Lagrange, who developed the technique of the calculus of variations. The famous problem of Dido, well known to the ancient geometers, was a variational problem which involved the minimization of an integral. Hero of Alexandria derived the law of reflection from the principle that a light ray which is emitted at the point A and proceeds to a certain point B after reflection from a mirror, reaches its destination in the shortest possible time. Fermat extended this principle to derive the law of refraction. All these problems were solved by employing essentially geometrical methods. The problem of the "brachistochrone"—the curve of quickest descent—was proposed by John Bernoulli and solved independently by him, Newton and

Leibniz. The basic differential equation of variational problems was discovered by Euler and Lagrange. A general method for the solution of variational problems was introduced by Lagrange in his *Mécanique Analytique*
(1788).

Before we proceed to develop the formal tools of the calculus
of variations, we shall consider a simple but typical example
which can well serve to elucidate the general nature of extremum
problems. Let us suppose that we wish to find the highest point
of a mountain. The height of the mountain can be analytically
represented by the equation

$$z = f(x, y), \tag{21.1}$$

where $f(x, y)$ will be considered as a continuous and differentiable
function of x and y.

Now the question of a maximum or minimum involves by its
very nature a *comparison*. If we claim to be on the top of a
mountain, we have to show that all nearby points are below us.
Here we encounter the first characteristic restriction of an extremum problem. There may be other peaks still higher in more
distant parts of the mountain. We are satisfied if our own height
is established as a maximum in comparison with the *immediate*
surroundings, even if it is not established as a maximum in relation to arbitrarily wide surroundings. We thus speak of a *local*
maximum (or minimum) in contrast to an *absolute* maximum (or
minimum).

However, even this search for a relative extremum can be
further restricted by first exploring the *infinitesimal* (i.e., arbitrarily small) neighbourhood of our position. It is obvious that
on the peak of a mountain all points of the infinitesimal neighbourhood must have the same height (to the first order), which
means that *the rate of change of the height must be zero in whatever
direction we proceed.* Indeed, a positive rate of change in some
direction would mean that there are points in our neighbourhood
whose height exceeds our own. On the other hand, if the rate of
change is negative in a certain direction, we can proceed in the
opposite direction—provided that every direction can be reversed
(a very essential condition)—and thus make the rate of change
positive.

We see that both positive and negative values of the rate of change of the function have to be excluded if the relative maximum of a function is required at a certain point. This leads to the important principle that the maximum of a given function at a certain point requires that the *rate of change of the function in every possible direction from that point must be zero.*

However, this condition in itself is not enough to guarantee a maximum. It is possible that we are at the level point of a saddle surface, which yields a minimum with respect to some directions and a maximum with respect to some others. We thus see that the vanishing of the rate of change in every possible direction is a necessary but by no means a sufficient condition for an extremum. Subject to this condition, an additional investigation is required which will decide whether a maximum or a minimum is realized or whether we are at a saddle point without any extremum value.

However, even the fact that we are at a point where the rate of change of the function vanishes in every possible direction is of interest in itself. Points where this happens are exceptional points, no matter whether the further condition of a maximum or minimum is satisfied or not. For this reason a special name has been coined for such points. We say that a *function has a "stationary value" at a certain point if the rate of change of the function in every possible direction from that point vanishes.*

It so happens that the problems of motion require only the *stationary value* of a certain integral. Under these circumstances there is a noticeable difference between the calculus of variations as a science of pure mathematics, and the application of that calculus to problems of mechanics. From the purely mathematical standpoint the problem of a stationary value is of minor interest. After establishing the criteria for a stationary value, we go on and seek the additional criteria for a true extremum. For the variational principles of mechanics, however, this latter investigation is of interest only if stability problems are involved —when we look for the actual minimum of the potential energy. Problems of motion are not influenced by the specific minimum conditions.

Summary. The problem of finding the position of a point at which a function has a relative maximum or minimum requires the exploration of the infinitesimal neighbourhood of that point. This exploration must show that the function has a stationary value at the point in question. Although this requirement cannot guarantee an extremum without further conditions, yet we need not go further for the general aims of dynamics because problems of motion require merely the stationary value, and not necessarily the minimum, of a certain definite integral.

2. The stationary value of a function. We consider a function of an arbitrary number of variables:

$$F = F(u_1, u_2, \ldots, u_n) \tag{22.1}$$

These variables can be pictured as the rectangular coordinates of a point P in a space of n dimensions. If we plot the value of the function along one additional dimension, we obtain a surface in a space of $n + 1$ dimensions. We assume that F is a continuous and differentiable function of the variables u_k.

Let us now translate the expression: "the exploration of the infinitesimal neighbourhood of a point" into exact mathematical language. It is this process to which the name "variation" is attached. A "variation" means an infinitesimal change, in analogy with the d-process of ordinary calculus. However, contrary to the ordinary d-process, this infinitesimal change is not caused by the *actual* change of an independent variable, but is imposed by us on a set of variables as a kind of *mathematical experiment.* Let us consider for example a marble which is at rest at the lowest point of a bowl. The actual displacement of the marble is zero. It is our desire, however, to bring the marble to a neighbouring position in order to see how the potential energy changes. A displacement of this nature is called a "virtual displacement." The term "virtual" indicates that the displacement was intentionally made in any kinematically admissible manner. Such a virtual and infinitesimal change of position is called briefly a "variation" of the position. The corresponding change of the given function F—which is in our example the potential energy of the marble—is determined by this variation. We have the variation of the position at our disposal, but the corresponding change of the function—called the "variation of the function"—is not at our disposal.

It was Lagrange's ingenious idea to introduce a special symbol for the process of variation, in order to emphasize its virtual character. This symbol is δ. The analogy to d brings to mind that both symbols refer to *infinitesimal changes*. However, d refers to an *actual*, δ to a *virtual change*. Since in problems involving the variation of definite integrals both types of change have to be considered simultaneously, the distinction is of vital importance.

In accordance with this notation we write the infinitesimal virtual changes of our coordinates in the form

$$\delta u_1, \delta u_2, \ldots, \delta u_n. \tag{22.2}$$

The corresponding change of the function F becomes by the rules of elementary calculus

$$\delta F = \frac{\partial F}{\partial u_1} \delta u_1 + \frac{\partial F}{\partial u_2} \delta u_2 + \ldots + \frac{\partial F}{\partial u_n} \delta u_n. \tag{22.3}$$

This expression is called the "first variation" of the function F.

In order to operate with finite rather than infinitesimal quantities, we put:

$$\delta u_1 = \epsilon a_1, \delta u_2 = \epsilon a_2, \ldots, \delta u_n = \epsilon a_n, \tag{22.4}$$

denoting by $a_1, a_2, \ldots, a_n$ the direction cosines of the virtual direction in which we have proceeded, while ϵ is a parameter which tends toward zero.

The rate of change of the function F in the specified direction now becomes

$$\frac{\delta F}{\epsilon} = \frac{\partial F}{\partial u_1} a_1 + \frac{\partial F}{\partial u_2} a_2 + \ldots + \frac{\partial F}{\partial u_n} a_n. \tag{22.5}$$

In order that F shall have a stationary value, this quantity has to vanish:

$$\sum_{k=1}^{n} \frac{\partial F}{\partial u_k} a_k = 0. \qquad (22.6)$$

But now the "virtual" nature of our displacement implies that we may proceed in any direction we like, so that the a_k are arbitrary and hence

$$\frac{\partial F}{\partial u_k} = 0, \quad (k=1, 2, \ldots, n). \qquad (22.7)$$

Conversely, if the equations (22.7) are satisfied, the quantity (22.5) vanishes and thus F has a stationary value. Hence the equations (22.7) are both necessary and sufficient.

Notice that the equations (22.7) determine the *position* of a stationary value of a function rather than the stationary value itself. If we have found n values $u_1, u_2, \ldots, u_n$ which satisfy the n equations (22.7) we can substitute these values in the expression for F and thus obtain the stationary value of the function.

Summary. The necessary and sufficient condition that a function F of n variables shall have a stationary value at a certain point P is that the n partial derivatives of F with respect to all the n variables shall vanish at that point P.

3. The second variation. Let us once more determine the infinitesimal change of a function due to a virtual variation of the coordinates. Writing the variation of the coordinates once more in the form (22.4), we have to evaluate the quantity

$$\Delta F = F(u_1 + \epsilon a_1, u_2 + \epsilon a_2, \ldots, u_n + \epsilon a_n)$$
$$- F(u_1, u_2, \ldots, u_n). \qquad (23.1)$$

We consider this quantity as a function of the parameter ϵ and expand it in powers of ϵ:

$$\Delta F = \epsilon \sum_{k=1}^{n} \frac{\partial F}{\partial u_k} a_k + \frac{1}{2} \epsilon^2 \sum_{i, k=1}^{n} \frac{\partial^2 F}{\partial u_i \partial u_k} a_i a_k + \ldots \qquad (23.2)$$

Now if we are at a point where F has a stationary value, the first sum vanishes and ΔF starts with the second order terms.

If ϵ is chosen sufficiently small, we can neglect the terms of order higher than the second and write

$$\Delta F = \frac{1}{2}\,\delta^2 F, \tag{23.3}$$

where

$$\delta^2 F = \epsilon^2 \sum_{i,\,k=1}^{n} \frac{\partial^2 F}{\partial u_i \partial u_k}\, a_i a_k. \tag{23.4}$$

This expression is called the "second variation" of F. If it so happens that $\delta^2 F$ is always positive no matter what values we choose for the direction cosines a_i—except that the sum of their squares has to equal 1—then F is *increasing* in every possible direction from P and we have a real minimum at that point. If $\delta^2 F$ is always negative, no matter what values we choose for the a_i, then F is *decreasing* in every possible direction from P and we have a real maximum at that point. If $\delta^2 F$ is positive in some directions and negative in others, then we have neither a maximum nor a minimum. Hence it is the *sign of the second variation* which enables us to determine the existence of an extremum value. We can write down the equation

$$\delta^2 F = 0, \tag{23.5}$$

and see whether we can find real values a_i which satisfy it. If this is the case, then $\delta^2 F$ can change sign and no extremum exists. If, on the other hand, the equation (23.5) has no real solution for the a_i, then we have an extremum value because $\delta^2 F$ cannot change sign. (The further decision whether we have a minimum or maximum is reached by evaluating $\delta^2 F$ for an *arbitrary* set of a_i and recording whether the result is positive or negative.)

We shall deal with the second variation of a function in greater detail later when we discuss the small vibrations near a state of equilibrium (cf. chap. v, section 10). We then find more exact criteria for deciding between a maximum or minimum. It is worth mentioning, however, that occasionally the investigation of the second variation is superfluous because the nature of the problem may be such that the existence of, say, a minimum is *a priori* established. If for example a function is to be minimized which is composed of purely positive terms, we know in advance that this function must have somewhere a smallest value. Hence, if it so happens that the condition for a stationary value has *one* definite solution, that solution must automatically yield a minimum (cf. chap. IV, section 8).

Summary. After satisfying the conditions for a stationary value, the further criterion for an extremum depends on the sign of the second variation. An extremum requires that the sign of the second variation shall remain the same for any possible infinitesimal virtual displacement—positive for a minimum, negative for a maximum. If the sign of the second variation is positive for some displacements and negative for others, then the stationary value of a function does not lead to an extremum.

4. Stationary value versus extremum value. We should keep well in mind the difference between stationary value and extremum and the mutual relation of these two problems. A stationary value requires solely the vanishing of the first variation, without any restriction on the second variation. An extremum requires the vanishing of the first variation *plus* further conditions on the second variation. Moreover, we have investigated the question of an extremum under the condition that we are *inside* the boundaries of the configuration space. A function which does not assume any extremum inside a certain region may well assume it on the boundary of that region. On the boundary of the configuration space the displacements are no longer reversible and hence our argument that the first variation must vanish—because otherwise it can be made positive as well as negative—no longer holds. For non-reversible displacements a function may well assume an extremum *without* having a stationary value at that point. In that case an extremum exists without the vanishing of the first variation. If a ball rolls down a groove, it will come to equilibrium at the lowest point of the groove where the tangent to the path is horizontal. But the ball may be stopped earlier by a peg which prevents its further descent. The ball is now in its lowest position although the tangent is not zero and the height has no stationary value. That condition is no longer required because the ball arrived at the limit of the available configuration space and there the variation of the position is not reversible.

Summary. The extremum of a function requires a stationary value only for reversible displacements. On the boundary of the configuration space, where the variation of the position is not reversible, an extremum is possible without a stationary value.

5. Auxiliary conditions. The Lagrangian λ-method. The problem of minimizing a function does not always present itself in the form considered above. The configuration space in which the point P can move may be restricted to less than n dimensions by certain kinematical relations which exist between the coordinates. Such kinematical conditions are called "auxiliary conditions" of the given variation problem. If such conditions do not exist and the variables $u_1, \ldots, u_n$ can be varied without restriction, we have a "free" variation problem, as considered previously (sections 1-3).

We shall now investigate the variation of the function

$$F = F(u_1, u_2, \ldots, u_n), \qquad (25.1)$$

with the auxiliary condition

$$f(u_1, u_2, \ldots, u_n) = 0. \qquad (25.2)$$

Our first thought would be to eliminate one of the u_k—for example u_n—from the auxiliary condition, expressing it in terms of the other u_k. Then our function would depend on the n-1 unrestricted variables $u_1, \ldots, u_{n-1}$ and could be handled as a free variation problem. This method is entirely justified and sometimes advisable. But frequently the elimination is a rather cumbersome procedure. Moreover, the condition (25.2) may be symmetric in the variables $u_1, \ldots, u_n$ and there would be no reason why one of the variables should be artificially designated as dependent, the others as independent variables.

Lagrange devised a beautiful method for handling auxiliary conditions, the "method of the undetermined multiplier," which preserves the symmetry of the variables without eliminations, and still reduces the problem to one of free variation. The method works quite generally for any number of auxiliary conditions and is applicable even to non-holonomic conditions which are given as non-integrable relations between the differentials of the variables, and not as relations between the variables themselves.

In order to understand the nature of the Lagrangian multiplier method, we start with a single auxiliary condition, given in the form (25.2). Taking the variation of this equation we obtain the following relation between the δu_k:

$$\delta f = \frac{\partial f}{\partial u_1} \delta u_1 + \ldots + \frac{\partial f}{\partial u_n} \delta u_n = 0 ; \qquad (25.3)$$

while the fact that the variation of F has to vanish at a stationary value, gives

$$\delta F = \frac{\partial F}{\partial u_1} \delta u_1 + \ldots + \frac{\partial F}{\partial u_n} \delta u_n = 0. \qquad (25.4)$$

We know from section 3 that the condition (25.4) would lead to the vanishing of each $\frac{\partial F}{\partial u_k}$ if the δu_k were all independent of each other. This, however, is not the case, because of the condition (25.3). We agree to eliminate δu_n in terms of the other variations—assuming that $\frac{\partial f}{\partial u_n}$ is not zero at the point P—and then consider the other δu_k as free variations. But before we do so, we shall modify the expression (25.4). It is obviously permissible to multiply the left-hand side of (25.3) by some undetermined factor λ, which is a function of $u_1, \ldots, u_n$, and add it to δF. This does not change the value of δF at all since we have added zero. Hence it is still true that:

$$\frac{\partial F}{\partial u_1} \delta u_1 + \ldots + \frac{\partial F}{\partial u_n} u \delta_n +$$
$$\lambda \left(\frac{\partial f}{\partial u_1} \delta u_1 + \ldots + \frac{\partial f}{\partial u_n} \delta u_n \right) = 0. \qquad (25.5)$$

This move is not trivial because, although we have added zero, we have actually added a *sum*; the individual terms of the sum are not zero, only the total sum is zero.

We write (25.5) in the form

$$\sum_{k=1}^{n} \left(\frac{\partial F}{\partial u_k} + \lambda \frac{\partial f}{\partial u_k} \right) \delta u_k = 0. \tag{25.6}$$

We wish to eliminate δu_n. But now we can choose λ so that *the factor multiplying δu_n shall vanish*:

$$\frac{\partial F}{\partial u_n} + \lambda \frac{\partial f}{\partial u_n} = 0. \tag{25.7}$$

This dispenses with the task of eliminating δu_n. After that our sum is reduced to only n-1 terms:

$$\sum_{k=1}^{n-1} \left(\frac{\partial F}{\partial u_k} + \lambda \frac{\partial f}{\partial u_k} \right) \delta u_k = 0, \tag{25.8}$$

and since only those δu_k remain which can be chosen arbitrarily, the conditions of a free variation problem are applicable. These require that the coefficient of each δu_k shall vanish:

$$\frac{\partial F}{\partial u_k} + \lambda \frac{\partial f}{\partial u_k} = 0, \ (k = 1, 2, \ldots, n-1). \tag{25.9}$$

The conditions (25.9), combined with the condition (25.7) on λ, lead to the conclusion that *each coefficient of the sum* (25.6) *vanishes, just as if all the variations δu_k were free variations*. The result of Lagrange's "method of the undetermined multiplier" can be formulated thus: instead of considering the vanishing of δF, consider the vanishing of

$$\delta F + \lambda \delta f, \tag{25.10}$$

and drop the auxiliary condition, handling the u_k as free, independent variables.

We have $\qquad \delta F + \lambda \delta f = \delta(F + \lambda f), \tag{25.11}$

since the term $f \delta \lambda$ vanishes on account of the auxiliary condition $f = 0$. Hence we can express the result of our deductions in an even more striking form. Instead of putting the first variation of F equal to zero, modify the function F to

$$\overline{F} = F + \lambda f, \tag{25.12}$$

and put its first variation equal to zero, for *arbitrary* variations of the u_k.

We generalize this λ-method for the case of an arbitrary number of auxiliary conditions. Let us assume once more that the stationary value of F is sought, but under m independent restricting conditions:

$$f_1\ (u_1, u_2, \ldots, u_n) = 0,$$
$$\cdot \quad \cdot \quad \cdot \quad \cdot \quad \cdot \quad \cdot \quad \cdot \quad \cdot$$
$$\cdot \quad \cdot \quad \cdot \quad \cdot \quad \cdot \quad \cdot \quad \cdot \quad \cdot \quad (m < n). \qquad (25.13)$$
$$\cdot \quad \cdot \quad \cdot \quad \cdot \quad \cdot \quad \cdot \quad \cdot \quad \cdot$$
$$f_m(u_1, u_2, \ldots, u_n) = 0.$$

These auxiliary conditions establish the following relations between the variations δu_k:

$$\delta f_1 = \frac{\partial f_1}{\partial u_1}\ \delta u_1 + \ldots + \frac{\partial f_1}{\partial u_n}\ \delta u_n = 0,$$
$$\cdot \quad \cdot \quad \cdot \quad \cdot \quad \cdot \quad \cdot \quad \cdot \quad \cdot \quad \cdot \quad \cdot$$
$$\cdot \quad \cdot \quad \cdot \quad \cdot \quad \cdot \quad \cdot \quad \cdot \quad \cdot \quad \cdot \quad \cdot \qquad (25.14)$$
$$\cdot \quad \cdot \quad \cdot \quad \cdot \quad \cdot \quad \cdot \quad \cdot \quad \cdot \quad \cdot$$
$$\delta f_m = \frac{\partial f_m}{\partial u_1}\ \delta u_1 + \ldots + \frac{\partial f_m}{\partial u_n}\ \delta u_n = 0.$$

Because of these conditions m of the δu_k can be designated as dependent variables and expressed in terms of the others. We shall consider the *last m* of the u_k as dependent, the first $n - m$ as independent, variables.

Now the given variational problem requires the vanishing of

$$\delta F = \sum_{k=1}^{n} \frac{\partial F}{\partial u_k}\ \delta u_k \qquad (25.15)$$

for all possible variations δu_k which satisfy the given auxiliary conditions. We should express the last m δu_k in terms of the independent δu_k. However, before doing so let us modify the expression (25.15) by adding the left-hand sides of the equations (25.14) after multiplying each one by some undetermined λ-factor. We thus get

$$\sum_{k=1}^{n} \left(\frac{\partial F}{\partial u_k} + \lambda_1 \frac{\partial f_1}{\partial u_k} + \ldots + \lambda_m \frac{\partial f_m}{\partial u_k} \right) \delta u_k = 0. \qquad (25.16)$$

Now the elimination of the last m δu_k can be accomplished by the proper choice of the λ-factors, so that

$$\frac{\partial F}{\partial u_k} + \lambda_1 \frac{\partial f_1}{\partial u_k} + \ldots + \lambda_m \frac{\partial f_m}{\partial u_k} = 0, \quad (k = n - m + 1, \ldots, n). \quad (25.17)$$

This leaves

$$\sum_{k=1}^{n-m} \left(\frac{\partial F}{\partial u_k} + \lambda_1 \frac{\partial f_1}{\partial u_k} + \ldots + \lambda_m \frac{\partial f_m}{\partial u_k} \right) \delta u_k = 0. \quad (25.18)$$

But all the δu_k which remain in (25.18) are *free* variations. Hence the coefficient of each δu_k must vanish separately. In the final analysis we have the equations

$$\frac{\partial F}{\partial u_k} + \lambda_1 \frac{\partial f_1}{\partial u_k} + \ldots + \lambda_m \frac{\partial f_m}{\partial u_k} = 0, \quad (k = 1, \ldots, n), \quad (25.19)$$

which can be considered as obtained from the variational principle

$$\delta F + \lambda_1 \delta f_1 + \ldots + \lambda_m \delta f_m = 0, \quad (25.20)$$

considering *all* the u_k as independent variables. Thus in the final result the distinction between dependent and independent variables disappears.

Equation (25.20) can be stated even more strikingly by writing it in the form

$$\delta(F + \lambda_1 f_1 + \ldots + \lambda_m f_m) = 0, \quad (25.21)$$

and interpreting this equation as follows: *Instead of asking for the stationary value of F, we ask for the stationary value of the modified function*

$$\overline{F} = F + \lambda f_1 + \ldots + \lambda_m f_m, \quad (25.22)$$

dropping the auxiliary conditions and handling this as a free variation problem. This yields n equations. In addition to these equations we have to satisfy the m auxiliary conditions (25.13). This gives $n + m$ equations for the $n + m$ unknowns

$$u_1, u_2, \ldots, u_n; \lambda_1, \lambda_2, \ldots, \lambda_m. \quad (25.23)$$

The ingenious multiplier-method of Lagrange changes a problem of $n - m$ degrees of freedom to a problem of $n + m$

degrees of freedom. If we add to the n variables u_k the m quantities λ_i as additional variables and ask for the stationary value of the function $\bar{F}$, this variation problem gives the same n equations as we had before if we vary with respect to the u_k, while the variations of the λ_i give the m additional conditions

$$f_1 = 0, \ldots, f_m = 0. \tag{25.24}$$

These are exactly the given auxiliary conditions, but now obtained *a posteriori,* on account of the variation problem.

The method of Lagrange permits the use of surplus coordinates—a great convenience in many considerations of mechanics. It preserves the full symmetry of all coordinates by making it unnecessary to distinguish between dependent and independent variables.

Summary. The Lagrangian-multiplier method reduces a variation problem with auxiliary conditions to a free variation problem without auxiliary conditions. We modify the given function F, which is to be made stationary, by adding the left-hand sides of the auxiliary conditions, after multiplying each by an undetermined factor λ. Then we handle the modified problem as a free variation problem. The resulting conditions, together with the given auxiliary conditions, determine the unknowns and the λ-factors.

6. Non-holonomic auxiliary conditions. As was pointed out in chap. I, section 6, the restrictions on the mechanical variables of a problem may be given in a differential instead of a finite form. We then have a variation problem with non-holonomic auxiliary conditions. The equations (25.13) do not exist in this case, but we have relations analogous to the *differentiated* forms (25.14) of the auxiliary conditions. The only difference is that the left-hand sides of these equations are no longer exact differentials but merely infinitesimal quantities. We can write the non-holonomic conditions in the following form:

$$\overline{\delta f}_1 = A_{11}\delta u_1 + A_{12}\delta u_2 + \ldots + A_{1n}\delta u_n = 0,$$

$$\cdots \cdots \cdots \cdots \cdots \cdots \quad (26.1)$$

$$\overline{\delta f}_m = A_{m1}\delta u_1 + A_{m2}\delta u_2 + \ldots + A_{mn}\delta u_n = 0.$$

Here the A_{ik} are given functions of the u_i which cannot be considered as the partial derivatives of a function f_i.

Non-holonomic conditions cannot be handled by the elimination method, because the equations for eliminating some variables as dependent variables do not exist. The Lagrangian λ-method, however, is again available. By exactly the same procedure as before, we can obtain an equation analogous to (25.20), namely:

$$\delta F + \lambda_1 \overline{\delta f}_1 + \ldots + \lambda_m \overline{\delta f}_m = 0 ; \quad (26.2)$$

and again all the δu_k are handled as free variations. The only difference lies in the fact that we cannot proceed to the equation (25.21) and have to be content with the differential formulation of the procedure. The reduction of a conditioned variation problem to a free variation problem is once more accomplished.

Summary. The Lagrangian λ-method is applicable even to non-holonomic conditions. We multiply the left sides of these conditions by some undetermined λ-factors and add them to the variation of the function F which is to be made stationary. This whole expression is put equal to zero, considering all the variations δu_k as free variations.

---

7. **The stationary value of a definite integral.** The analytical problems of motion involve a special type of extremum problem: the stationary value of a *definite integral.* The branch of mathematics dealing with problems of this nature is called the Calculus of Variations. A typical problem of this kind is that of the brachistochrone (the curve of quickest descent), first formulated and solved by John Bernoulli (1696); it is one of the earliest instances of a variational problem. We wish to find a

suitable plane curve along which a particle descends in the shortest possible time, starting from A and arriving at B. If

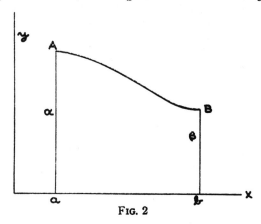

FIG. 2

the unknown curve is analytically characterized by the equation

$$y = f(x), \tag{27.1}$$

the time which must be minimized, is given by the following definite integral:

$$t = \frac{1}{\sqrt{2g}} \int_a^b \frac{\sqrt{1 + y'^2}}{\sqrt{a - y}}\, dx. \tag{27.2}$$

Here y is an unknown function of x which is to be determined. Amongst all possible $y = f(x)$ we want to find that particular $f(x)$ which yields the smallest possible value of t. The general conditions are that y must be a continuous and differentiable function of x, with a continuous tangent. Moreover, since the two end points of the unknown curve $y = f(x)$ are given, y has to satisfy the following boundary conditions:

$$f(a) = a, \quad f(b) = \beta. \tag{27.3}$$

There is one and only one function $f(x)$ which satisfies these conditions.

The type of problem we encounter here can be more generally characterized as follows: we are given a function F of three variables:

$$F = F(y, y', x) \tag{27.4}$$

(in the above example F happens to be independent of x, but it is unnecessary to make that restriction); and given the definite integral

$$I = \int_a^b F(y, y', x)dx; \qquad (27.5)$$

we are given also the boundary conditions

$$f(a) = \alpha, \quad f(b) = \beta. \qquad (27.6)$$

The problem is to find a function

$$y = f(x) \qquad (27.7)$$

—restricted by the customary regularity conditions—which will make the integral I an extremum, or at least give it a stationary value.

At first sight this problem appears utterly different from the previous problem where we dealt with the extremum or stationary value of a function $F(u_1, \ldots, u_n)$ of a set of variables. Instead of a *function*, a *definite integral* must be minimized. Moreover, instead of a *set of variables* $u_1, \ldots, u_n$ we have a certain *unknown function* $y = f(x)$ at our disposal. Yet closer inspection reveals that the mathematical nature of this new problem is not substantially different from that of the previous problem.

Let us make use of the concept of the "function space," originated by Hilbert. The arbitrary function $y = f(x)$—provided that it satisfies certain general continuity requirements—can be expanded in an infinite Fourier series throughout the given range between a and b:

$$f(x) = \tfrac{1}{2} a_0 + a_1 \cos \xi + a_2 \cos 2\xi + \ldots$$
$$+ b_1 \sin \xi + b_2 \sin 2\xi + \ldots, \qquad (27.8)$$

with
$$\xi = \frac{2\pi}{b-a} \left(x - \frac{a+b}{2} \right).$$

The coefficients of this expansion are uniquely determined. Then we can associate a definite set of coefficients $a_0, a_1, \ldots, a_n; b_1, \ldots, b_n$ with any function $y = f(x)$, provided we choose n large enough to make the remainder of the expansion sufficiently small. We now plot these coefficients as rectangular coordinates of a point P in a $(2n + 1)$-dimensional space. In this picture an arbitrary function is mapped as a definite point of a space of many dimensions; the value of the integral I associated with this $f(x)$ can be plotted perpendicularly as an additional dimension. We thus return to the picture of a *surface in a many-dimensional space*. A small change of the function $f(x)$

means a small variation of the position of the point P. The problem of finding the function $f(x)$ which minimizes the definite integral I is translated into the problem of finding the coordinates of the deepest point of a surface in a space of $2n + 2$ dimensions. This is exactly the problem we have studied in the previous sections of this chapter.

Euler has shown how the problem can be solved by elementary means, without resorting to the tools of a specific calculus. We make use of the fact that a definite integral can be replaced by a sum of an increasing number of terms. Moreover, the derivative can be replaced by a difference coefficient. The errors we commit in this procedure can be made as small as we wish.

In accordance with the customary procedure of the calculus, we divide the interval between $x = a$ and $x = b$ into many equal small intervals, obtaining a set of abscissa values:

$$x_0 = a, \ x_1, x_2, \ldots, x_n, x_{n+1} = b, \tag{27.9}$$

and the corresponding ordinates

$$y_0 = a, \ y_1, y_2, \ldots, y_n, y_{n+1} = \beta, \tag{27.10}$$

where

$$y_k = f(x_k). \tag{27.11}$$

We then replace the derivative $f'(x_k)$ by the difference coefficient

$$z_k = \left(\frac{\Delta y}{\Delta x}\right)_{x = x_k} = \frac{y_{k+1} - y_k}{x_{k+1} - x_k}, \tag{27.12}$$

and the integral (27.5) by the sum

$$S = \sum_{k=0}^{n} F(y_k, z_k, x_k) \ (x_{k+1} - x_k). \tag{27.13}$$

This involves a certain error which, however, tends toward zero as each one of the intervals $\Delta x_k = x_{k+1} - x_k$ tends toward zero.

We replace the original integral by the sum (27.13) and ask for the stationary value of this sum. This new problem is of the customary type. We have a given function S of the n variables $y_1, \ldots, y_n$ which take the place of our earlier $u_1, \ldots, u_n$. We know that this problem is solved by setting the partial derivatives of S with respect to y_k equal to zero. Then we investigate what happens to these conditions as Δx_k tends toward zero.

Before carrying out this program we alter the expression (27.13) to an extent which becomes negligible in the limit. Since

y_k and y_{k+1} are arbitrarily near to each other, it is permissible to change $F(y_k, z_k, x_k)$ to $F(y_{k+1}, z_k, x_k)$. Hence the function we want to minimize is finally defined as follows:

$$S' = \sum_{j=0}^{n} F(y_{j+1}, z_j, x_j)(x_{j+1} - x_j). \qquad (27.14)$$

Now, in forming the partial derivative of S' with respect to, say, y_{k+1}, we have to bear in mind that y_{k+1} appears in the sum S' in *two* neighbouring terms, namely, those for which $j = k$ and $j = k + 1$, in view of the definition (27.12) of z_{k+1}. Partial differentiation with respect to y_{k+1} gives therefore

$$\frac{\partial S'}{\partial y_{k+1}} = \left(\frac{\partial F}{\partial y}\right)_{x=x_k} (x_{k+1} - x_k) +$$
$$\left(\frac{\partial F}{\partial y'}\right)_{x=x_k} - \left(\frac{\partial F}{\partial y'}\right)_{x=x_{k+1}} \qquad (27.15)$$

Dividing by $\Delta x_k = x_{k+1} - x_k$, this equation can be written as follows:

$$\left[\frac{\partial F}{\partial y} - \frac{\Delta}{\Delta x}\left(\frac{\partial F}{\partial y'}\right)\right]_{x=x_k} = 0, \ (k = 0, 1, 2, \ldots, n - 1). \qquad (27.16)$$

Here we have the necessary, and also sufficient conditions for the sum S' to be stationary. It is important to note that the two limiting ordinates y_0 and y_{n+1} are *given quantities* and thus remain unvaried. Should they also be unknown, we would get two boundary conditions in addition to the equations (27.16).

In the limit, when Δx decreases toward zero, the difference equation (27.16) becomes a *differential equation*. Moreover, since the points x_k come arbitrarily near to any point of the interval (a, b), that differential equation has to hold for the entire interval:

$$\frac{\partial F}{\partial y} - \frac{d}{dx}\left(\frac{\partial F}{\partial y'}\right) = 0, \ (a \leqq x \leqq b). \qquad (27.17)$$

This fundamental equation was discovered independently by Euler and Lagrange and is usually called the Euler-Lagrange differential equation. We notice that this differential equation is derivable by elementary means, as the condition that the sum which replaces the integral originally given shall be stationary.

This method of deriving the basic differential equation of the calculus of variations, essentially due to Euler, is objectionable from the rigorous point of view since it makes use of a double limit process in a fashion which is not necessarily admissible. The direct method of Lagrange which we shall discuss in the next chapter is free from this objection.

Summary. The problem of minimizing a definite integral which contains an unknown function and its derivative can be reduced to the elementary problem of minimizing a function of many variables. For this purpose the integral is replaced by a sum and the derivative by a difference coefficient. The condition for the vanishing of the first variation takes the form of a difference equation which in the limit goes over into the differential equation of Euler and Lagrange.

8. The fundamental processes of the calculus of variations. Lagrange realized that the problem of minimizing a definite integral requires specific tools, different from those of the ordinary calculus. With the help of these tools we can attack the problem directly, instead of reverting to the limiting process by which Euler obtained the solution of the problem.

Consider the function $y = f(x)$ which by hypothesis gives a stationary value to the definite integral (27.5). In order to prove that we *do* have a stationary value, we have to evaluate the same integral for a slightly *modified* function $y = \overline{f(x)}$ and show that the rate of change of the integral due to the change in the function becomes zero.

Now the modified function $\overline{f(x)}$ can obviously be written in the form

$$\overline{f(x)} = f(x) + \epsilon\phi(x), \tag{28.1}$$

where $\phi(x)$ is some arbitrary new function which satisfies the same general continuity conditions as $\overline{f(x)}$. Hence $\phi(x)$ has to be continuous and differentiable.

It is obvious that the selection of the hypothetical function $f(x)$ which solves our variation problem has to be made from the class of continuous functions which can be differentiated at least once, because otherwise the integral $F(y, y', x)$ would have no meaning. However, we want to restrict our $f(x)$ by the further condition that even $f''(x)$ exists throughout the range. (A similar assumption for $\phi(x)$ is not required.)

Making use of the variable parameter ϵ we have it in our power to modify the function $f(x)$ by arbitrarily small amounts. For that purpose we let ϵ decrease toward zero.

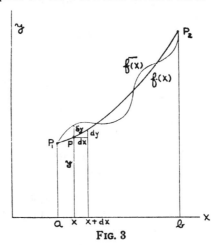

FIG. 3

We now compare the values of the modified function $\overline{f(x)}$ with the values of the original function $f(x)$ *at a certain definite point x* of the independent variable, by forming the difference between $\overline{f(x)}$ and $f(x)$. This difference is called the "variation" of the function $f(x)$ and is denoted by δy:

$$\delta y = \overline{f(x)} - f(x) = \epsilon \phi(x). \tag{28.2}$$

The variation of a function—in full analogy with the variation of the position of a point, studied before—is characterized by two fundamental features. It is an *infinitesimal* change, since the parameter ϵ decreases toward zero. Moreover, it is a *virtual* change, that means we make it in any arbitrary manner we please. Hence $\phi(x)$ is a function chosen arbitrarily, so long as the general continuity conditions are satisfied.

Note the fundamental difference between δy and dy. Both are infinitesimal changes of the function y. However the dy

refers to the infinitesimal change of the given function $f(x)$ caused by the infinitesimal change dx of the independent variable, while δy is an infinitesimal change of y which produces a *new function* $y + \delta y$.

It is in the nature of the process of variation that only the dependent function y should be varied while the variation of x serves no useful purposes. Hence we agree that we shall always put

$$\delta x = 0. \tag{28.3}$$

Moreover, if the two limiting ordinates $f(a)$ and $f(b)$ of the function $f(x)$ are prescribed, these two ordinates cannot be varied, which means

$$\begin{aligned}
[\delta f(x)]_{x=a} &= 0, \\
[\delta f(x)]_{x=b} &= 0.
\end{aligned} \tag{28.4}$$

We then speak of a "variation between definite limits."

Summary. The calculus of variations considers a virtual infinitesimal change of a function $y = f(x)$. The variation δy refers to an arbitrary infinitesimal change of the value of y, *at the point x*. The independent variable x does not participate in the process of variation.

9. The commutative properties of the δ-process. The variation of the function $f(x)$ defines an entirely new function $\epsilon\phi(x)$. We can take the derivative of this function. On the other hand, we can take the derivative of the new function $\overline{f(x)}$ and the derivative of the original function $f(x)$. The difference of these two derivatives can naturally be called the variation of the derivative $f'(x)$. In the first case we have "the derivative of the variation":

$$\frac{d}{dx}\,\delta y = \frac{d}{dx}\,[\overline{f(x)} - f(x)] = \frac{d}{dx}\,\epsilon\phi(x) = \epsilon\phi'(x). \tag{29.1}$$

In the second case we have "the variation of the derivative":

$$\delta \frac{d}{dx} f(x) = \overline{f'(x)} - f'(x) = (y' + \epsilon\phi') - y' = \epsilon\phi'(x). \qquad (29.2)$$

This gives
$$\frac{d}{dx} \delta y = \delta \frac{d}{dx} y. \qquad (29.3)$$

This shows that *the derivative of the variation is equal to the variation of the derivative.*

In a similar way we may be interested in the variation of a definite integral. This means that we take the definite integral evaluated for the modified integrand minus the definite integral evaluated for the original integrand:

$$\delta \int_a^b F(x)dx = \int_a^b \overline{F(x)}dx - \int_a^b F(x)dx$$
$$= \int_a^b [\overline{F(x)} - F(x)]\,dx = \int_a^b \delta F(x)dx. \qquad (29.4)$$

Hence
$$\delta \int_a^b F(x)dx = \int_a^b \delta F(x)dx. \qquad (29.5)$$

This shows that *the variation of a definite integral is equal to the definite integral of the variation.*

Summary. The δ-process reveals two characteristic properties:

(*a*) Variation and differentiation are permutable processes.

(*b*) Variation and integration are permutable processes.

---

10. The stationary value of a definite integral treated by the calculus of variations. We consider once more the problem of section 7, but this time treated by the direct methods of the calculus of variations. Given the definite integral (27.5), with the boundary conditions (27.6), the stationary value of this integral is to be found.

In order to solve this problem we investigate the rate of change of the given integral, caused by the variation of the function $y = f(x)$. We start out with the variation of the integrand $F(y, y', x)$ itself, caused by the variation of y (remembering that F is a *given function* of the three variables y, y', x and this functional dependence is *not altered* by the process of variation):

$$\delta F(y, y', x) = F(y + \epsilon\phi, y' + \epsilon\phi', x) - F(y, y', x)$$

$$= \epsilon\left(\frac{\partial F}{\partial y}\,\phi + \frac{\partial F}{\partial y'}\,\phi'\right). \qquad (210.1)$$

The higher order terms of the Taylor development can be neglected since ϵ approaches zero.

We are now in a position to evaluate the variation of the definite integral (27.5):

$$\delta\int_a^b F\,dx = \int_a^b \delta F\,dx = \epsilon\int_a^b\left(\frac{\partial F}{\partial y}\,\phi + \frac{\partial F}{\partial y'}\,\phi'\right)dx. \qquad (210.2)$$

In order to get the "rate of change" of our integral, we have to divide by the infinitesimal parameter ϵ, as we did in section 2 when the stationary value of an ordinary function $F(u_1, \ldots, u_n)$ was involved. Hence the quantity which has to vanish for a stationary value of the integral I is

$$\frac{\delta I}{\epsilon} = \int_a^b\left(\frac{\partial F}{\partial y}\,\phi + \frac{\partial F}{\partial y'}\,\phi'\right)dx. \qquad (210.3)$$

This expression is not accessible to further analysis in its present form, because $\phi(x)$ and $\phi'(x)$ are not independent of each other, although their relation cannot be formulated in any algebraic form. The difficulty can be removed by an ingenious application of the method of integration by parts. We can transform the second part of (210.3) as follows:

$$\int_a^b\frac{\partial F}{\partial y'}\,\phi'dx = \left[\frac{\partial F}{\partial y'}\,\phi\right]_a^b - \int_a^b\frac{d}{dx}\left(\frac{\partial F}{\partial y'}\right)\cdot\phi dx. \qquad (210.4)$$

The first term drops out since we vary between definite limits so that $\phi(x)$ vanishes at the two end-points a and b. Making use of this transformation, equation (210.3) becomes

$$\frac{\delta I}{\epsilon} = \int_a^b\left(\frac{\partial F}{\partial y} - \frac{d}{dx}\frac{\partial F}{\partial y'}\right)\phi dx. \qquad (210.5)$$

For the sake of brevity we introduce the following notation:

$$E(x) = \frac{\partial F}{\partial y} - \frac{d}{dx}\frac{\partial F}{\partial y'},\qquad(210.6)$$

and write the condition for a stationary value of I in the form

$$\int_a^b E(x)\phi(x)dx = 0.\qquad(210.7)$$

Now it is not difficult to see that this integral can vanish for *arbitrary* functions $\phi(x)$ only if $E(x)$ vanishes everywhere between a and b. Indeed, we may arrange that the function $\phi(x)$ shall vanish everywhere with the exception of an arbitrarily small interval around the point $x = \xi$. But within this interval $E(x)$ is practically constant and can be put in front of the integral sign[1]

$$\frac{\delta I}{\epsilon} = E(\xi)\int_{\xi-\rho}^{\xi+\rho}\phi(x)dx.\qquad(210.8)$$

The error we have made tends to zero as ρ tends to zero. Since the second integral is at our disposal and need not vanish, the vanishing of $\delta I/\epsilon$ requires the vanishing of the first factor. The point $x = \xi$ may be chosen as *any* point of the interval between a and b. Hence we obtain for the entire interval between a and b the differential equation

$$\frac{\partial F}{\partial y} - \frac{d}{dx}\frac{\partial F}{\partial y'} = 0.\qquad(210.9)$$

This condition is *necessary* for the vanishing of δI. On the other hand, it is also *sufficient*, because, if the integrand of (210.5) vanishes the integral also vanishes. *The differential equation (210.9) is thus the necessary and sufficient condition that the definite integral I shall be stationary under the given boundary conditions (27.6).*

Problem. Consider the definite integral

$$I = \int_a^b F(y, y', y'', x)\, dx,\qquad(210.10)$$

which contains the first *and* second derivatives of y. Find the condition for a stationary value by forming the variation of the integral and applying the method of integration by parts *twice*. Derive the differential equation

$$\frac{\partial F}{\partial y} - \frac{d}{dx}\frac{\partial F}{\partial y'} + \frac{d^2}{dx^2}\frac{\partial F}{\partial y''} = 0,\qquad(210.11)$$

[1]We assume that $E(x)$ is a *continuous* function of x.

and the boundary term

$$\delta I = \left[\left(\frac{\partial F}{\partial y'} - \frac{d}{dx} \frac{\partial F}{\partial y''} \right) \delta y + \frac{\partial F}{\partial y''} \delta y' \right]_a^b, \qquad (210.12)$$

which vanishes if y and y' are prescribed at the two end-points.

Summary. The necessary and sufficient condition for the integral

$$\int_a^b F(y, y', x) dx$$

to be stationary, with the boundary conditions

$$y(a) = a, \quad y(b) = \beta,$$

is that the differential equation of Euler-Lagrange

$$\frac{\partial F}{\partial y} - \frac{d}{dx} \frac{\partial F}{\partial y'} = 0.$$

shall be satisfied.

---

11.   The Euler-Lagrange differential equations for n degrees of freedom.   In mechanics the problem of variation presents itself in the following form.   Find the stationary value of a definite integral

$$I = \int_{t_1}^{t_2} L(q_1, \ldots, q_n; \dot{q}_1, \ldots, \dot{q}_n; t) dt, \qquad (211.1)$$

with the boundary conditions that the q_k are given (and thus their variation is zero) at the two end-points t_1 and t_2:

$$[\delta q_k(t)]_{t=t_1} = 0, \quad [\delta q_k(t)]_{t=t_2} = 0. \qquad (211.2)$$

The $q_1, \ldots, q_n$ are unknown functions of t, to be determined by the condition that the actual motion shall make the integral I stationary:

$$\delta I = 0, \qquad (211.3)$$

for arbitrary independent variations of the q_k, subject only to the boundary conditions (211.2).

Now we can obviously select one definite q_k and vary it all by itself, leaving the other q_i unchanged.   Hence we can apply the differential equation (210.9) to our present problem, after adapting it to the present notation. The y corresponds to q_k, the y' to $\dot{q}_k$, the independent variable x is now the time t. The function F is denoted by L and the limits of integration are t_1 and t_2. Hence we get

$$\frac{\partial L}{\partial q_k} - \frac{d}{dt} \frac{\partial L}{\partial \dot{q}_k} = 0, \ (t_1 \leqq t \leqq t_2). \tag{211.4}$$

These equations have to hold for each separate q_k, k running from 1 to n.

The variations we have employed so far are but *special* variations, and the question arises whether a simultaneous variation of *all* the q_k would not bring in additional conditions. This is actually not the case, on account of the superposition principle of infinitesimal processes. Let us denote by $\delta_k I$ the variation of I produced by varying q_k alone. Then the simultaneous variation of all the q_k produces the following resultant variation:

$$\delta I = \delta_1 I + \delta_2 I + \ldots + \delta_n I. \tag{211.5}$$

Now the differential equation (211.4) guarantees the vanishing of $\delta_k I$. If that differential equation holds for *all* indices k the sum (211.5) vanishes, and thus δI is zero for *arbitrary* variations of the q_k.

The problem of finding the stationary value of I for arbitrary variations of the q_k between definite limits is thus solved. The conditions for the stationary value of I come out in the form of the following *system of simultaneous differential equations*:

$$\frac{d}{dt} \frac{\partial L}{\partial \dot{q}_k} - \frac{\partial L}{\partial q_k} = 0, \ (k = 1, 2, \ldots, n). \tag{211.6}$$

They are called "the differential equations of Euler and Lagrange," or also, if applied to problems of mechanics, "the Lagrangian equations of motion."

With the exception of the singular case in which the function L depends on some or all the $\dot{q}_k$ in a *linear* way, the partial derivatives $\dfrac{\partial L}{\partial \dot{q}_k}$ will contain all the $\dot{q}_k$, so that differentiation with respect to t brings in all the second derivatives $\ddot{q}_k$. We can solve the equations for the $\ddot{q}_k$ algebraically and thus rewrite the differential equations (211.6) in the following explicit form:

$$\ddot{q}_k = \phi_k(q_1 \ldots, q_n, \dot{q}_1, \ldots, \dot{q}_n, t). \qquad (211.7)$$

The integration of such a system of differential equations of the second order involves $2n$ constants of integration, so that the complete solution of the equations (211.7) may be written as follows:

$$q_k = q_k(A_1, \ldots, A_n; B_1, \ldots, B_n; t). \qquad (211.8)$$

The constants of integration A_k and B_k can be adjusted to the given boundary conditions. Our variational problem requires variation between *definite limits*, which means that the coordinates q_k are given at $t = t_1$ and $t = t_2$. These are $2n$ boundary conditions which can be satisfied by the proper choice of the constants A_k and B_k. The nature of mechanical problems is such that more frequently *initial conditions* take the place of boundary conditions. The freedom of $2n$ constants of integration allows all the initial position coordinates and velocities to be prescribed arbitrarily.

Summary. If a definite integral is given which contains not one but n unknown functions, to be determined by the condition that the integral be stationary, we can vary these functions independently of each other. Thus the Euler-Lagrange differential equation can be formed for each function separately. This gives n simultaneous differential equations of the second order. The solution of these differential equations determines the n unknown functions in terms of the independent variable and $2n$ constants of integration.

12. Variation with auxiliary conditions. We consider once more the problem of the previous paragraph, with the modification that the variables $q_1, \ldots, q_n$ shall not be independent, but restricted by given auxiliary conditions. These conditions will take the form of certain functional relations between the q_k:

$$f_1(q_1, \ldots, q_n, t) = 0,$$

. . . .

. . . .   (212.1)

. . . .

$$f_m(q_1, \ldots, q_n, t) = 0.$$

It is possible to eliminate m of the q_k in terms of the other variables and thus reduce the problem to $n-m$ degrees of freedom. After the reduction the differential equations of Euler-Lagrange come into play. However, this elimination may be rather cumbersome; moreover, the conditions between the variables may be of a form which makes the distinction between dependent and independent variables artificial. Here again the method of the Lagrangian multiplier, studied before in section 5, gives an adequate solution of the problem.

Variation of the equations (212.1) gives:

$$\delta f_1 = \frac{\partial f_1}{\partial q_1} \delta q_1 + \ldots + \frac{\partial f_1}{\partial q_n} \delta q_n = 0,$$

. . . .

. . . .   (212.2)

. . . .

$$\delta f_m = \frac{\partial f_m}{\partial q_1} \delta q_1 + \ldots + \frac{\partial f_m}{\partial q_n} \delta q_n = 0.$$

These equations hold at any time t. According to the principle of the Lagrangian multiplier we multiply each one of these equations by an undetermined factor λ_i. Since the auxiliary conditions are prescribed for every value of the independent variable t, the λ-factors have also to be applied for every value of t, which makes them functions of t. Moreover, the summation over all the auxiliary conditions—after multiplication by λ_i—amounts to an integration with respect to the time t. Thus the method of the Lagrangian multiplier appears here in the following form: instead of putting the variation of the given integral I equal to zero, modify it in the following fashion:

$$\delta I' = \delta \int_{t_1}^{t_2} L dt + \int_{t_1}^{t_2} (\lambda_1 \delta f_1 + \ldots + \lambda_m \delta f_m) dt = 0. \qquad (212.3)$$

This is obviously permissible since what we have actually added to the variation of I is zero. Now in the first term we perform the standard integration by parts and reduce it to the form

$$\delta \int_{t_1}^{t_2} L dt = \int_{t_1}^{t_2} [E_1(t)\delta q_1 + \ldots + E_n(t)\delta q_n] \, dt. \qquad (212.4)$$

We now unite the two integrands of (212.3) and collect the terms which belong to a certain δq_k. We should eliminate the last $(n - m)\delta q_k$ with the help of the equations (212.2), but we can avoid this step by choosing the λ_i in such a manner that the coefficients of these δq_k shall vanish. The remaining δq_i, on the other hand, can be chosen arbitrarily and hence our previous conclusion that the coefficients of these δq_i must vanish individually throughout the range becomes valid again. But then we see that in the final analysis the coefficient of *each* individual δq_k vanishes, irrespective of whether a certain q_k was designated as a dependent or an independent variable. The resulting equations are:

$$\frac{\partial L}{\partial q_k} - \frac{d}{dt} \frac{\partial L}{\partial \dot{q}_k} + \lambda_1 \frac{\partial f_1}{\partial q_k} + \ldots + \lambda_m \frac{\partial f_m}{\partial q_k} = 0. \qquad (212.5)$$

This is equivalent to the following variational problem. Instead of considering the variation of the definite integral

$$I = \int_{t_1}^{t_2} L dt, \qquad (212.6)$$

with the auxiliary conditions (212.1), consider the variation of the modified integral

$$I' = \int_{t_1}^{t_2} L' dt, \qquad (212.7)$$

where

$$L' = L + \lambda_1 f_1 + \ldots + \lambda_m f_m, \qquad (212.8)$$

and drop the auxiliary conditions.

Here again, as in the general procedure, the λ_i have to be considered as constants relative to the process of variation. But we can also include them in the variational problem as additional unknown functions of t. The variation of the λ_i gives then the auxiliary conditions (212.1) *a posteriori*.

The solution of the differential equations (212.5) has to be completed by satisfying the m auxiliary conditions (212.1). This determines the λ_i as functions of t. In the matter of initial conditions we can choose arbitrarily only $n - m$ position coordinates and $n - m$ velocities, because the remaining q_i and $\dot{q}_i$ are determined by the auxiliary conditions. This is in accord with the fact that the given system represents a mechanical system of $n - m$ degrees of freedom and the use of surplus coordinates is merely a matter of mathematical convenience.

Summary. The Lagrangian λ-method in treating auxiliary conditions dispenses with the elimination of surplus variables and takes the auxiliary conditions into account without reducing the number of variables. The integrand L of the given variational problem is modified by adding the left-hand sides of the given auxiliary conditions, each one multiplied by an undetermined factor λ. Then the problem is treated as a problem of free variation. The λ-factors are determined afterwards as functions of t, by satisfying the given auxiliary conditions.

13. Non-holonomic conditions. If the auxiliary conditions of the variational problem are not given as algebraic relations between the variables, but as differential relations—cf. chap. I, section 6 and chap. II, section 6—the Lagrangian λ-method is still applicable. We again get the equations (212.5) with the only modification that the $\dfrac{\partial f_i}{\partial q_k}$ are replaced by the coefficients A_{ik} of the non-holonomic conditions (26.1). A difference exists in the matter of initial conditions. The coordinates q_i are at present not restricted by any conditions, only their differentials. Hence the initial values of all the q_i can now be prescribed arbitrarily. The velocities, however, are restricted on account of the given conditions, according to the equations

$$A_{i1}\dot{q}_1 + \ldots + A_{in}\dot{q}_n = 0, \; (i = 1, \ldots, m). \tag{213.1}$$

We can thus assign arbitrarily n initial position coordinates and $n-m$ initial velocities.

The m equations (213.1) serve not only the purpose of eliminating m of the initial velocities. They have the further function of determining the λ-factors which enter the equations of motion as undetermined multipliers.

Non-holonomic auxiliary conditions which are rheonomic, i.e. time-dependent, require particular care. Here it is necessary to know what conditions exist between the δq_k if the variation is not performed instantaneously but during the infinitesimal time δt. The auxiliary conditions now take the form

$$A_{i1}\delta q_1 + \ldots + A_{in}\delta q_n + B_i\delta t = 0, \qquad (213.2)$$

with coefficients A_{ik} and B_i which are in general functions of the q_i and the time t. The quantities B_i do not enter into the equations of motion since the virtual displacements δq_k are performed *without* varying the time; but they do enter into the relations which exist between the velocities:

$$A_{i1}\dot{q}_1 + \ldots + A_{in}\dot{q}_n + B_i = 0. \qquad (213.3)$$

Summary. A variational problem with non-holonomic auxiliary conditions cannot be reduced to a form where the variation of a certain quantity is put equal to zero. However, the equations of motion are once more derivable, with the help of the multiplier method, in a fashion analogous to the case of holonomic conditions.

14. Isoperimetric conditions. It can happen that an auxiliary condition does not appear as an algebraic relation between the q_k, but in the form of a definite integral which must have a prescribed value C:

$$\int_{t_1}^{t_2} f(q_1, \ldots, q_n, t)dt = C. \qquad (214.1)$$

Auxiliary conditions of this form are called "isoperimetric conditions," since the first historically recorded extremum problem, that of finding the maximum area bounded by a perimeter of given length (Dido's problem) prescribes a condition of this nature.

Generally speaking we might have an *arbitrary* number of

conditions of the form (214.1) prescribed as auxiliary conditions, but it will suffice to treat the case of a single condition since the generalization to m conditions is obvious.

The condition (214.1) must hold not only for the actual functions $q_1, \ldots, q_n$ which make the given integral I of (211.1) stationary, but also for the varied functions $\bar{q}_1, \ldots, \bar{q}_n$; hence we can take the variation of (214.1), which gives the following integral relation between the $\delta q_1, \ldots, \delta q_n$:

$$\int_{t_1}^{t_2} \left(\frac{\partial f}{\partial q_1} \delta q_1 + \ldots + \frac{\partial f}{\partial q_n} \delta q_n \right) dt = 0. \qquad (214.2)$$

Hence our present problem is to satisfy the equation

$$\delta I = \int_{t_1}^{t_2} [E_1(t)\delta q_1 + \ldots + E_n(t)\delta q_n] \, dt = 0, \qquad (214.3)$$

with the auxiliary condition (214.2). Again, it is allowable to multiply the left-hand side of the equation (214.2) by some undetermined constant λ and add it to δI. Hence we consider the equation

$$\int_{t_1}^{t_2} \left[\left(E_1 + \lambda \frac{\partial f}{\partial q_1} \right) \delta q_1 + \ldots + \left(E_n + \lambda \frac{\partial f}{\partial q_n} \right) \delta q_n \right] dt = 0. \qquad (214.4)$$

Let us now replace dt by Δt, and the definite integral by a sum, making Δt smaller and smaller and thus approaching the given definite integral more and more. We can see by similar reasoning as before that during this process the λ can always be determined in such a way that the integral (214.4) shall vanish for *arbitrary* variations of the δq_k. Since this holds independently of Δt, it must hold even in the limit; which gives the principle that with a proper choice of λ we have

$$\delta \int_{t_1}^{t_2} (L + \lambda f) \, dt = 0, \qquad (214.5)$$

for arbitrary variations of the δq_k. But this means that we once more have succeeded in transforming our conditioned variation problem into a free variation problem by changing the original L into

$$L' = L + \lambda f, \qquad (214.6)$$

where λ is an undetermined constant.

From here on, the conclusions are the same as in section 12. We once more arrive at the equations (212.5) with the only difference that the λ_i are now *constants* and not functions of t. These constants can again be determined by satisfying the given integral conditions.

The same λ-method applies if the isoperimetric condition (214.1) depends not only on the q_k but also on their derivatives with respect to t:

$$\int_{t_1}^{t_2} f(q_1, \ldots, q_n; \dot{q}_1, \ldots, \dot{q}_n; t) \, dt = C. \qquad (214.7)$$

For a problem of this nature, see the problem of the flexible chain, treated in chap. III, section 4.

Summary. The Lagrangian λ-method can be extended to the case of auxiliary conditions which are such that some given definite integrals must not change their value during the process of variation. The resulting equations have the same form as those which hold for algebraic auxiliary conditions. The only difference is that now the λ-factors are constants and not functions of t.

15. The calculus of variations and boundary conditions The problem of the elastic bar.[1] In all our previous considerations we were essentially interested in the *differential equations* which can be deduced as the solution of the problem that a given definite integral shall assume a stationary value. The derivation of these equations by the method of integration by parts revealed that the variation of a definite integral is composed of two parts, namely, an integral extended over the given range, *plus a boun-*

[1]The credit for the material discussed in this section goes to the various brief but fundamental references to this subject in *Courant-Hilbert* (see Bibliography), particularly pp. 206-207 of volume I. The connection between the calculus of variations and boundary conditions is not only interesting for its own sake but quite important for many applications, particularly to stability problems of elasticity.

dary term. We did not pay much attention to this boundary term, since we assumed boundary conditions which made this term vanish. But actually there are situations where this boundary term plays a more active role. We shall see later, in the Hamiltonian theory of solving the differential equations of motion by means of a partial differential equation, that in the advanced mathematical theory of dynamics the previously neglected boundary term is important. Here, however, we wish to discuss another aspect of the boundary term which has a more direct physical significance.

We know that any problem which involves the solution of differential equations requires a proper number of *boundary conditions* in order to make the solution unique. These conditions might be determined by the given physical circumstances. But it is also possible that the physical conditions under consideration give no boundary conditions or not enough boundary conditions to make the solution unique.

As an illustrative example we consider a problem from the mechanics of continua. We consider a horizontal elastic bar, loaded by weights and supported at the two ends. The differential equation which gives the transverse displacement of the bar is of the fourth order and thus requires the prescription of four boundary conditions. Indeed, we can *clamp* the bar at both ends, which means that not only the displacement, but also its first derivative, is prescribed as zero at both ends. This gives four boundary conditions.

However, instead of being clamped, the bar may be merely *supported* at both ends. In this case the physical conditions prescribe but *two* boundary conditions, namely, the vanishing of the displacement at the two ends while the derivative remains undetermined. From where will the additional two boundary conditions come?

It is a particularly beautiful feature of variational problems that *they always furnish automatically the right number of boundary conditions.* Those boundary conditions which are not imposed on the system by the given external circumstances, follow from the nature of the variational problem. The condition for a stationary value requires these surplus boundary conditions, no less than it requires the fulfillment of the Euler-Lagrange differential equations. It is the boundary term of δI which is responsible for these surplus boundary conditions. *"Imposed" and "natural" boundary conditions, taken together, provide a unique solution.*

This principle is illustrated by the problem of the loaded elastic bar already referred to. The equilibrium of the bar is determined by the condition that its potential energy shall be a minimum. We are not concerned here with the derivation of the expression for the elastic energy of a bar, which can be found in any textbook on elasticity. We denote by l the length of the bar and by x the independent variable which runs from 0 to l, characterizing the position of a point of the bar. The small vertical displacement caused by the loading of the bar will be denoted by $y(x)$, while $\rho(x)$ denotes the load per unit length. We assume that the bar has everywhere the same cross-section. Then the potential energy due to the elastic forces is given by

$$V_1 = \frac{k}{2} \int_0^l (y'')^2 dx, \qquad (215.1)$$

where k is a constant, while the potential energy due to gravity is

$$V_2 = - \int_0^l \rho y \, dx. \qquad (215.2)$$

Hence the integral which must be minimized is

$$I = \int_0^l \left(\frac{k}{2} y''^2 - \rho y \right) dx. \qquad (215.3)$$

The Euler-Lagrange differential equation of this variational problem is—see (210.11):

$$ky''''(x) = \rho(x). \qquad (215.4)$$

If this equation is satisfied, the variation δI is reduced to a boundary term which in our problem appears as follows—see (210.12):

$$\delta I = k \, [- y'''(l)\delta y(l) + y'''(0)\delta y(0)$$
$$+ y''(l)\delta y'(l) - y''(0)\delta y'(0) \,] \qquad (215.5)$$

1. *Clamped ends.* The boundary conditions imposed on the bar are here:

$$y(0) = 0, \qquad\qquad y(l) = 0,$$
$$y'(0) = 0, \qquad\qquad y'(l) = 0. \qquad (215.6)$$

Since the variation of fixed quantities is zero, we notice that all the terms of (215.5) vanish and the only condition for the vanishing of δI is the differential equation (215.4). All the boundary conditions are here provided by the external circumstances of the problem and no natural boundary conditions have to be added.

2. *Supported ends.* The only imposed conditions are now

$$y(0) = 0, \qquad\qquad y(l) = 0. \qquad (215.7)$$

Hence the variations $\delta y'(0)$ and $\delta y'(l)$ are arbitrary and the vanishing of (215.5) now requires the two additional boundary conditions:

$$y''(0) = 0, \qquad\qquad y''(l) = 0, \qquad (215.8)$$

which supplement the "imposed" boundary conditions (215.7).

3. *One end clamped, the other free.* Let the bar be clamped at the point $x = 0$, but free at the other end. The given boundary conditions are thus:

$$y(0) = 0, \qquad\qquad y'(0) = 0, \qquad (215.9)$$

while $y(l)$ and $y'(l)$ are not prescribed. Hence the variation of these two quantities is arbitrary and the vanishing of (215.5) requires the following two conditions:

$$y''(l) = 0, \qquad\qquad y'''(l) = 0, \qquad (215.10)$$

as a consequence of the variational problem, supplementing the imposed conditions (215.9).

4. *Both ends free.* The natural boundary conditions (215.10) come into play at both ends:

$$y''(0) = 0, \qquad\qquad y'''(0) = 0,$$
$$y''(l) = 0, \qquad\qquad y'''(l) = 0. \qquad (215.11)$$

Notice that now *all* the boundary conditions are furnished by the variational problem. We have no "imposed" conditions, but the "natural" boundary conditions are present in sufficient number.

Yet we feel that this problem must have its own difficulties, since obviously the bar cannot be in equilibrium without proper support from two upward forces. We shall regard these forces as a limiting case of forces continuously distributed along the bar, so that they can be included in $\rho(x)$, counting an upward force with a negative sign. It is shown in the general theory of differential equations that the right-hand side of (215.4) can be prescribed arbitrarily only if the corresponding homogeneous differential equation—i.e., the differential equation without any load distribution—has no solution except the trivial solution $y = 0$. In all the previous cases the boundary conditions were such that the differential equation

$$y''''(x) = 0 \qquad (215.12)$$

had no solution under the given boundary conditions. Here, however, *two* such independent solutions exist, namely

$$y = 1, \qquad (215.13)$$

and
$$y = x. \qquad (215.14)$$

In such cases our boundary value problem is not solvable unless $\rho(x)$ is "orthogonal" to the homogeneous solutions (215.13), (215.14)—i.e.

$$\int_0^l \rho(x)dx = 0, \qquad (215.15)$$

and
$$\int_0^l \rho(x)x\,dx = 0. \qquad (215.16)$$

The physical significance of these conditions is that *the sum of the forces and the sum of the moments of the forces acting on the bar are zero.* These conditions determine the two upward forces which are necessary to support the bar. They are derivable as the integrability conditions of the boundary value problem which is here completely furnished by the calculus of variations.

Problem 1. Investigate the case where the end $x = 0$ is supported (but not clamped), the end $x = l$ free. Show that there is now the single integrability condition "sum of all moments equals zero."

Problem 2. Consider the case of two bars of different cross-sections rigidly joined together. Here the constant k of the differential equation (215.4) makes a sudden *jump* at the point $x = \xi$ where the two bars are joined. Show that now the boundary term (215.5) yields two continuity conditions at the point $x = \xi$, namely, that ky'' and ky''' *are continuous at the point* $x = \xi$.

Summary. The solution of a differential equation —or a set of differential equations—is not unique without the addition of the proper number of boundary conditions. Problems arising from finding the stationary value of a definite integral have always the right number of boundary conditions. If the given external circumstances do not provide the sufficient number of boundary conditions, the remainder is always made up by the variational problem itself, because the boundary term in the expression for δI furnishes some "natural" boundary conditions in addition to the "imposed" boundary conditions. The problem of the elastic bar with various end-conditions provides an excellent illustration of this principle.

CHAPTER III

THE PRINCIPLE OF VIRTUAL WORK

1. The principle of virtual work for reversible displacements. The first variational principle we encounter in the science of mechanics is the principle of virtual work. It controls the equilibrium of a mechanical system and is fundamental for the later development of analytical mechanics.

In the Newtonian form of mechanics a particle is in equilibrium if the resulting force acting on that particle is zero. This form of mechanics isolates the particle and replaces all constraints by forces. The inconvenience of this procedure is obvious if we think of such a simple problem as the equilibrium of a lever. The lever is composed of an infinity of particles and an infinity of inner forces acting between them. The analytical treatment can dispense with all these forces and take only the external force—i.e. in this case the force of gravity—into account. This is accomplished by performing only such virtual displacements as are in harmony with the given constraints. In the case of the lever, for example, we let the lever rotate around its fulcrum *as a rigid body*, thus preserving the mutual distance of any two particles. By this procedure the inner forces which produce the constraints need not be considered.

The mechanical behavior of a rigid body is certainly very different from that of a sandpile. There are strong inner forces acting between the particles of a rigid body which keep these particles together and which do not act between the particles of a sandpile. But how can we prove the presence of these forces? By trying to *break* the rigid body, i.e. by trying to move the particles relative to each other in a manner which is *not* in harmony with the given constraints. If we merely move a rigid body or a sandpile by rotation and translation, the mechanical difference between the two systems disappears, because now the

strong inner forces which characterize the rigid body in contrast
to the sandpile do not come into action.   This is the reason why
in the variational treatment of mechanics the "forces of constraint" which maintain certain given kinematical conditions are
neglected, and only the work of the "impressed forces" needs to be
taken into account.   We may eliminate the action of the inner
forces, since the virtual displacements applied to the system are
in harmony with the given kinematical conditions. The number
of equations obtained by this procedure is smaller than the
number of particles.   But it is exactly equal to the number of
degrees of freedom which characterize the system.

Let us use at first the language of vectorial mechanics.   We
assume that the given external forces F_1, F_2, ..., F_n act at
the points P_1, P_2, ..., P_n of the system.   The virtual displacements of these points will be denoted by

$$\delta R_1, \ \delta R_2, \ \ldots, \ \delta R_n. \tag{31.1}$$

These virtual displacements must be in harmony with the given
kinematical constraints, and we shall assume that they are *reversible*, i.e. the given constraints do not prevent us from changing
an arbitrary δR_i into $-\delta R_i$.

Now the principle of virtual work asserts that *the given mechanical system will be in equilibrium if, and only if, the total virtual
work of all the impressed forces vanishes*:

$$\overline{\delta w} = F_1 . \ \delta R_1 + F_2 . \ \delta R_2 + \ldots + F_n . \ \delta R_n = 0. \tag{31.2}$$

Let us translate this equation into analytical language.   For
this purpose we express the rectangular coordinates x_i, y_i, z_i as
functions of the generalized coordinates q_1, q_2, ..., q_n, exactly
as we have done in chap. I, section 7.   The differential form
(31.2) is then transformed into the new differential form

$$\overline{\delta w} = F_1 \delta q_1 + F_2 \delta q_2 + \ldots + F_n \delta q_n, \tag{31.3}$$

where F_1, F_2, ..., F_n are called the components of the generalized force.   They form a vector of the n-dimensional configuration space.

The principle of virtual work requires that

$$F_1 \delta q_1 + F_2 \delta q_2 + \ldots + F_n \delta q_n = 0. \tag{31.4}$$

We can give a striking geometrical interpretation of this equation. The left-hand side of the equation is nothing but the "scalar product" of force and virtual displacement. The vanishing of this scalar product means that *the force F_i is perpendicular to any possible virtual displacement.*

Let us assume at first that the given mechanical system is free of any constraints. In that case the C-point of the configuration space can be displaced in an arbitrary direction. Then the principle (31.4) requires that the force F_i shall vanish, because there is no vector which can be perpendicular to all directions in space.

Let us assume that the C-point has to stay within a certain $(n - m)$-dimensional *subspace* of the configuration space, on account of m given kinematical constraints. Then the condition (31.2) no longer requires the vanishing of the force F_i, but only its *perpendicularity* to that subspace. This amounts to $n - m$ equations, in conformity with the $n - m$ degrees of freedom of the mechanical system.

We now come to the *physical* interpretation of the principle of virtual work. According to Newtonian mechanics, the state of equilibrium requires that the resultant force acting on *any* particle of the system shall vanish. This resultant force is the sum of the impressed force and the forces which maintain the given constraints. These latter forces are usually called "forces of reaction." Since the principle of equilibrium requires that "impressed force plus resultant force of reaction equals zero," we see that the virtual work of the impressed forces can be replaced by the negative virtual work of the forces of reaction. Hence the principle of virtual work can be formulated in the following form, which we shall call Postulate A:

"The virtual work of the forces of reaction is always zero for any virtual displacement which is in harmony with the given kinematic constraints."

This postulate is not restricted to the realm of statics. It applies equally to dynamics, when the principle of virtual work is suitably generalized by means of d'Alembert's principle. Since all the fundamental variational principles of mechanics, the principles of Euler, Lagrange, Jacobi, Hamilton, are but alternative mathematical formulations of d'Alembert's principle, Postulate A is actually the *only* postulate of analytical mechanics, and is thus of fundamental importance.[1]

The principle of least action assumes a special significance in the particularly important case where the impressed force F_i is monogenic, i.e. derivable from a single scalar function, the work function. In this case the virtual work is equal to the variation of the work function $U(q_1, \ldots, q_n)$. Since the work function can be replaced by the negative of the potential energy, we can say that the state of equilibrium of a mechanical system is distinguished by the stationary value of the potential energy, i.e. by the condition

$$\delta V = 0. \tag{31.5}$$

If the equilibrium is *stable*, the potential energy must assume its minimum value—the minimum understood in the local sense—while in general, equilibrium does not require the minimum, but only the stationary value, of V.

Summary. The principle of virtual work demands that for the state of equilibrium the work of the impressed forces is zero for any infinitesimal variation of the configuration of the system which is in harmony with the given kinematical constraints. For monogenic forces, this leads to the condition that, for equilibrium, the potential energy shall be stationary with respect to all kinematically permissible variations.

The following two sections will be devoted to the application of the principle of virtual work to the statics of a rigid body. The results are well known

---

[1]Those scientists who claim that analytical mechanics is nothing but a mathematically different formulation of the laws of Newton must assume that Postulate A is deducible from the Newtonian laws of motion. The author is unable to see how this can be done. Certainly the third law of motion, "action equals reaction," is not wide enough to replace Postulate A.

from elementary vectorial mechanics, but their deduction from one funda-mental principle is a valuable experience.

2. The equilibrium of a rigid body. A rigid body which can move freely in space has six degrees of freedom: three on account of translation and three on account of rotation. Making use of the superposition principle of infinitesimal quantities, we can apply these two types of displacements independently of one another.

(A) *Translation.* An infinitesimal translation produces at each point of the rigid body the same displacement. Let ϵ be the extent of the infinitesimal displacement, and B a vector of unit length. We then have for the virtual displacement δR_k of the particle P_k

$$\delta R_k = \epsilon B, \tag{32.1}$$

and the resultant work becomes

$$\overline{\delta w} = \Sigma(F_k \cdot \epsilon B) = \epsilon B \cdot \Sigma F_k. \tag{32.2}$$

Since the vector B can be chosen in any direction, the vanishing of (32.2) requires:

$$\overline{F} = \Sigma F_k = 0, \tag{32.3}$$

which means that the *resultant force* F *of all the impressed forces vanishes.*

A piston may be forced to move up and down in its tube. The vector **B** has then a definite direction, and the vanishing of (32.3) requires only that the *component of* $\overline{F}$ *in the direction of motion shall vanish.*

(B) *Rotation.* Let ϵ be the angle of an infinitesimal rotation, and Ω a vector of unit length along the axis of rotation. The displacement of the point P_k due to the rotation can be written as follows:

$$\delta R_k = \epsilon \Omega \times R_k, \tag{32.4}$$

where R_k denotes the position vector of P_k with respect to an origin on the axis of rotation.

The work of the force F_k becomes

$$\overline{\delta w}_k = F_k \cdot \epsilon \Omega \times R_k = \epsilon \Omega \cdot (R_k \times F_k) = \epsilon \Omega \cdot M_k, \tag{32.5}$$

where we have introduced the vector

$$M_k = R_k \times F_k \tag{32.6}$$

to denote the "moment of the force" about the origin.

Hence the total work of all acting forces becomes

$$\overline{\delta w} = \Sigma \epsilon \Omega \cdot \mathbf{M}_k = \epsilon \Omega \cdot \Sigma \mathbf{M}_k. \tag{32.7}$$

Notice the complete analogy of (32.7) with (32.2), the vector Ω taking over the role of the translation vector $\mathbf{B}$ and the moment $\mathbf{M}$ the role of the force $\mathbf{F}$.

Since a free body may rotate about any axis, and thus Ω has an arbitrary direction, the vanishing of (32.7) requires that

$$\overline{\mathbf{M}} = \Sigma \mathbf{M}_k = 0. \tag{32.8}$$

The resultant moment of all the impressed forces vanishes.

Here we have the second condition for the equilibrium of a rigid body. Since an arbitrary virtual displacement of a rigid body can always be produced by the superposition of an infinitesimal translation and rotation, the conditions (32.3) and (32.8) together determine its equilibrium.

Note 1. The body may be fixed at the origin 0 and thus its freedom of translation destroyed. In that case the first condition, "the sum of all forces vanishes," no longer holds; but the second condition, the vanishing of the resultant moment, still holds. The radius vectors $\mathbf{R}_k$ are now measured from the fixed point; hence it is the resulting moment *about that point* which must vanish.

Note 2. Not only one point of the body, but a whole axis may be fixed. In this case Ω is no longer an arbitrary vector but has the direction of the given axis. The vanishing of (32.7) requires only

$$\Omega \cdot \overline{\mathbf{M}} = 0, \tag{32.9}$$

which means that *only the component of the resultant moment in the direction of the axis of rotation has to vanish.*

Summary. The general kinematical possibilities of a rigid body are translation and rotation. The possibility of translation requires the sum of all forces to vanish, and the possibility of rotation requires the sum of all moments to vanish for equilibrium.

---

3. Equivalence of two systems of forces. As we shall see later, the virtual work of the impressed forces determines not only the equilibrium but also the dynamical behaviour of a mechanical system. *Two systems of forces which produce the same virtual work are dynamically equivalent.*

Since the virtual work of an arbitrary system of forces acting on a rigid body depends solely on two quantities, viz. the resultant force $\overline{\mathbf{F}}$ and the resultant moment $\overline{\mathbf{M}}$, we get at once the important theorem that *two systems of forces which have the same resultant force and resultant moment are mechanically equivalent.*

Problem 1. Show that any given system of forces acting on a rigid body can be replaced by a single force if, and only if, the resultant moment $\overline{\mathbf{M}}$ and the resultant force $\overline{\mathbf{F}}$ are perpendicular to each other:

$$\overline{\mathbf{F}} \cdot \overline{\mathbf{M}} = 0. \tag{33.1}$$

Problem 2. Show that two forces can be replaced by a single force if, and only if, the two forces are coplanar and do not form a couple.

Problem 3. Show that an arbitrary *parallel* system of forces

$$\mathbf{F}_i = m_i \mathbf{G}, \tag{33.2}$$

can always be replaced by a single force

$$\overline{\mathbf{F}} = \overline{m} \mathbf{G} \tag{33.3}$$

applied at the point

$$\mathbf{R}_0 = \frac{\Sigma m \mathbf{R}}{\Sigma m}, \tag{33.4}$$

provided that $\overline{m} = \Sigma m$ is not zero.

Problem 4. Show that an arbitrary system of forces can be replaced by a single force $\overline{\mathbf{F}} = \Sigma \mathbf{F}$ plus a couple whose axis is parallel to $\overline{\mathbf{F}}$.

4. Equilibrium problems with auxiliary conditions. Occasionally, equilibrium problems occur which involve one or more auxiliary conditions. In such problems the infinitesimal virtual work $\overline{\delta w}$, before being equated to zero, must be augmented by the variation of the auxiliary conditions, each one multiplied by an undetermined Lagrangian factor λ, according to the general method discussed in chap. II, sections 5 and 12. As an illustration of the general method we consider here two problems of statics. One requires the minimizing of an ordinary function, the other of a definite integral.

1. *The problem of jointed bars.* Consider a system of uniform rigid bars of constant cross-section, freely jointed at their end-points. The two free ends of the chain are suspended. Find the position of equilibrium of the system.

In this problem the impressed force is the force of gravity, which is a monogenic force. We determine the potential energy V of the system which has to be minimized. Let us denote by x_k, y_k the rectangular coordinates of the end points of the bars—the x-axis being chosen horizontally, the y-axis vertically downwards—while the given lengths of the bar will be denoted by l_k:

$$l_k^2 = (x_{k+1} - x_k)^2 + (y_{k+1} - y_k)^2 \tag{34.1}$$
$$= (\Delta x_k)^2 + (\Delta y_k)^2, \qquad (k = 0, 1, \ldots, n - 1).$$

If σ is the mass per unit length, we obtain the following expression for the potential energy of the system:

$$V = \frac{\sigma g}{2} \sum_{k=0}^{n-1} (y_k + y_{k+1}) l_k. \tag{34.2}$$

This is the potential energy to be minimized, considering the equations (34.1) as n auxiliary conditions.

According to the general Lagrangian procedure we form the *modified* potential energy:

$$\overline{V} = \frac{1}{2} \sum_{k=0}^{n-1} (y_k + y_{k+1}) l_k + \sum_{k=0}^{n-1} \lambda_k [(\Delta x_k)^2 + (\Delta y_k)^2] \tag{34.3}$$

(omitting the irrelevant constant factor σg in V), and this is the new function which has to be minimized. The variables of the problem are the x_k and y_k ($k = 1, 2, \ldots, n - 1$), while x_0, y_0 and x_n, y_n are given as the two points of suspension.

Variation with respect to the x_k yields

$$\lambda_k \Delta x_k - \lambda_{k-1} \Delta x_{k-1} = 0. \tag{34.4}$$

This difference equation is soluble in the form

$$\lambda_k = \frac{C}{\Delta x_k}. \tag{34.5}$$

Variation with respect to the y_k yields

$$\frac{1}{2} (l_k + l_{k-1}) - \lambda_k \Delta y_k + \lambda_{k-1} \Delta y_{k-1} = 0, \tag{34.6}$$

and if we substitute for λ_k its value from (34.5), we obtain the following difference equation as the solution of our equilibrium problem:

$$c \left(\frac{\Delta y_k}{\Delta x_k} - \frac{\Delta y_{k-1}}{\Delta x_{k-1}} \right) = \frac{1}{2} (l_k + l_{k-1}). \tag{34.7}$$

2. *The problem of the uniform chain (catenary).* If the number of jointed bars increases more and more, while their lengths decrease (irrespective of whether these lengths are equal or not), the bars approach a smooth, continuous and differentiable *curve*. In the limit we get the problem of the uniform *chain*. Now the difference equation (34.7) obviously approaches in the limit the differential equation

$$c \frac{d}{ds} \left(\frac{dy}{dx} \right) = 1, \tag{34.8}$$

if we denote by ds the line element of the curve. Let us characterize the curve by the equation $y = f(x)$; then (34.8) may be written in the form

$$\frac{y''}{\sqrt{1 + y'^2}} = a, \tag{34.9}$$

where we have put $a = \dfrac{1}{c}$. This equation can be integrated, and we obtain

$$y = \frac{1}{2a}\left(e^{ax} + e^{-ax}\right), \tag{34.10}$$

which is the well-known equation of the catenary.

Instead of using this limiting process, let us now apply the direct methods of the calculus of variations to our problem, translating the previous problem of the jointed bars directly to the problem of the chain. We assume that the resulting curve is given in parametric form by

$$\begin{aligned} x &= x(\tau), \\ y &= y(\tau). \end{aligned} \tag{34.11}$$

The definite integral to be minimized is now

$$V = \int_{\tau_1}^{\tau_2} y\sqrt{x'^2 + y'^2}\,d\tau. \tag{34.12}$$

As far as auxiliary conditions are concerned, we can follow two different procedures. If we proceed as in the previous problem, we must keep the length of any arbitrarily small part of the chain constant during the process of variation. This amounts to the auxiliary condition

$$x'^2 + y'^2 = 1, \tag{34.13}$$

if we agree that our parameter τ shall coincide with the arc length s of the actual curve. In this case the method of the Lagrangian multiplier requires that we minimize the modified integral

$$\overline{V} = \int_{\tau_1}^{\tau_2} \left[y + \lambda(x'^2 + y'^2)\right]\,d\tau. \tag{34.14}$$

However, since our chain is perfectly flexible, we can perform our variation in a different manner. We can allow any arbitrary variations of the curve (34.11) which will not change the *total* length of the curve. Here the problem is to minimize the integral (34.12) with the auxiliary condition

$$\int_{\tau_1}^{\tau_2} \sqrt{x'^2 + y'^2}\,d\tau = \text{Constant.} \tag{34.15}$$

This problem is of the isoperimetrical type (see chap. II, section 14). The modified integral here becomes

$$\overline{V} = \int_{\tau_1}^{\tau_2} (y + \lambda)\sqrt{x'^2 + y'^2}\,d\tau. \tag{34.16}$$

The essential difference between the two treatments is that in the first case— where the auxiliary condition is of the local type—the Lagrangian factor λ will be a function of τ; while in the second case—where the auxiliary condition is an integral extended over the entire range—λ will be a constant.

Problem 1.   Apply the Euler-Lagrange differential equations to the integral (34.14). The first equation (connected with the variable x) is directly integrable. Eliminating λ and substituting in the second differential equation shows that the differential equation (34.8) is obtained. Verify that λ comes out as a function of τ.

Problem 2.   Apply the Euler-Lagrange differential equation to the integral (34.16). If we identify τ with y, the differential equation for x becomes directly integrable. Show that the results agree in both problems. In the second problem λ is a constant and entirely different from the λ of the first problem.

5. Physical interpretation of the Lagrangian multiplier method.

Let us assume that we have a mechanical system of n degrees of freedom, characterized by the generalized coordinates $q_1, \ldots, q_n$, and that there is a kinematical condition given in the form

$$f(q_1, \ldots, q_n) = 0. \tag{35.1}$$

Now the Lagrangian multiplier method requires that

$$\delta V + \lambda \delta f = 0. \tag{35.2}$$

This equation, however, can be expressed in the form

$$\delta \overline{V} = 0, \tag{35.3}$$

where

$$\overline{V} = V + \lambda f. \tag{35.4}$$

Once more we have the problem of finding the stationary value of a function, but that function is no longer the original potential energy V but a modified potential energy $\overline{V}$. This, however, is physically very plausible. If we do not restrict the variation of the configuration of the system by the condition (35.1) but permit *arbitrary* variations of the q_i, then not only the impressed forces will act but also the forces which maintain the given kinematical condition. They too have a potential energy which has to be added to the potential energy of the impressed forces. And thus the modification of the potential energy by the additional term λf is not merely a matter of mathematical method but has a very real physical significance. *The modification of the potential energy on account of the Lagrangian λ-method represents the potential energy of the forces which are responsible for the maintenance of the given auxiliary conditions.*

It is entirely in the nature of the problem that we cannot expect too much information about the mechanism which maintains a given kinematical condition if we know nothing but that condition itself. The term λf contains the factor λ, which can be calculated only for the *actual* configuration of the mechanical system. Hence λ is known only at a point C of the configuration space which lies on the surface (35.1). And yet this scanty knowledge about the work function of the force which maintains a given kinematical condition is quite sufficient to give us this force. The reason is that in forming the gradient of the additional potential energy

$$V_1 = \lambda f, \tag{35.5}$$

we obtain

$$F_{1i} = -\frac{\partial V_1}{\partial x_i} = -\lambda \frac{\partial f}{\partial x_i} - \frac{\partial \lambda}{\partial x_i} f. \tag{35.6}$$

At the point C the second term drops out because it is multiplied by f which vanishes at C; and thus the "force of reaction" F_{1i} is reduced to

$$F_{1i} = -\lambda \frac{\partial f}{\partial x_i}. \tag{35.7}$$

We see that the Lagrangian λ-term has the remarkable property of giving us the force of reaction connected with a given kinematical constraint. The same holds not only in the state of equilibrium but also in the state of motion, as we shall see later (cf. chap. V, section 8).

Problem 1. Let us assume that the work function of an assembly of free particles, subject to mutual forces between these particles, depends only on the "relative coordinates"

$$\xi_{ik} = x_i - x_k, \quad \eta_{ik} = y_i - y_k, \quad \zeta_{ik} = z_i - z_k, \tag{35.8}$$

of any pair of particles P_i and P_k:

$$U = U(\xi_{ik}, \eta_{ik}, \zeta_{ik}). \tag{35.9}$$

Let the coordinate x_i be varied by δx_i, thus obtaining the x-component of the force acting at P_i. Show that the quantity

$$X_{ik} = \frac{\partial U}{\partial \xi_{ik}} \qquad (35.10)$$

can be interpreted as the x-component of "the force on P_i due to P_k." We see directly that

$$X_{ik} = - X_{ki}, \qquad (35.11)$$

which shows that these forces satisfy Newton's law of "action equals reaction." The forces appear in pairs of equal magnitude but opposite direction and the sum of all forces is zero.

Problem 2. Assume now more specifically that U depends solely on the combination

$$r_{ik}^2 = \xi_{ik}^2 + \eta_{ik}^2 + \zeta_{ik}^2, \qquad (35.12)$$

i.e. on the *mutual distance* of any two particles. Show that the inner forces are now of the nature of central forces which act along the straight line between P_i and P_k. Such forces produce no resultant moments.

Problem 3. Consider a rigid body as an assembly of free particles, restricted by the auxiliary conditions

$$(x_i - x_k)^2 + (y_i - y_k)^2 + (z_i - z_k)^2 = c_{ik}^2. \qquad (35.13)$$

Show with the help of the Lagrangian multiplier method, and making use of the results of the previous two problems, that the forces which keep a rigid body together produce no resultant force $\overline{F}$ and no resultant moment $\overline{M}$.

If a given kinematical condition is non-holonomic, we can no longer modify the function which is to be minimized. But the λ-terms still appear in the conditions of equilibrium. This again has a direct physical significance. The λ-terms augment the impressed forces by those forces which maintain the given kinematical constraints. No work function exists in this case, *but once more the forces of reaction are provided.*

Hence the ingenious method of the Lagrangian multiplier elucidates the nature of holonomic and non-holonomic kinematical conditions, and shows that holonomic conditions and monogenic forces are mechanically equivalent; on the other hand, non-holonomic conditions and polygenic forces are also mechanically equivalent. *A holonomic condition is maintained by monogenic forces; a non-holonomic condition by polygenic forces.*

Summary. The Lagrangian multiplier method has the physical significance that it provides the forces of reaction which are exerted on account of given kinematical constraints. In the case of holonomic conditions these forces are deducible from a work function, while in the case of non-holonomic conditions such a work function does not exist. The forces of reaction are provided in both cases.

6. Fourier's inequality. All our previous conclusions were made under the tacit assumption that the virtual displacements are reversible. This is the case if we are somewhere *inside* the configuration space, so that motion can occur in every direction. The situation is quite different, however, if we reach the *boundaries* of the configuration space. Here a virtual displacement has to be directed *inward* and the opposite displacement is not possible because it would lead out of the region. Consider a ball hanging on the end of a flexible string. That ball can move upwards, which merely relaxes the string. But it cannot move downwards because the string does not permit it. Again, a ball on the table can move on the surface of the table but it can also move upwards in any direction we like; however, it cannot move downwards. The virtual displacement is reversible if the motion occurs in the horizontal direction, but it is irreversible in all other directions.

Fourier pointed out that the ordinary formulation of the principle of virtual work

$$\delta V = 0, \tag{36.1}$$

is restricted to *reversible* displacements; for irreversible displacements it has to be replaced by the inequality

$$\delta V \geqq 0. \tag{36.2}$$

Indeed, our final goal is to minimize the potential energy. This requires for small but finite displacements that V shall always increase:

$$\Delta V > 0, \tag{36.3}$$

For infinitesimal displacements this must be replaced by

$$\Delta V \geqq 0. \tag{36.4}$$

There is now no objection to the "greater than" sign. The only reason we were able to remove it before was that "greater than" would become "smaller than" if applied to the reversed displacement, and "smaller than zero" contradicts the hypothesis of a minimum. For irreversible displacements, however, the argument is no longer valid, and we have to leave (36.4) in its original form because there is no objection to positive changes of the potential energy. Since the change of the potential energy is the same as the negative work of the forces, we can express the inequality (36.4) in the form

$$\overline{\delta w} \leqq 0. \tag{36.5}$$

In this form Fourier's inequality includes even polygenic forces which have no potential energy. *If the work of the forces for any virtual displacement is zero or negative, the system is in equilibrium.*

If the ball at the end of a string is moved upwards, the work of the force of gravity is negative. For a horizontal (and reversible) displacement it is zero. Similarly, moving a ball on the horizontal table (reversible displacements) the work of the force of gravity is zero, but moving the ball upwards the work becomes negative. Any mechanical system which cannot come to equilibrium *within* the configuration space will move to the *boundary* of that space and find its equilibrium there. It is easier to satisfy the inequality (36.5) on the boundary of a region than the equality (36.1) inside the region. Notice that on the boundary of the configuration space equilibrium does not require a stationary value of the potential energy.

Summary. The customary formulation of the principle of virtual work, "the sum of all virtual works is equal to zero," is true for reversible displacements only. For irreversible displacements which occur on the boundary of the configuration space, "equal to zero" has to be replaced by "equal to or less than zero."

CHAPTER IV

D'ALEMBERT'S PRINCIPLE

Introduction: With d'Alembert's principle we leave the realm of statics and enter the realm of dynamics. Here the problems are much more complicated and their solution requires more elaborate methods. While the problems of statics of systems of a finite number of degrees of freedom lead to algebraic equations which may be solved by eliminations and substitutions, the problems of dynamics lead to differential equations. The present treatise is primarily devoted to the formulation and interpretation of the basic differential equations of motion, rather than to their final integration. D'Alembert's principle, which we shall discuss in the present chapter, did not contribute directly to the problem of integration. Yet it is an important landmark in the history of theoretical mechanics, since it gives an interpretation of the force of inertia which is fundamental for the later development of variational methods.

1. The force of inertia. With a stroke of genius the eminent French mathematician and philosopher d'Alembert (1717-1783) succeeded in extending the applicability of the principle of virtual work from statics to dynamics. The simple but far-reaching idea of d'Alembert can be approached as follows. We start with the fundamental Newtonian law of motion: "mass times acceleration equals moving force":

$$m\mathbf{A} = \mathbf{F}, \tag{41.1}$$

and rewrite this equation in the form

$$\mathbf{F} - m\mathbf{A} = 0. \tag{41.2}$$

We now define a vector $\mathbf{I}$ by the equation

$$\mathbf{I} = -m\mathbf{A}. \tag{41.3}$$

This vector $\mathbf{I}$ can be considered as a force, created by the motion. We call it the "force of inertia." With this concept the equation of Newton can be formulated as follows:

$$\mathbf{F} + \mathbf{I} = 0. \tag{41.4}$$

Apparently nothing is gained, since the intermediate step (41.3) gives merely a new name to the negative product of mass times acceleration.   It is exactly this apparent triviality which makes d'Alembert's principle such an ingenious invention and at the same time so open to distortion and misunderstanding.

The importance of the equation (41.4) lies in the fact that it is *more* than a reformulation of Newton's equation.   It is the expression of a *principle*.   We know that the vanishing of a force in Newtonian mechanics means equilibrium.   Hence the equation (41.4) says that the addition of the force of inertia to the other acting forces produces equilibrium.   But this means that if we have any criterion for the equilibrium of a mechanical system, we can immediately extend that criterion to a system which is in motion.   All we have to do is to add the new "force of inertia" to the previous forces.   By this device *dynamics is reduced to statics*.

This does not mean that we can actually *solve* a dynamical problem by statical methods.   The resulting equations are *differential equations* which have to be solved.   We have merely *deduced* these differential equations by statical considerations. The addition of the force of inertia I to the acting force **F** changes the problem of motion to a problem of equilibrium.

The objection can be raised that since the mass m is actually in motion, why should we treat it as if it were in equilibrium? Two answers can be given to this objection.   Firstly, motion is a relative phenomenon.   We can introduce a reference system which moves with the body and we can observe the body in that system.   The body is then actually at rest. Secondly, d'Alembert's principle focuses attention on the forces, not on the moving body, and the equilibrium of a given system of forces can be treated without referring to the state of motion of the body on which they act.   The criterion for the equilibrium of an arbitrary system of forces is that the total virtual work of all forces vanishes.   This criterion involves *virtual*, not actual displacements, and is thus equally applicable to masses at rest and to masses in motion.   Since the virtual displacement involves a possible, but purely mathematical experiment, it can be applied *at a certain definite time* (even if such a displacement would involve physically infinite velocities). At that instant the actual motion of the body does not enter into account.

D'Alembert generalized his equilibrium consideration from a single particle to any arbitrary mechanical system. His principle states that *any system of forces is in equilibrium if we add to the impressed forces the forces of inertia.* This means that *the total virtual work of the impressed forces, augmented by the inertial forces, vanishes for reversible displacements.* It seems appropriate to have a special name for the force which results if we add the force of inertia I_k to the given impressed force F_k which acts on a particle. We shall call this force the "effective force"[1] and denote it by F_k^e:

$$F_k^e = F_k + I_k. \tag{41.5}$$

D'Alembert's principle can now be formulated as follows: *The total virtual work of the effective forces is zero for all reversible variations which satisfy the given kinematical conditions*:

$$\sum_{k=1}^{N} F_k^e \cdot \delta R_k \equiv \sum_{k=1}^{N} (F_k - m_k A_k) \cdot \delta R_k = 0. \tag{41.6}$$

Notice that the impressed forces F_k may act at just a few points, while the effective forces F_k^e are present wherever a mass is in accelerated motion.

A given system of impressed forces will generally not be in equilibrium. This requires the fulfilling of special conditions. The total virtual work of the impressed forces will usually be different from zero. In that case the *motion* of the system makes up for the deficiency. The body moves in such a way that the additional inertial forces, produced by the motion, bring the balance up to zero. In this way d'Alembert's principle gives *the equations of motion of an arbitrary mechanical system.*

In contrast to the impressed forces, which are usually derivable from a single work function by differentiation (monogenic forces), the forces of inertia are of a polygenic character. While

[1]Synge and Griffith, *The Principles of Mechanics* (see Bibliography), and a few other authors use the name "effective force" for the product of mass times acceleration, i.e., for the negative of the force of inertia.

the virtual work of the impressed forces can be written as the complete variation of the work function:

$$\overline{\delta w} = \delta U, \tag{41.7}$$

the virtual work of the forces of inertia:

$$\overline{\delta w^i} = -\Sigma m_k \mathbf{A}_k \cdot \delta \mathbf{R}_k \tag{41.8}$$

is merely a differential form, not reducible to the variation of a scalar function. (We shall see later how this shortcoming can be remedied by an integration with respect to the time.)

The Newtonian equation (41.1) holds only if the mass is a constant. If m changes during the motion, the fundamental equation of motion has to be written in the form:

$$\frac{d}{dt}(m\mathbf{v}) = \mathbf{F}, \tag{41.9}$$

i.e., "rate of change of momentum equals moving force." Accordingly, the force of inertia I has to be defined as the negative rate of change of momentum:

$$\mathbf{I} = -\frac{d}{dt}(m\mathbf{v}). \tag{41.10}$$

For the customary case of constant m the general definition can be replaced by (41.3).

The definition of the force of inertia requires an "absolute reference system" in which the acceleration is measured. This is an inherent difficulty of Newtonian mechanics, keenly felt by Newton and his contemporaries. The solution of this difficulty came in recent times through Einstein's great achievement, the Theory of General Relativity.

We can ask the question, what is the *physical* significance of d'Alembert's principle? From the definition of the "effective force" by (41.5) it follows that this force is zero in the case of a free particle, while it is equal to the negative force of reaction if the particle is subject to constraints. Hence the application of the principle of virtual work to the effective force $\mathbf{F}^e$ is equivalent to the assumption that "the virtual work of the forces of reaction is zero for any virtual displacement which is in harmony with the given constraints." We thus come back to the same "Postulate A" that we have encountered before in chap. III, section 1. D'Alembert's principle generalizes this postulate from the field of statics to the field of dynamics, without any alteration.

Summary. D'Alembert's principle introduces a new force, the force of inertia, defined as the negative of the product of mass times acceleration. If this force is added to the impressed forces we have equilibrium, which means that the principle of virtual work is satisfied. The principle of virtual work is thus extended from the realm of statics to the realm of dynamics.

2. The place of d'Alembert's principle in mechanics. D'Alembert's principle gives a complete solution of problems of mechanics. All the different principles of mechanics are merely mathematically different formulations of d'Alembert's principle. The most advanced variational principle of mechanics, Hamilton's principle, can be obtained from d'Alembert's principle by a mathematical transformation. It is equivalent to d'Alembert's principle where both are applicable. In fact, Hamilton's principle is restricted to holonomic systems, while d'Alembert's principle can equally well be applied to holonomic and non-holonomic systems.

This principle is more elementary than the later variational principles because it requires no integration with respect to the time. But the disadvantage of the principle is that the virtual work of the inertial forces is a polygenic quantity and thus not reducible to a single scalar function. This makes the principle unsuited to the use of curvilinear coordinates. However, in many elementary problems of dynamics which can be adequately treated with the help of rectangular coordinates, or by the vector methods without the use of any coordinates, d'Alembert's principle is of great use.

For certain problems d'Alembert's principle is even more flexible than the more advanced principle of least action. The differential equations of motion are of the second order, determining the accelerations of the moving system. These accelerations are the second derivatives $\ddot{q}_i$ of the position coordinates q_i or the first derivatives of the velocities $\dot{q}_i$. Now it may happen—and such a situation arises particularly in the dynamics of rigid bodies—that it is more convenient to characterize the motion with the help of certain velocities which are *not* the derivatives of actual position coordinates. Such quantities are called "kinematical variables." A good example is the spin of a top about its axis of symmetry. This spin is the angular velocity of rotation, $\omega = \overline{d\phi}/dt$, but the $\overline{d\phi}$ is here merely an infinitesimal angle of rotation and not the differential of an angle ϕ. This angle exists only if the axis is fixed but not if the axis is changing. And yet it is convenient to use the spin as one of the quantities which characterize the motion of a top. In the principle of least action we cannot use kinematical variables, whereas in d'Alembert's principle we can.

D'Alembert's principle is fundamental in still another respect: it makes possible the use of moving reference systems, and is thus a forerunner of Einstein's revolutionary ideas concerning the relativity of motion, explaining—within the scope of Newtonian physics—the nature of those "apparent forces" which are present in a moving frame of reference.

Summary. D'Alembert's principle requires a polygenic quantity in forming the virtual work of the forces of inertia; hence it cannot provide the same facilities in the analytical use of curvilinear coordinates as the principle of least action. However, in problems which involve the use of kinematical variables (non-holonomic velocities) and the transformation to moving reference systems, d'Alembert's principle is eminently useful.

In the following sections we show the application of d'Alembert's principle to a number of typical problems, particularly connected with the use of moving reference systems. The first application concerns the deduction of the law of the conservation of energy, well known from elementary mechanics. This deduction is of interest, however, because it shows the limits within which the law holds.

3. **The conservation of energy as a consequence of d'Alembert's principle.** Although d'Alembert's principle is generally of a polygenic nature, in one special case it becomes monogenic and integrable. This special case leads to one of the most fundamental laws of mechanics, the law of the conservation of energy.

We consider the general formulation of d'Alembert's principle, given in equation (41.6). Let us assume that the impressed forces are monogenic, derivable from a potential energy function. Then the work of the impressed forces is equal to the negative variation of the potential energy, and d'Alembert's principle may be written as follows:

$$\delta V + \Sigma m_k \mathbf{A}_k \cdot \delta \mathbf{R}_k = 0. \tag{43.1}$$

The second term, the negative work of the forces of inertia, cannot in general be written as the variation of something. But now let us dispose of the $\delta \mathbf{R}_k$—which mean arbitrary tentative variations of the radius vector $\mathbf{R}_k$—in a special way. Let these tentative displacements coincide with the *actual* displacements as they occur during the time dt. This means that we replace $\delta \mathbf{R}_k$ by $d\mathbf{R}_k$, which is merely a special application of the variation principle (43.1).

For this special variation the virtual change δV of the potential energy will coincide with the actual change dV that occurs during the time dt. Moreover, the second term too becomes the perfect differential of a quantity, as can be seen if the acceleration $\mathbf{A}_k$ is replaced by the second derivative of the radius vector $\mathbf{R}_k$:

$$\Sigma m_k \ddot{\mathbf{R}}_k \cdot d\mathbf{R}_k = \Sigma m_k \ddot{\mathbf{R}}_k \cdot \dot{\mathbf{R}}_k \, dt = \frac{d}{dt} \left(\tfrac{1}{2} \Sigma m_k \dot{\mathbf{R}}_k^2 \right) dt = dT, \tag{43.2}$$

where we put

$$\tfrac{1}{2} \Sigma m_k \dot{\mathbf{R}}_k^2 = \tfrac{1}{2} \Sigma m_k \mathbf{v}_k^2 = T. \tag{43.3}$$

T is the kinetic energy of the mechanical system.

Equation (43.1) now takes the form

$$dV + dT = d(V + T) = 0, \tag{43.4}$$

which can be integrated to give

$$T + V = \text{constant} = E.$$

The sum of the kinetic and potential energies remains unchanged during the motion. This fundamental theorem is called the "law of the conservation of energy." It is a scalar equation and only *one* of the integrals of the equations of motion. Although in itself insufficient for the complete solution of the problem of motion, except in the case of a single degree of freedom, it is one of the most fundamental and universal laws of nature and, properly modified, holds not only for mechanical but for all physical phenomena. The constant E is called the "energy constant."

At first sight we might think that it is *always* permissible to identify the $\delta \mathbf{R}_k$ with the $d\mathbf{R}_k$ because the actual displacement must surely be among the kinematically possible displacements. And yet this apparently obvious reasoning is not always flawless. It holds in the case of free particles, but not always in the case of mechanical systems with constraints. It is certainly permissible to vary the position of the C-point in the configuration space arbitrarily, and we can certainly let the δq_i coincide with the dq_i. This, however, will not always mean that the variations $\delta \mathbf{R}_i$ of the positions of the particles coincide with the actual displacements $d\mathbf{R}_i$. We have to bear in mind that the variation is applied *suddenly*, at a certain instant, which means with infinite velocity, while the actual motion occurs with finite velocity. If we compare the equations (12.8) with the equations (18.3), we notice that in the first case the identification of δq_i with dq_i leads to $\delta \mathbf{R}_i = d\mathbf{R}_i$; in the second case, not. Now equations of the first type hold in the "scleronomic" case, where the given kinematical conditions do not contain the time explicitly, while equations of the second type hold in the "rheonomic" case, where the kinematical conditions involve the time explicitly.

A similar remark holds with regard to the potential energy V. If V is a function of the q_i only, then the choice $\delta q_i = dq_i$ will lead to $\delta V = dV$. But this would no longer be true if V were rheonomic through depending on t explicitly. Since V was defined as $-U$, our result implies that the work function U must not depend on t explicitly if the conservation of energy is to hold.

We see that the law of the conservation of energy has to be restricted to systems which are *scleronomic in both work function and kinematical conditions.* Moreover, our deduction assumed implicitly that the masses m_k are *constants*.

Summary. If the variations of position in d'Alembert's principle are made to coincide with the actual displacements which occur during the time dt, the resulting equation can be integrated. It yields the law of the conservation of energy, which states that the sum of kinetic and potential energies remains constant during the motion. The conditions for the validity of this procedure are that the masses of the particles shall be constant, and that both the work function of the system and the given kinematical conditions must be scleronomic, i.e., independent of the time.

4. Apparent forces in an accelerated reference system. Einstein's equivalence hypothesis. The fundamental soundness of d'Alembert's principle can be tested by the most exact physical experiments. The conclusions from the principle are in agreement with observation. D'Alembert's principle assumes that the force of inertia is just one more additional force which acts exactly like all the other forces. Now for the definition of the force of inertia we have to make use of an "absolute reference system." If the reference system in which the accelerations are measured is in motion relative to the absolute system, the force of inertia measured in the moving system is no longer the true force of inertia but has to be corrected by additional terms. According to d'Alembert's principle, the correction terms which enter the force of inertia on account of the motion of the reference system can equally well be interpreted as belonging to the impressed forces produced by an external field of force. This gives rise to the phenomenon of "apparent forces" which occur in reference systems that are in motion.

In the following discussion a comparison between two reference systems will be necessary, one of which is at "absolute rest," the other in motion. We shall distinguish these two reference systems as "the system S" and "the system S'," and we agree to the following convention: the quantities measured in the moving system will be denoted by the same letters as the quantities measured in the system at rest, but with a "prime." We are particularly interested in the comparison of three fundamental kinematical quantities: the radius vector **R**, the velocity **v** and the acceleration **A** of a moving particle, measured in the systems S and S'.

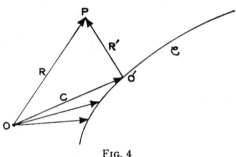

Fig. 4

First we consider a reference system which is in purely *translatory* motion, such as an elevator which can move up and down with arbitrary accelerations.

Such a translatory motion of the reference system can be characterized as follows. The origin O' of a rigid reference system S' moves along some given curve C, traced out by the vector **C** which is some function of the time t. The radius vector **R'** measured in the moving system and the radius vector **R** measured in the absolute system are in the following relation to each other (see Figure 4):

$$\mathbf{R} = \mathbf{C} + \mathbf{R}'. \tag{44.1}$$

Differentiating twice with respect to the time we obtain

$$\ddot{\mathbf{R}} = \ddot{\mathbf{C}} + \ddot{\mathbf{R}}'. \tag{44.2}$$

This equation may be written

$$\mathbf{A} = \ddot{\mathbf{C}} + \mathbf{A}', \tag{44.3}$$

and multiplication by $-m$ gives

$$I = -m\ddot{C} + I'. \tag{44.4}$$

We see that the force of inertia which, according to d'Alembert's principle, has to be added to the impressed forces $\mathbf{F}$, consists of two parts. The quantity $-m\ddot{C}$ is part of the true force of inertia and is due to the fact that the reference system in which the measurements are made is in motion relative to the absolute system. Whenever the motion of the reference system generates a force which has to be added to the relative force of inertia $\mathbf{I}'$, measured in that system, we call that force an "apparent force." The name is well chosen, inasmuch as that force does not exist in the absolute system and is created solely by the fact that our reference system moves relative to the absolute system. The name is misleading, however, if it is interpreted as a force which is not as "real" as any given physical force. In the moving reference system the apparent force is a perfectly real force which is not distinguishable in its nature from any other impressed force. Let us suppose that the observer is not aware of the fact that his reference system is in accelerated motion. Then purely mechanical observations cannot reveal to him that fact. According to d'Alembert's principle the physical action is determined by the sum of the impressed force $\mathbf{F}$ and the force of inertia $\mathbf{I}$. There is no way of separating these two forces in any fundamental way. If the physicist who is unaware of his own motion interprets the apparent force $-m\ddot{C}$ as an external force, he comes into no conflict with the facts.

This is indeed a logical consequence of d'Alembert's principle. Let us assume that the true force of inertia $\mathbf{I}$ and the force of inertia $\mathbf{I}'$ measured in a moving system are related to each other as follows:

$$\mathbf{I} = \mathbf{J} + \mathbf{I}'. \tag{44.5}$$

Now the effective force which acts in the moving system is

$$\mathbf{F}^e = \mathbf{F} + \mathbf{J} + \mathbf{I}'. \tag{44.6}$$

Since it is this effective force alone which determines the mechanical action, it makes no difference whether we interpret $\mathbf{F}^e$ as the sum of $\mathbf{F}$ and $\mathbf{J} + \mathbf{I}'$, or as the sum of $\mathbf{F} + \mathbf{J}$ and $\mathbf{I}'$. In the first case, we take notice of the fact that our reference system is in motion, and that that motion creates the force **J** as part of the force of inertia. In the second case, we forget about the motion of our reference system and interpret the force **J** as an additional impressed force which acts on the particle.

Let us assume in particular that we are in a closed laboratory which is pulled upwards with constant acceleration **G**, while at the same time the field of gravity suddenly disappears (Einstein's "box-experiment"). Then d'Alembert's principle implies that no mechanical experiment can detect this change, because it is impossible to distinguish between the following two hypotheses:

1. Our reference system is moving upwards with constant acceleration **G**; no field of gravity exists.

2. Our reference system is at rest but there is a field of gravity present which pulls every mass m downwards with the force m**G**.

This "equivalence hypothesis" was introduced by Einstein in his early attempts to solve the riddle of gravity, before establishing the general principle of relativity. For purely mechanical phenomena the equivalence hypothesis is a direct consequence of d'Alembert's principle. Einstein raised it to a general principle of nature. It seems justifiable to call the apparent force

$$\mathbf{E} = - m\ddot{\mathbf{C}}, \tag{44.7}$$

which acts in an accelerated reference system on account of d'Alembert's principle, the "Einstein force." Note that *all apparent forces, being generated by the force of inertia* $- m\mathbf{A}$, *are proportional to the inertial mass* m. The empirical fact that the gravitational mass is always proportional to the inertial mass was the decisive clue which enabled Einstein to establish the force of gravity as an apparent force.

(a) The pilot of an airplane bailing out from his plane has a peculiar sensation during the first moments of his fall, caused by the fact that he is entirely weightless. The apparent force generated by his fall compensates the force of gravity. As his parachute opens, his weight comes gradually back. (b) A similar sensation of becoming lighter is felt in a lesser degree in the first moments after an elevator starts downwards. When the elevator decelerates before it stops, our apparent weight increases. (c) When a bus begins to move, we are pushed backwards, when it decelerates, forwards. (d) The acceleration and deceleration of a train can be measured by a sensitive spring balance fastened to a wall which is perpendicular to the tracks.—All these phenomena are illustrations of d'Alembert's principle.

Summary. If a reference system is in accelerated motion, this motion produces the same effect as if external forces were present which have to be added to the impressed forces. These forces are called the "d'Alembert forces" or "apparent forces" because they are generated by the motion of the reference system but are not present in an absolute system. If the reference system is in a state of uniformly accelerated translation, the apparent force generated is that of a uniform gravitational field.

5. Apparent forces in a rotating reference system. Any motion of a rigid reference system can be considered as a translation plus a rotation. After studying a pure translation of the system we now deal with the case of pure rotation. The general problem is a superposition of these two special problems.

If the origin O of the reference system is kept fixed, the radius vectors R and R' in the absolute system S and in the moving system S' are the same:

$$R = R'. \tag{45.1}$$

Nevertheless, the velocities and accelerations measured in both systems differ from each other because of the fact that rates of change observed in the two systems are different. If a certain vector B is constant in S' it rotates with the system and thus, if observed in S, undergoes in the time dt an infinitesimal change

$$dB = (\Omega \times B)\, dt \tag{45.2}$$

(see 32.4), where Ω is the angular velocity vector which gives the rotation of S' with respect to S. Hence

$$\frac{dB}{dt} = \Omega \times B, \tag{45.3}$$

while at the same time

$$\frac{d'\mathbf{B}}{dt} = 0. \tag{45.4}$$

We have introduced here the notation d'/dt which refers to the operation of observing the rate of change of a quantity in the moving system S'.

If, on the other hand, the vector $\mathbf{B}$ is not constant in S', so that $d'\mathbf{B}/dt \neq 0$, we find, by making use of the superposition principle of infinitesimal processes, that

$$\frac{d\mathbf{B}}{dt} = \frac{d'\mathbf{B}}{dt} + \Omega \times \mathbf{B}. \tag{45.5}$$

Applying this equation to the radius vector $\mathbf{R} = \mathbf{R}'$ we get

$$\frac{d\mathbf{R}}{dt} = \frac{d'\mathbf{R}'}{dt} + \Omega \times \mathbf{R}'. \tag{45.6}$$

This yields the following relation between the velocities $\mathbf{v}$ and $\mathbf{v}'$:

$$\mathbf{v} = \mathbf{v}' + \Omega \times \mathbf{R}'. \tag{45.7}$$

In order to obtain the relation between the absolute and relative accelerations $\mathbf{A}$ and $\mathbf{A}'$, we differentiate the equation (45.7):

$$\frac{d\mathbf{v}}{dt} = \frac{d\mathbf{v}'}{dt} + \dot{\Omega} \times \mathbf{R}' + \Omega \times \frac{d\mathbf{R}'}{dt}. \tag{45.8}$$

On the right-hand side, the relation (45.5) can be employed in order to express everything in terms of quantities which belong to S'. The result is:

$$\frac{d\mathbf{v}}{dt} = \frac{d'\mathbf{v}'}{dt} + 2\Omega \times \mathbf{v}' + \Omega \times (\Omega \times \mathbf{R}') + \dot{\Omega} \times \mathbf{R}'. \tag{45.9}$$

Multiplication by $-m$ gives the relation between the two forces of inertia $\mathbf{I}$ and $\mathbf{I}'$,

$$\mathbf{I} = \mathbf{I}' - 2m\,\Omega \times \mathbf{v}' - m\Omega \times (\Omega \times \mathbf{R}') - m\dot{\Omega} \times \mathbf{R}'. \tag{45.10}$$

From now on, we abandon the comparison between absolute and rotating systems. *We agree to stay consistently in the rotating system S'.* Hence we may omit the primes on the understanding that all quantities will refer to S'. We notice that the motion of the reference system gives rise to three apparent forces of a different character. The third term on the right-hand side of (45.10) can be transformed into

$$C = m\omega^2 \, R_\perp, \tag{45.11}$$

where $R_\perp$ denotes that component of the radius vector R which is perpendicular to the vector Ω. This force is of a monogenic nature, being derivable from the potential energy

$$\phi = -\frac{m}{2} \omega^2 r_\perp^2. \tag{45.12}$$

It is called the "centrifugal force." On the earth, which rotates around its axis, the plumb-line indicates the direction of the resultant of the force of gravity and the centrifugal force. D'Alembert's principle shows that these two forces cannot be separated by any experiment.

Of a different nature is the second term:

$$B = -2m \, \Omega \times v = 2m \, v \times \Omega, \tag{45.13}$$

which depends on the velocity of the moving particle. It is always perpendicular to the velocity and thus does no work. This force is called the "Coriolis force." Its horizontal component can be shown to exist by Foucault's celebrated experiment of the precessing pendulum,—the first mechanical demonstration of the rotation of the earth.

The existence of the vertical component can be demonstrated by the ingenious experiment of the Hungarian physicist Eötvös who took a chemical balance and rotated it with constant angular velocity about a vertical axis, after taking away the two pans of the balance. The beam of the balance rotates in a horizontal plane, a particle of the right beam and a symmetrically situated particle of the left beam having equal velocities in opposite directions. The Coriolis force acts up and down and produces a periodic torque on the beam which brings the balance into forced vibration (the maximum of the torque occurs when the beam is parallel to the meridian, while in passing the east-west direction the torque becomes zero). Although the effect is very small, it is measurable by resorting to the resonance principle and letting the period of rotation coincide with the period of oscillation of the free balance.

The last term of (45.10) is present only if the angular velocity vector changes in either direction or magnitude. It vanishes for uniform rotation about a fixed axis. This third apparent force has no universally accepted name. The author likes to call it the "Euler force," in view of the outstanding investigations of Euler in this subject. We put

$$\mathbf{K} = - m\, \dot{\Omega} \times \mathbf{R}, \qquad (45.14)$$

and finally obtain for the effective force $\mathbf{F}^e$ which acts in a rotating system:

$$\mathbf{F}^e = \mathbf{F} + \mathbf{C} + \mathbf{B} + \mathbf{K} + \mathbf{I} \qquad (45.15)$$

(impressed force + centrifugal force + Coriolis force + Euler force + force of inertia).

Summary. An observer who is attached to a rotating reference system experiences in his system the centrifugal force and the Coriolis force as apparent forces. If in addition the rotation is not uniform, a third apparent force is generated, the "Euler force."

6. Dynamics of a rigid body. The motion of the centre of mass. As a last illustration of the action of apparent forces and d'Alembert's principle we consider the dynamics of a rigid body which can move freely in space. We assume an observer who makes his measurements in a reference system attached to the moving body. In this reference system the body is at rest and is thus in equilibrium.

Now we have seen in chap. III, section 2 that the equilibrium of a rigid body requires two vector conditions: the sum of all forces and the sum of all moments must vanish. In the present section we consider the first condition only.

In forming the resultant force $\overline{\mathbf{F}}^e$ we have to keep in mind that in addition to the impressed forces $\mathbf{F}$ we have the apparent forces acting in our system. We have firstly the "Einstein force" (44.7), resulting from the translatory motion of the rigid body. Then we have the centrifugal force (45.11) and the "Euler force" (45.14) on account of the rotatory motion of the rigid body (the Coriolis force $\mathbf{B}$ and the force of inertia $\mathbf{I}$ drop out since there are no velocities and accelerations in our reference system). We shall assume that the origin O' of our reference system coincides with the centre of mass of the rigid body. This means

$$\Sigma m \mathbf{R} = 0. \qquad (46.1)$$

Problem 1. Show that in our reference system the vector sum of the centrifugal forces vanishes.

Problem 2. Show that in our reference system the vector sum of the Euler forces vanishes.

Since the resultant centrifugal force and the resultant Euler force vanish in our reference system, we must have equilibrium between the impressed forces and the Einstein forces. This gives the condition

$$\bar{\mathbf{F}} - \Sigma m \ddot{\mathbf{C}} = 0, \qquad (46.2)$$

or

$$\bar{m}\, \ddot{\mathbf{C}} = \bar{\mathbf{F}}, \qquad (46.3)$$

where we have put

$$\bar{m} = \Sigma m. \qquad (46.4)$$

If we keep in mind that $\ddot{\mathbf{C}}$ is the acceleration of the centre of mass, we see that the equation (46.3) expresses the well-known theorem of the centre of mass: *the centre of mass of a rigid body moves like a particle in which the total mass of the body is concentrated and on which the resultant of all the forces impressed on the rigid body acts.*

Summary. The law which governs the motion of the centre of mass of a rigid body can be conceived as a consequence of the fact that the impressed forces and the Einstein forces which act in a reference system attached to the rigid body are in equilibrium.

---

7. Dynamics of a rigid body. Euler's equations. We come now to the *second* condition for the equilibrium of a rigid body, freely movable in space: the sum of all moments must vanish.

Problem 1. Show that the resultant moment of the Einstein forces vanishes in our reference system.

Since the Einstein forces have no resultant moment, the sum of the moments of the following three forces must vanish: the impressed forces, the centrifugal forces and the Euler forces.

We set up a rectangular system of axes x, y, z, with origin at the centre of mass. We shall assume that these axes are rigidly connected with the moving body, and coincide with the "principal axes." This means that in our reference system the "products of inertia" are zero:

$$\Sigma myz = \Sigma mzx = \Sigma mxy = 0. \tag{47.1}$$

Moreover, we introduce the customary notations:

$$\Sigma m(y^2 + z^2) = A,$$
$$\Sigma m(z^2 + x^2) = B,$$
$$\Sigma m(x^2 + y^2) = C. \tag{47.2}$$

Problem 2. Show that the resultant moment of the centrifugal forces (45.11) has components

$$(B - C)\omega_y\omega_z, \ (C - A)\omega_z\omega_x, \ (A - B)\omega_x\omega_y. \tag{47.3}$$

Problem 3. Show that the resultant moment of the Euler forces (45.14) has components

$$- A\dot{\omega}_x, \ - B\dot{\omega}_y, \ - C\dot{\omega}_z. \tag{47.4}$$

Let us denote the components of the resultant moment $\overline{\mathbf{M}} = \Sigma(\mathbf{R} \times \mathbf{F})$ of the impressed forces by

$$\overline{M}_x, \ \overline{M}_y, \ \overline{M}_z. \tag{47.5}$$

The vanishing of the resultant of the moments (47.3), (47.4), and (47.5) gives Euler's equations:

$$A\dot{\omega}_x - (B - C)\omega_y\omega_z = \overline{M}_x,$$
$$B\dot{\omega}_y - (C - A)\omega_z\omega_x = \overline{M}_y, \tag{47.6}$$
$$C\dot{\omega}_z - (A - B)\omega_x\omega_y = \overline{M}_z.$$

These famous equations describe how the instantaneous axis of rotation Ω changes with the time relative to a body-centered reference system. They give a partial solution of the dynamical problem of a freely rotating rigid body, but they have to be completed by fixing the position of the body-centered reference system relative to a space-centered system. This problem, and many others in the dynamics of rigid bodies, is outside the scope of the present book, which is devoted to the general principles of mechanics and deals with applications only as illustrations of the operation of the general principles. For further developments the reader is referred to the textbooks given in the Bibliography.

Summary. Euler's equations, which describe the rate of change of the instantaneous axis of rotation of a rigid body relative to the body-centered reference system of the principal axes, can be interpreted as the conditions for the vanishing of the resultant of three moments: those of the Euler forces, the centrifugal forces, and the external forces.

8. Gauss' principle of least constraint. D'Alembert's principle is not a minimum principle. It makes use of an infinitesimal quantity—the virtual work of the impressed forces, augmented by the virtual work of the forces of inertia—but the latter quantity is not the variation of some function. Gauss (1777-1855) gave an ingenious reinterpretation of d'Alembert's principle which changes it into a minimum principle. The procedure of Gauss can be described as follows.

The differential equations of motion determine the *accelerations* of the position coordinates q_i. Let us suppose that we are confronted with the following problem. At a certain time t the position and velocity of all the particles of a mechanical system are given. However, we may dispose of the accelerations at will, though subject to the given constraints. If the actual acceleration of a particle at time t is $\mathbf{A}(t)$, the change of that acceleration to $\mathbf{A} + \delta\mathbf{A}$ has the following effect. The position of the particle at time $t + \tau$ is given by Taylor's theorem:

$$\mathbf{R}(t + \tau) = \mathbf{R}(t) + \mathbf{v}(t)\tau + \tfrac{1}{2}\,\mathbf{A}(t)\tau^2 + \dots \qquad (48.1)$$

Now the variation of the acceleration will cause a deviation in the path of the particle. The variation of the first two terms of the right-hand side of (48.1) is zero, in accordance with our convention that we shall not change the initial position and velocity of the system. Hence, if we consider τ as an arbitrarily small time interval and thus neglect terms of higher order, we obtain

$$\delta\mathbf{R}(t + \tau) = \tfrac{1}{2}\,\delta\mathbf{A}(t) \cdot \tau^2. \qquad (48.2)$$

This consideration is entirely independent of any constraints; they enter only in the choice of the $\delta\mathbf{A}(t)$.

Now we know that d'Alembert's principle requires that

$$\sum_{i=1}^{N} (\mathbf{F}_i - m_i\mathbf{A}_i) \cdot \delta\mathbf{R}_i = 0, \tag{48.3}$$

where the $\delta\mathbf{R}_i$ may be identified with any virtual displacements which are in accord with the given constraints. We can apply this principle at time $t + \tau$, choosing our $\delta\mathbf{R}_i(t + \tau)$ according to (48.2). This gives

$$\sum_{i=1}^{N} (\mathbf{F}_i - m_i\mathbf{A}_i) \cdot \delta\mathbf{A}_i = 0. \tag{48.4}$$

Since the impressed forces $\mathbf{F}_i$ are given and cannot be varied, the condition (48.4) may be rewritten as follows:

$$\sum_{i=1}^{N} (\mathbf{F}_i - m_i\mathbf{A}_i) \cdot \delta\left(\frac{\mathbf{F}_i - m_i\mathbf{A}_i}{m_i}\right) = 0, \tag{48.5}$$

but this means that

$$\delta \sum_{i=1}^{N} \frac{1}{2m_i} (\mathbf{F}_i - m_i\mathbf{A}_i)^2 = 0. \tag{48.6}$$

Gauss defined the quantity

$$Z = \sum_{i=1}^{N} \frac{1}{2m_i} (\mathbf{F}_i - m_i\mathbf{A}_i)^2 \tag{48.7}$$

as the "constraint" of the motion and expressed the equation (48.6) in the form of the "principle of least constraint": the actual motion occurring in nature is such that under the given kinematical conditions the constraint becomes as small as possible. If the particles are free from constraints, Z can assume its absolute minimum, which is zero. We get then

$$m_i\mathbf{A}_i = \mathbf{F}_i \tag{48.8}$$

which expresses Newton's law of motion. If constraints prevent the free choice of the $\mathbf{A}_i$, we can still minimize Z under the given auxiliary conditions. The solution obtained yields the actual motion of the system realized in nature.

Example: A particle is forced to stay on the surface

$$z = f(x, y), \tag{48.9}$$

and is acted upon by the force $\mathbf{F}$. Find the equations of motion. We now have

$$\ddot{z} = f_x\ddot{x} + f_y\ddot{y} + f_{xx}\dot{x}^2 + \ldots. \tag{48.10}$$

The quantity to be minimized is

$$(F_1 - m\ddot{x})^2 + (F_2 - m\ddot{y})^2 + (F_3 - m\ddot{z})^2, \tag{48.11}$$

where $\ddot{z}$ is given by (48.10). The independent variables are $\ddot{x}, \ddot{y}$. This yields:

$$F_1 - m\ddot{x} + (F_3 - m\ddot{z})f_x = 0,$$
$$F_1 - m\ddot{y} + (F_3 - m\ddot{z})f_y = 0, \tag{48.12}$$

and these are the equations of motion.

Gauss was much attached to this principle because it represented a perfect physical analogy to the "method of least squares" (discovered by him and independently by Legendre), in the adjustment of errors. If a functional relation involves certain parameters which have to be determined by observations, the calculation is straight-forward so long as the number of observations agrees with the number of unknown parameters. But if the number of observations exceeds the number of parameters, the equations become contradictory on account of the errors of observation. The hypothetical value of the function minus the observed value is the "error." The sum of the squares of all the individual errors is now formed, and the parameters of the problem are determined by the principle that this sum shall be a minimum.

The principle of minimizing the quantity Z is completely analogous to the procedure sketched above. The $3N$ terms of the sum (48.7) correspond to $3N$ observations. This number is in excess of the number of unknowns $\ddot{q}_i$, on account of the m given kinematical conditions. The "error" is represented by the deviation of the impressed force from the force of inertia "mass times acceleration." Even the factor $1/m_i$ in the expression for Z can be interpreted as a "weight factor," in analogy with the case of observations of different quality which are weighted according to their estimated reliability.

Although the condition (48.6) establishes at first only the stationary value of Z, we can easily prove that in the present case the stationary value leads automatically to a minimum, without any further conditions. This follows from the fact that Z—being a sum of essentially positive terms—must have a minimum somewhere. Hence, if the condition for the stationary value gives a unique solution, that solution must give the minimum of Z. Now the uniqueness of the solution provided by the Gaussian principle can be established as follows. We have seen in the above example that $\ddot{z}$ came out as a *linear* function of $\ddot{x}$ and $\ddot{y}$. Generally, no matter what the given kinematical conditions are, differentiating twice we always get *linear* relations between the accelerations. If we eliminate the surplus accelerations from these conditions, the resulting expression for Z will again be a quadratic function of the remaining accelerations which can now be varied freely. Hence we get a system of *linear* equations which has a unique solution.

The Gaussian principle of least constraint is thus a true minimum principle, comparable with the principle of least action, but simpler in not requiring an integration with respect to the time. However, this advantage is more than offset by the disadvantage that the *accelerations* are required, while the principle of least action needs the velocities only. This puts the principle of Gauss in a notably inferior position. On the other hand, it can be used for non-holonomic forces no less than for holonomic ones, without losing its minimum character, while the principle of least action cannot be formulated as a minimum principle if the forces or the auxiliary conditions are non-holonomic.

Hertz[1] gave a striking geometrical interpretation of the Gaussian principle for the special case where the impressed forces vanish. He showed that in this case the constraint Z can be interpreted as the "geodesic curvature" of the C-point which represents the position of the mechanical system in a $3N$-dimensional Euclidean space with the rectangular coordinates $\sqrt{m_i}\, x_i$, $\sqrt{m_i}\, y_i$, $\sqrt{m_i}\, z_i$ (cf. chap. I, sections 4 and 5). Our point is bound to stay within a certain subspace of that $3N$-dimensional space, owing to the given constraints. The principle that Z has to become a minimum can now be formulated as the principle that the moving C-point tries to reduce the curvature of its path at any point to the smallest value compatible with the given constraints. This means that the path of the C-point tends to be *as straight as possible*. This Hertzian "principle of the straightest

[1] Cf. H. Hertz, *Principles of Mechanics* (London, Macmillan, 1900).

path" brings the Gaussian principle into close relation with Jacobi's principle which achieves the same aim much more directly, by minimizing the arc length of the configuration space. The Hertzian "paths of least curvature" can be interpreted as *the geodesic lines of that curved configuration space which is imbedded in the flat $3N$-dimensional space of Hertz.* (cf. chap. I, section 5).

Appell[1] gave a modified formulation of the Gaussian principle which made it particularly suitable for the derivation of the equations of motion in the case of non-holonomic conditions and in cases where it is desirable to employ kinematical variables of the type considered in section 2.

Summary. By a special form of variation Gauss succeeded in changing d'Alembert's principle into a genuine minimum principle in which a scalar quantity—called by Gauss the "constraint"—is minimized, considering the accelerations as the variables of the minimum problem. As a minimum principle, the principle of least constraint parallels the principle of least action. It is simpler than the latter principle in not requiring the calculus of variations because no definite integral, but an ordinary function, has to be minimized. However, the great disadvantage of the principle is that it requires the evaluation of the accelerations; this, in general, involves rather cumbersome and elaborate calculations, while in the principle of least action everything is deduced from a scalar function which contains no derivatives higher than the first.

[1]P. E. Appell, *"Sur une forme générale des équations de la dynamique"* (*Mémorial des sciences mathém.,* fasc. 1, Paris, Gauthier-Villars, 1925).

CHAPTER V

THE LAGRANGIAN EQUATIONS OF MOTION

Introduction. We now arrive at the typical "variational principles," namely, those principles which operate with the minimum—or more generally the stationary value—of a definite integral. The polygenic character of the force of inertia can be overcome if we integrate with respect to the time. By this procedure the problem of dynamics is reduced to the investigation of a scalar integral. The condition for the stationary value of this integral gives all the equations of motion.

In spite of the various names connected with these principles—Euler, Lagrange, Jacobi, Hamilton—they are all closely related to each other, and the name "principle of least action," if taken in the broader sense, applies to them all.

In our exposition of these principles we shall not follow the historical development, but shall start with "Hamilton's principle," which is the most direct and most natural transformation of d'Alembert's into a minimum principle. From this we can obtain by specialization the older form of the principle used by Euler and Lagrange, and likewise Jacobi's principle.

1. Hamilton's principle. D'Alembert's principle operates with a non-integrable differential. A certain infinitesimal quantity $\overline{\delta w^e}$—the total virtual work done by impressed and inertial forces—is equated to zero. The two parts of the work done are very different in their nature. While the virtual work of the impressed forces is a monogenic differential, deducible from a single function, the work function, the virtual work of the inertial forces cannot be deduced from a single function but has to be formed for each particle separately. This puts the inertial forces at a great disadvantage compared with the impressed forces. It is of the greatest theoretical and practical importance that this situation can be remedied by a transformation which brings d'Alembert's principle into a monogenic form. Although implicitly used by Euler and Lagrange, it was Hamilton who first transformed d'Alembert's principle, showing that an integration with respect to the time brings the work done by the inertial forces into a monogenic form.

111

Let us multiply $\overline{\delta w^e}$ by dt and integrate between the limits $t = t_1$ and $t = t_2$:

$$\int_{t_1}^{t_2} \overline{\delta w^e} \, dt \equiv \int_{t_1}^{t_2} \Sigma \left[\mathbf{F}_i - \frac{d}{dt}(m_i \mathbf{v}_i) \right] \cdot \delta \mathbf{R}_i \, dt. \qquad (51.1)$$

We separate the right-hand side into two parts. The first part can be written

$$\int_{t_1}^{t_2} \Sigma \, \mathbf{F}_i \cdot \delta \mathbf{R}_i \, dt = -\int_{t_1}^{t_2} \delta V \, dt = -\delta \int_{t_1}^{t_2} V \, dt. \qquad (51.2)$$

(We assume that the work function is independent of the velocities and we put $V = -U$). In the second part an integration by parts can be performed:

$$-\int_{t_1}^{t_2} \frac{d}{dt}(m_i \mathbf{v}_i) \cdot \delta \mathbf{R}_i \, dt = -\int_{t_1}^{t_2} \frac{d}{dt}(m_i \mathbf{v}_i \cdot \delta \mathbf{R}_i) \, dt$$

$$+ \int_{t_1}^{t_2} m_i \mathbf{v}_i \cdot \frac{d}{dt}(\delta \mathbf{R}_i) \, dt. \qquad (51.3)$$

The first part of this sum is integrable and we get a boundary term:

$$- [m_i \mathbf{v}_i \cdot \delta \mathbf{R}_i]_{t_1}^{t_2}, \qquad (51.4)$$

while the second part, making use of the interchangeable nature of variation and differentiation (see eq. (29.3)), may be written

$$\int_{t_1}^{t_2} m_i \mathbf{v}_i \cdot \frac{d}{dt} \delta \mathbf{R}_i \, dt = \int_{t_1}^{t_2} m_i \mathbf{v}_i \cdot \delta \mathbf{v}_i \, dt$$

$$= \tfrac{1}{2} \int_{t_1}^{t_2} m_i \, \delta(\mathbf{v}_i \cdot \mathbf{v}_i) \, dt = \tfrac{1}{2} \delta \int_{t_1}^{t_2} m_i v_i^2 \, dt. \qquad (51.5)$$

Summing over all particles we finally get:

$$\int_{t_1}^{t_2} \overline{\delta w^e} \, dt = \delta \int_{t_1}^{t_2} \tfrac{1}{2} \Sigma \, m_i v_i^2 \, dt - \delta \int_{t_1}^{t_2} V \, dt$$

$$- [\Sigma m_i \mathbf{v}_i \cdot \delta \mathbf{R}_i]_{t_1}^{t_2}. \qquad (51.6)$$

We introduce the kinetic energy T of the mechanical system according to the definition (15.9) and we furthermore set

$$L = T - V. \qquad (51.7)$$

This function L, defined as the excess of kinetic energy over potential energy, is the most fundamental quantity in the mathematical analysis of mechanical problems. It is frequently referred to as the "Lagrangian function."

With these definitions we can write (51.6) in the form:

$$\int_{t_1}^{t_2} \overline{\delta w^e}\, dt = \delta \int_{t_1}^{t_2} L\, dt - [\Sigma\, m_i \mathbf{V}_i \cdot \delta \mathbf{R}_i]_{t_1}^{t_2}. \qquad (51.8)$$

So far the variations $\delta \mathbf{R}_i$ are arbitrary virtual changes of the radius vectors $\mathbf{R}_i$. We now require that the $\delta \mathbf{R}_i$ *shall vanish at the two limits* t_1 *and* t_2:

$$\delta \mathbf{R}_i\,(t_1) = 0,$$
$$\delta \mathbf{R}_i\,(t_2) = 0. \qquad (51.9)$$

This means that the position of the mechanical system is supposed to be given for $t = t_1$ and $t = t_2$, and no variations are allowed at these limits. We say that we "vary between definite limits", because the limiting positions of the system are precribed. In that case the boundary term on the right-hand side of (51.8) vanishes and *the integral with respect to the time of the virtual work done by the effective forces becomes the variation of a definite integral*:

$$\int_{t_1}^{t_2} \overline{\delta w^e}\, dt = \delta \int_{t_1}^{t_2} L\, dt = \delta A, \qquad (51.10)$$

with

$$A = \int_{t_1}^{t_2} L\, dt. \qquad (51.11)$$

Since d'Alembert's principle requires the vanishing of $\overline{\delta w^e}$ at any time, the left-hand side of (51.10) vanishes also. Hence d'Alembert's principle can be reformulated as follows:

$$\delta A = 0. \qquad (51.12)$$

This is "Hamilton's principle" which states that the motion of an arbitrary mechanical system occurs in such a way that *the definite integral A becomes stationary for arbitrary possible variations of the configuration of the system, provided the initial and final configurations of the system are prescribed.*

Hamilton gave an improved mathematical formulation of a principle which was well established by the fundamental investigations of Euler and Lagrange; the integration process employed by him was likewise known to Lagrange. The name "Hamilton's principle," coined by Jacobi, was not adopted by the scientists of the last century. It came into use, however, through the text-books of more recent date.

Our reasoning which led to Hamilton's principle can be pursued in the opposite order. We can start out with the postulate that δA vanishes for arbitrary variations of position, transform δA to the left-hand side of (51.10) and deduce the vanishing of $\overline{\delta w^e}$, which is d'Alembert's principle. This shows that Hamilton's principle and d'Alembert's principle are mathematically equivalent and their scopes are the same as long as the impressed forces which act on the mechanical system are monogenic. For polygenic forces the transformation of d'Alembert's principle into a minimum—or strictly speaking stationary value—principle is not possible. Since holonomic kinematical conditions are mechanically equivalent to monogenic forces, non-holonomic conditions to polygenic forces, we can say that *Hamilton's principle holds for arbitrary mechanical systems which are characterized by monogenic forces and holonomic auxiliary conditions.* The *conservative* nature of these forces and auxiliary conditions—i.e. their independence of the time—is *not* required.

While d'Alembert's principle makes an independent statement at each instant of time during the motion, Hamilton's principle includes all these in one single statement. The motion is taken *as a whole.*

The unifying quality of a variational principle is truly remarkable. Although the modern development of physics deviates essentially from the older course on account of relativity and quantum theory, yet the idea of deriving the basic equations of nature from a variational principle has never been abandoned, and both the equations of relativity and the equations of wave mechanics share with the older equations of physics the common feature that they are derivable from a "principle of least action." It is only the Lagrangian function L which has to be defined in a different manner.

Summary. By an integration with respect to the time the virtual work done by the force of inertia can be transformed into a true variation. D'Alembert's principle can thus be mathematically reformulated as Hamilton's principle, which requires that a definite integral shall be stationary, namely the time-integral of the Lagrangian function L, where L is the difference between the kinetic and potential energies. The variation has to occur between definite end-positions.

2. The Lagrangian equations of motion and their invariance relative to point transformations. In the derivation of the principle of least action rectangular coordinates are used. However, the mechanical system may be subject to kinematical conditions; if these conditions are holonomic, the $3N$ rectangular position coordinates of the system will be expressible in terms of n generalized coordinates $q_1, \ldots, q_n$, as in (12.8) and (18.3). Then both the potential and kinetic energies are functions of the generalized coordinates and the generalized velocities $\dot{q}_1, \ldots, \dot{q}_n$. The motion of the system can now be pictured as that of the single C-point in the n-dimensional configuration space of the q_i. However, we can go one step further in the geometrization of dynamics by plotting the time t as one more additional dimension and thus operate with an $n + 1$-dimensional "extended configuration space" which comprises the generalized coordinates *and the time t* as independent variables.[1] In this space the successive phases of the motion show up as successive points of a *curve.* This curve, the "world-line" of the C-point, contains in geometrical form the entire physical history of the mechanical system.

Now in this picture the variation of the position of the system at any time between t_1 and t_2 becomes a variation of the curve C

[1]The idea of adding the time t as an additional dimension for the geometrization of mechanics was first introduced by Lagrange in his *Mécanique Analytique* (1788).

i.e. of the world-line of the mechanical system. Since the two end-positions of the system at the times t_1 and t_2 are given, the variation occurs between definite limits, which means that the varied curve C' has the same end-points A and B. The time t does not play any special role in this representation and need not be considered even as the independent variable. One could characterize the curve C in parametric form equally well, giving all the q_i *and the time t* as functions of a parameter τ.

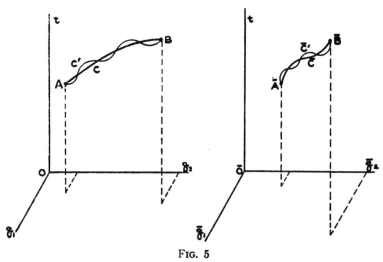

Fig. 5

The necessary and sufficient conditions for the action integral (51.11) to be stationary are (see chap. II):

$$\frac{d}{dt}\frac{\partial L}{\partial \dot{q}_i} - \frac{\partial L}{\partial q_i} = 0, \; (i = 1, \ldots, n). \tag{52.1}$$

It is a remarkable fact that the problem of minimizing a definite integral is quite independent of any special reference system. Let us assume that the original set of coordinates q_i is changed to a new set of coordinates by a point-transformation (14.3). This point-transformation can be pictured as a mapping of the n-dimensional q-space on itself: (cf. chap. I, section 4). In the (q, t)-space the curve C is transformed into some new curve

$\bar{C}$. The varied curve C' is transformed into a corresponding variation $\bar{C}'$ of $\bar{C}$. Variation between definite limits in the (q, t)-space means variation between definite limits in the $(\bar{q}, t)$-space; (cf. Fig. 5). The vanishing of the variation of the integral A requires the vanishing of this variation when expressed in terms of the new coordinates. Hence *the differential equations of Euler-Lagrange remain valid in the new reference system.* The Lagrangian function L and the action integral A are invariants of the transformation. We merely substitute for q_i the functions which express them in terms of the new variables $\bar{q}_i$. The *form* of L in the new variables is quite different from what it was before, but its *value* remains the same. Similarly, the value of the action integral A, taken along the curve C, and then along the curve $\bar{C}$, remains the same. The Lagrangian equations of motion, if taken separately for a certain index i, are not transformed into the corresponding equations in the new reference system. But the *complete set of equations* is transformed into the corresponding set in the new coordinate system,[1] because the complete set of Lagrangian equations expresses the vanishing of the variation of the action integral A, and this statement is independent of any special system of coordinates.

Since t is merely one more additional variable, these considerations remain valid even if the relations between the old and the new q_i depend on the time t (see chap. I, section 8). This is the case if the mechanical phenomena are described in a reference system which is in motion. *The Lagrangian equations of motion remain valid even in arbitrarily moving reference systems.*

The remarkable *invariance of the Lagrangian equations with respect to arbitrary point-transformations* gives these equations a unique position in the development of mathematical thought. These equations stand out as the first example of that " principle of invariance" which was one of the leading ideas of 19th century mathematics, and which has become of dominant importance in contemporary physics.

The invariance of the Lagrangian equations of motion is one of their most important features. It makes it possible to adjust

---

[1]For this reason we use the word "covariant" rather than "invariant" if a whole system of equations is involved.

the type of coordinates employed to the nature of the problem. There is no general method known by which the Lagrangian equations can be solved. The best that we can do is to try to find a system of coordinates in which the equations are at least partially integrable.

Another great advantage of the principle of least action, compared with d'Alembert's principle, is the use of the *single scalar function L*. It is no longer necessary to find the acceleration of each particle and the virtual work done by all the inertial forces. *The scalar function L = T − V determines the entire dynamics of the given system.*

Problem 1. Characterize the position of a compound pendulum by the angle which the plane passing through the axis of suspension and the centre of mass makes with the vertical. Show that

$$T = \tfrac{1}{2} I \dot{\theta}^2,$$

where I is the moment of inertia about the axis of suspension, and

$$V = M g l (1 - \cos \theta),$$

where M is the mass of the pendulum and l the distance of the centre of mass from the axis of suspension.

Form the Lagrangian equation of motion and integrate it assuming θ to be small.

Problem 2. Characterize the position of a planet by the polar coordinates r, θ. Show that:

$$T = \frac{m}{2} (\dot{r}^2 + r^2 \dot{\theta}^2),$$

$$V = -\frac{fmM}{r},$$

assuming the sun fixed at the origin.

Form the Lagrangian equations of motion.

Problem 3. Characterize the position of a spherical pendulum of length l by spherical polar coordinates r, θ, ϕ and obtain:

$$T = \frac{m}{2} l^2 (\dot{\theta}^2 + \sin^2 \theta \, \dot{\phi}^2),$$

$$V = mgl (1 - \cos \theta).$$

Form the Lagrangian equations of motion.

The two basic functions T and V correspond to the two sides of the Newtonian equation: "Mass times acceleration equals force". This equation can be interpreted as the balance between the force of inertia and the moving force. A similar balance can be established in the analytical treatment by separating the two basic scalars of analytical mechanics: the kinetic energy T and the work function U. The Lagrangian equations can be written

$$\frac{d}{dt}\frac{\partial T}{\partial \dot{q}_i} - \frac{\partial T}{\partial q_i} = -\frac{d}{dt}\frac{\partial U}{\partial \dot{q}_i} + \frac{\partial U}{\partial q_i}. \tag{52.2}$$

The n quantities

$$\frac{d}{dt}\frac{\partial T}{\partial \dot{q}_i} - \frac{\partial T}{\partial q_i} \tag{52.3}$$

represent the n components of the force of inertia in the configuration space. The n quantities

$$-\frac{d}{dt}\frac{\partial U}{\partial \dot{q}_i} + \frac{\partial U}{\partial q_i} \tag{52.4}$$

similarly represent the n components of the moving force.

Ordinarily, the first term of (52.4) is omitted on the ground that U is a function of the q_i alone. Moreover, U is replaced by $-V$. However, there is no inherent reason for excluding the possibility that the work function may depend on the velocities $\dot{q}_i$. In the relativistic form of mechanics this is actually the case; the electromagnetic force which acts on a charged particle can be derived from a work function dependent on velocity.

Summary. Hamilton's principle leads to a system of simultaneous differential equations of the second order, the Lagrangian equations of motion. These equations have the remarkable property of remaining invariant with respect to arbitrary transformations of the coordinates.

---

3. **The energy theorem as a consequence of Hamilton's principle.** The fundamental law of the conservation of energy, derived previously as a consequence of d'Alembert's principle (cf.

chap. IV, section 3), can now be established as a consequence of Hamilton's principle. This new derivation will bring into evidence the general relation which exists between the total energy of a mechanical system and the Lagrangian function L.

For this purpose we use once more that special variational process which we considered before when we related d'Alembert's principle to the energy theorem. We let the virtual displacement δq_i at each instant coincide with the actual displacement dq_i which takes place during the infinitesimal time $dt = \epsilon$. This means that we put

$$\delta q_i = dq_i = \epsilon \dot{q}_i. \tag{53.1}$$

Problem 1. Make a graph of this variation in the q, t diagram and show that it amounts to a *vertical shift of the actual curve downward* by the amount ϵ. Do not forget that the time t is *not* varied, in accordance with our general convention; (cf. chap. II, section 8).

This variation alters the coordinates $q_i(t)$ even at the two end points t_1 and t_2 and thus we cannot speak of a "variation between definite limits." We know that the equations of motion were derived from the principle

$$\delta \int_{t_1}^{t_2} L \, dt = 0, \tag{53.2}$$

provided that the q_i are not varied at the two end-points. But now we wish to see what happens to the variation of the action-integral if we do *not* adhere to the condition of varying between definite limits, but vary the position even at the limits. The method of integrating by parts in transforming the variation of a definite integral [cf. (210.4)] shows that the variation at the limits causes the appearance of a *boundary term*. We obtain:

$$\delta \int_{t_1}^{t_2} L \, dt = \left[\sum_{i=1}^{n} \frac{\partial L}{\partial \dot{q}_i} \delta q_i \right]_{t_1}^{t_2}. \tag{53.3}$$

Before we continue our deductions, we shall introduce a notation of fundamental importance. The partial derivations of the Lagrangian function with respect to the velocities play such a fundamental role in all considerations of analytical mechanics that a special name and notation has been created for them. We put

$$p_i = \frac{\partial L}{\partial \dot{q}_i},$$ (53.4)

and call these p_i the components of the "generalized momentum," because if we deal with a single free particle and use rectangular coordinates, the three quantities p_1, p_2, p_3 become identical with the rectangular components of the momentum $m\mathbf{v}$. In the general case, the quantities p_i do not even resemble the elementary definition and have to be considered as the components of a vector in the configuration space. The fundamental importance of the p_i as a new set of independent variables was recognized by Hamilton, who developed his entire theory from the realization that the momenta can be handled as a new set of mechanical variables. But even the Lagrangian equations gain in simplicity if we introduce the p_i as a set of intermediate variables and write the equations of motion in the form

$$\dot{p}_i = \frac{\partial L}{\partial q_i}.$$ (53.5)

With these momenta, the equation (53.3) becomes

$$\delta \int_{t_1}^{t_2} L \, dt = \left[\sum_{i=1}^{n} p_i \, \delta q_i \right]_{t_1}^{t_2}$$ (53.6)

Now for the special variation (53.1) of the q_i we can establish a certain relation between the variation of L and the actual infinitesimal change of L as it occurs during the interval dt. Let us assume that our system is *scleronomic*, i.e. L does not contain the time explicitly (cf. chap. I, section 8):

$$L = L(q_1, \ldots, q_n; \dot{q}_1, \ldots, \dot{q}_n).$$ (53.7)

Then we see at once that the condition (53.1) leads to a corresponding equation for the variation of L:

$$\delta L = dL = \epsilon \dot{L},$$ (53.8)

and we obtain

$$\delta \int_{t_1}^{t_2} L \, dt = \int_{t_1}^{t_2} \delta L \, dt = \int_{t_1}^{t_2} \epsilon \dot{L} \, dt = \epsilon \, [L]_{t_1}^{t_2},$$ (53.9)

while the right-hand side of (53.6) becomes

$$\left[\Sigma p_i \delta q_i \right]_{t_1}^{t_2} = \epsilon \left[\sum_{i=1}^{n} p_i \dot{q}_i \right]_{t_1}^{t_2}.$$ (53.10)

The equation (53.6) thus yields

$$\left[\sum_{i=1}^{n} p_i \dot{q}_i - L \right]_{t_1}^{t_2} = 0. \tag{53.11}$$

This means that the quantity $\Sigma p_i \dot{q}_i - L$ has the same value at the time $t = t_1$ as at the time $t = t_2$. Since, however, the limits t_1 and t_2 do not enter into the equations of motion and t_2 may be chosen arbitrarily, we obtain

$$\sum_{i=1}^{n} p_i \dot{q}_i - L = \text{const.} \tag{53.12}$$

This fundamental theorem does not assume any specific form for L, except its independence of the time t. Now in the mechanical applications of the calculus of variations the function L has usually the form $T - V$ and both T and V are functions of a definite character. T is a quadratic form in the velocities $\dot{q}_i$ [cf. (15.16)]:

$$T = \tfrac{1}{2} \sum_{i,\,k=1}^{n} a_{ik} \dot{q}_i \dot{q}_k, \tag{53.13}$$

— the a_{ik} being functions of the q_i — while V is ordinarily independent of the velocities. Under these circumstances we have

$$p_i = \frac{\partial T}{\partial \dot{q}_i} = \sum_{k=1}^{n} a_{ik} \dot{q}_k, \tag{53.14}$$

and thus

$$\sum_{i=1}^{n} p_i \dot{q}_i = \sum_{i,\,k=1}^{n} a_{ik} \dot{q}_i \dot{q}_k = 2T. \tag{53.15}$$

Equation (53.12) now assumes the form

$$2T - (T - V) = T + V = \text{const.} = E. \tag{53.16}$$

This is the well known *law of the conservation of energy.*

Our deduction shows that the customary form of the energy theorem: "The sum of kinetic and potential energies is constant during the motion," holds only under definite restricting conditions. It is not enough that the system shall be scleronomic. It is further necessary that the kinetic energy shall be a quadratic form in the velocities while the potential energy shall not contain the velocities. We encounter mechanical systems with "gyroscopic terms" which are linear and not quadratic in the velocities. Moreover, in relativistic mechanics the kinetic part of the Lagrangian function depends on the velocities in a more complicated way than in Newtonian mechanics. Such systems, if they are scleronomic, *still satisfy a conservation law.* But that law takes the more general form (53.12) which holds for *any* form of the Lagrangian function.

Notice that in the case of mechanical systems of *one* degree of freedom the energy theorem (53.12) leads to a *complete integration* of the problem. The equation (53.12) is then of the form

$$f(q, \dot{q}) = E, \tag{53.17}$$

from which $\dot{q}$ can be found:

$$\dot{q} = \varphi(E, q). \tag{53.18}$$

We then have

$$t = \int \frac{dq}{\varphi(E, q)} + \tau, \tag{53.19}$$

which gives t as a function of q. The inverse function gives q as a function of t, with the two constants of integration E and τ.

Problem 1. Solve the problem of the linear oscillator:
$$T = \tfrac{1}{2} m\dot{x}^2, \quad V = \tfrac{1}{2} kx^2,$$
with the help of the energy theorem.

Problem 2. Do the same for the problem of the compound pendulum, treated in Problem 1 of section 2. (Here the quadrature leads to an elliptic integral.)

We can interpret the quantity

$$\Lambda = \sum_{i=1}^{n} p_i \dot{q}_i - L \tag{53.20}$$

as the "total energy" of the mechanical system. It is, together with the Lagrangian function L, the most important scalar associated with a mechanical system. In fact, we shall see later that in the Hamiltonian form of dynamics this function Λ takes precedence over the Lagrangian function L, and, expressed in the proper variables, becomes the "Hamiltonian function" H of the mechanical system which completely replaces the original Lagrangian function L.

It is of interest to see what happens to the total energy Λ if the system is rheonomic, i.e. if L depends on the time t explicitly:

$$L = L(q_1, \ldots, q_n; \dot{q}_1, \ldots, \dot{q}_n; t). \tag{53.21}$$

In that case δL and dL are no longer equal because the variation δL occurs *at a definite time*, while the change dL occurs during the interval dt. Instead of (53.8) we now get the more general relation

$$\delta L = dL - \frac{\partial L}{\partial t}\, dt = \epsilon\left(\dot{L} - \frac{\partial L}{\partial t}\right), \tag{53.22}$$

and our previous equation (53.11) has now to be generalized as follows:

$$\left[\sum_{i=1}^{n} p_i \dot{q}_i - L\right]_{t_1}^{t_2} = -\int_{t_1}^{t_2} \frac{\partial L}{\partial t}\, dt. \tag{53.23}$$

This shows that the total energy Λ is no longer a constant, but changes according to the following law:

$$\Delta\Lambda = -\int_{t_1}^{t_2} \frac{\partial L}{\partial t}\, dt. \tag{53.24}$$

Problem 3. A simple pendulum hangs from a fixed pulley. The other end of the string is in the hand of an observer who pulls up the string slowly, thus shortening the length of the pendulum with uniform velocity. (Ehrenfest's pendulum.) Show that, neglecting friction, the amplitude of the oscillations increases in the following manner. The change of the total energy from the position $\theta = 0$ to the next position $\theta = 0$ is given by

$$\Delta E = -\tfrac{1}{2}\frac{\Delta l}{l}\, E,$$

where E is the energy constant of the undisturbed oscillations.

Summary. For scleronomic systems the Lagrangian equations of motion yield a first integral in the form $\Sigma p_i \dot{q}_i - L = E$, which can be interpreted as the law of the conservation of energy if we define the left-hand side of the equation as the total energy of the system. For the customary problems of classical mechanics the sum $\Sigma p_i \dot{q}_i$ is equal to twice the kinetic energy, in which case the energy theorem assumes the form $T + V = E$.

4. Kinosthenic or ignorable variables and their elimination.
We have mentioned before that a general method of integration
of the Lagrangian equations cannot be given. It is possible,
however, that under special circumstances at least a *partial*
integration can be accomplished. A particularly important example of such a situation is the case of "ignorable" or "kinosthenic" variables.

The Lagrangian function L is generally a function of all the
position coordinates q_i and the velocities $\dot{q}_i$. Now it may happen
that a certain variable q_k does not appear in the Lagrangian
function, although $\dot{q}_k$ is present. The fundamental importance
of these variables for the integration of the Lagrangian equations
was first recognized by Routh[1] and somewhat later by Helmholtz.[2]
Routh called these variables "absent coordinates" while J. J.
Thomson refers to them as "kinosthenic" or "speed coordinates."
Helmholtz calls the same coordinates "cyclic variables"
while the textbook of Whittaker (cf. Bibliography) uses the
name "ignorable variables." The present book adopts the latter
expression, but in a few instances the name "kinosthenic" seems
more appropriate.

We have seen that, after introducing the "momenta" p_k
according to the definition (53.4), we can write the Lagrangian
equations of motion in the following form:

$$\dot{p}_k = \frac{\partial L}{\partial q_k}. \tag{54.1}$$

Now, if q_k is an ignorable variable,

$$\frac{\partial L}{\partial q_k} = 0, \tag{54.2}$$

and (54.1) can be integrated at once to give

$$p_k = \text{const.} = c_k. \tag{54.3}$$

*The momentum connected with an ignorable variable is a constant
during the motion.* This important theorem has many physical
applications.

[1]E. J. Routh, *Dynamics of Rigid Bodies* (Macmillan, 1877).
[2]H. V. Helmholtz, *Journal of Math.*, 97 (1884), p. 111.

Problem 1. Consider the problem of planetary motion, treated before in Problem 2 of section 2. Show that the constancy of the kinosthenic momentum p_θ expresses Kepler's law of areas.

Problem 2. Examine in a similar way the problem of the spherical pendulum considered in Problem 3, section 2, interpreting the constancy of the kinosthenic momentum.

Since q_k is not present in the partial derivative $p_k = \partial L/\partial \dot{q}_k$, but $\dot{q}_k$ is present, we can obtain $\dot{q}_k$ from (54.3) as a function of the non-ignorable variables and velocities. For the sake of simpler formalism we restrict our considerations to one single ignorable coordinate, the generalization to any number being obvious. We assume that we have arranged the subscripts of the coordinates q_i in such a manner that the *last* coordinate q_n is the ignorable coordinate, so that

$$\frac{\partial L}{\partial \dot{q}_n} = c_n; \tag{54.4}$$

and hence

$$\dot{q}_n = f(q_1, \ldots, q_{n-1}; \dot{q}_1, \ldots, \dot{q}_{n-1}; c_n; t). \tag{54.5}$$

Whenever we encounter $\dot{q}_n$ in the Lagrangian equations (q_n itself is not present), we can replace it by the right-hand side of (54.5). Hence the problem of integration of the Lagrangian equations can be reduced to one involving the non-ignorable variables alone. After obtaining the solution of the reduced problem, we finally put the q_i and $\dot{q}_i$—which are now given functions of t— in the relation (54.5) and obtain q_n by a quadrature.

If this reduction process can be carried through after writing down the Lagrangian equations of motion, the question can be raised whether we could not eliminate the ignorable variables *before* writing down these equations, i.e. in the formulation of the variation problem itself. We know that the first variation of the integral

$$A = \int_{t_1}^{t_2} L(q_1, \ldots, q_{n-1}; \dot{q}_1, \ldots, \dot{q}_{n-1}, \dot{q}_n; t) \, dt \tag{54.6}$$

vanishes for arbitrary variations of the q_k between definite limits. If we eliminate $\dot{q}_n$ by means of the equation (54.5), our problem is reduced from n to $n - 1$ degrees of freedom *from the start*. Hence we can actually simplify the given variation problem by eliminating in advance all of the kinosthenic coordinates. However, we have to keep in mind that the relation (54.5) makes q_n a function of the non-ignorable variables in the sense that we assume the relation (54.4) to hold not only for the actual motion but for the varied motion as well. Now there is no objection to this restriction of the variation of q_n, since the variation of A vanishes for *arbitrary* variations of all the q_k. However, the condition that the variation of q_n must vanish at the two end-points of the range is violated because q_n is obtained from (54.5) by a quadrature. Hence, from equation (53.3) we now have

$$\delta \int_{t_1}^{t_2} L \, dt = \left[p_n \, \delta q_n \right]_{t_1}^{t_2}. \tag{54.7}$$

But p_n, the kinosthenic momentum, is constant everywhere along the C-curve. Hence we can write

$$\left[p_n \delta q_n \right]_{t_1}^{t_2} = p_n \, \delta \int_{t_1}^{t_2} \dot{q}_n \, dt = \delta \int_{t_1}^{t_2} p_n \dot{q}_n dt, \tag{54.8}$$

so that (54.7) becomes

$$\delta \int_{t_1}^{t_2} (L - p_n \dot{q}_n) \, dt = 0. \tag{54.9}$$

Let us put

$$L - c_n \dot{q}_n = \overline{L}, \tag{54.10}$$

and call $\overline{L}$ the "modified Lagrangian function." The result of our deduction can now be stated as follows: The minimizing of the action integral A *with* the ignorable variable q_n can be reduced to the minimizing of the modified action integral

$$\overline{A} = \int_{t_1}^{t_2} \overline{L} \, dt \tag{54.11}$$

without the ignorable variable q_n, after eliminating $\dot{q}_n$ with the help of the momentum integral $p_n = c_n$.

The process of eliminating an ignorable variable may be divided into three steps.

1. Write down the equation for the kinosthenic momentum

$$\frac{\partial L}{\partial \dot{q}_n} = c_n. \tag{a}$$

2.   Modify the given Lagrangian function to

$$\bar{L} = L - c_n \dot{q}_n. \tag{β}$$

3.   Eliminate the ignorable velocity $\dot{q}_n$ by solving the equation (a) for $\dot{q}_n$ and substituting in (β).   Then the new Lagrangian function $\bar{L}$ does not depend on the ignorable variable, and the original variation problem of n degrees of freedom is reduced to a new variation problem of $n - 1$ degrees of freedom.

This reduction process remains the same if the given problem contains more than one ignorable coordinate.   The modified Lagrangian function has now to be defined by

$$\bar{L} = L - \Sigma\, c_k \dot{q}_k, \tag{54.12}$$

the summation in the second term being extended over all the ignorable coordinates.

Problem 1. Apply the general reduction process to the problem of planetary motion studied before (cf. Problem 2 of section 2), eliminating $\dot{\theta}$. The reduced problem has but one degree of freedom and can be integrated by the energy theorem.

Problem 2.   Treat in the same way the problem of the spherical pendulum (cf. Problem 3 of section 2).

Problem 3.   Again, consider those ignorable coordinates which are present themselves but whose *velocities* do not appear in the Lagrangian function. Show that a variable of this nature can be eliminated algebraically with the help of the equation

$$\frac{\partial L}{\partial q_n} = 0,$$

without modifying the Lagrangian function L.

The elimination of an ignorable variable has interesting consequences for the reduced problem.   The kinetic energy T of the original system may be written as follows [cf. (53.13)] :

$$T = \tfrac{1}{2} \sum_{i,k=1}^{n-1} a_{ik} \dot{q}_i \dot{q}_k + \sum_{i=1}^{n-1} a_{in} q_i \dot{q}_n + \tfrac{1}{2} a_{nn} \dot{q}_n{}^2. \tag{54.13}$$

We have separated here the kinosthenic velocity from the other velocities.   Now the kinosthenic momentum becomes

$$p_n = \frac{\partial T}{\partial \dot{q}_n} = \sum_{i=1}^{n-1} a_{in} \dot{q}_i + a_{nn} \dot{q}_n = c_n, \tag{54.14}$$

and we obtain for the modified Lagrangian function

$$\bar{L} = L - c_n\dot{q}_n = \frac{1}{2} \sum_{i,k=1}^{n-1} a_{ik}\dot{q}_i\dot{q}_k - \frac{1}{2} a_{nn}\dot{q}_n^2 - V. \qquad (54.15)$$

The reduction process is not yet accomplished since we have to eliminate $\dot{q}_n$ with the help of (54.14). We obtain one term in $\bar{L}$:

$$- \frac{1}{2} \frac{c_n^2}{a_{nn}}, \qquad (54.16)$$

which is independent of any velocities. We further obtain $n-1$ terms which are *linear* in the non-kinosthenic velocities.

The term (54.16) behaves like an apparent potential energy and can be combined with V to give

$$\bar{V} = V + \frac{1}{2}\frac{c_n^2}{a_{nn}}. \qquad (54.17)$$

The remaining terms, however, are of an unusual nature. They can be considered as part of the kinetic energy T, but instead of being quadratic they are *linear* in the velocities. Such expressions in the Lagrangian function are called "gyroscopic terms" because they are responsible for the paradoxical behavior of a gyroscope. The Coriolis-force in a rotating reference system and the magnetic force due to electric currents are further examples of the effect of gyroscopic terms in the Lagrangian function.

Whether or not we get gyroscopic terms depends on the a_{in}-coefficients ($i \neq n$) between ignorable and non-ignorable velocities. We say that there is a "kinetic coupling" between ignorable and non-ignorable velocities if the a_{in} do not all vanish. If these coefficients vanish, there is no coupling and the reduced problem is free from gyroscopic terms.

Example. In the Kepler problem the elimination of θ yields an apparent potential energy of the form c^2/r^2. This means an apparent repulsive force which is proportional to $1/r^3$ while the attractive force is proportional to $1/r^2$. These two forces balance each other at a certain point, which is a point of stable equilibrium. The oscillation of r around that point explains the pulsation of the radius vector between perihelion and aphelion. If the attractive force decreased as $1/r^3$ or any higher inverse power of r, no stable equilibrium of the two forces would exist and the radius vector could not oscillate between finite limits. The paths of the planets would be either of a hyperbolic type, or of the nature of spirals approaching the sun—depending on the magnitude of the angular momentum constant. (The kinetic coupling is here zero.)

Summary. If a variable is not present explicitly in the Lagrangian function, but only its velocity, that variable is called kinosthenic or ignorable. The momentum connected with an ignorable variable is a constant during the motion. Ignorable variables can be eliminated from the Lagrangian function by suitably modifying it. The elimination produces an apparent potential energy; in addition, we may get an apparent kinetic energy which is not quadratic but linear in the velocities.

5. The forceless mechanics of Hertz. The properties of ignorable variables formed the basis of an ingenious theory of Hertz designed to explain the deeper significance of the potential energy. A mechanical system is characterized by a definite number of position coordinates, but it is not certain that all these coordinates are observable. A mechanical system may contain "microscopic parameters" which are not directly evident. On account of these "microscopic parameters" the number of degrees of freedom of a mechanical system may seem much smaller than it actually is. For example, a solid body is approximately rigid; but actually the molecules of the solid body oscillate about their average positions. The six degrees of freedom of a rigid body describe the mechanical behaviour of a solid only *macroscopically*.

Now these microscopic motions inside a mechanical system can be put into two categories: those expressible by kinosthenic and those expressible by non-kinosthenic variables. The motions associated with non-kinosthenic variables produce apparently polygenic forces in the macroscopic motion. A good example is the force of friction which acts macroscopically like a polygenic force, but is actually merely a substitute for the non-observable microscopic motion of the molecules.

Quite different in nature are those microscopic motions which are associated with kinosthenic variables. These can be eliminated, and it is possible to have a mechanical system with hidden motions which are not brought into evidence by any non-holonomic behavior. The reduced system is entirely holonomic and satisfies the principle of least action. The presence of hidden motions cannot be detected.

Now if there is a kinetic coupling between the macroscopic and the hidden ignorable coordinates, the Lagrangian function of the macroscopic system will contain gyroscopic terms which are linear in the observable velocities. If, however, this coupling is lacking, the presence of hidden motions will show up only in the form of an additional apparent potential energy of the macroscopic variables.

This consideration led Hertz[1] to the speculation that possibly the whole potential energy of the impressed forces may be caused by hidden motions expressible by kinosthenic variables. The dualism of kinetic and potential energy is a puzzling problem to philosophical thinking. We have on the one hand the inertial property of matter, on the other the force. The inertial property of matter is something derivable from the mere existence of mass. Pure inertia causes matter to move along a straight line, and the same holds true in the Riemannian space which pictures even the most complicated dynamical motion as the motion of a single point. One gets the impression that inertia is an inborn quality of matter which can hardly be reduced to something still simpler. Hence, from the philosophical point of view, we can be reconciled to the expression of the inertial property of matter by means of the kinetic energy. But no similar explanation can be offered for the "force." If the kinetic energy is the ultimate moving power of mechanics, could we not dispense somehow with the potential energy and thus eliminate the inexplicable dualism which creeps into mechanics on account of the two widely different forms of energy, kinetic and potential. Hertz thought to explain the potential energy as actually of *kinetic* origin, caused by the hidden motions of ignorable variables. The kinematical conditions which are imposed on the motion of the

[1] H. Hertz, *Die Prinzipien der Mechanik* (Barth, 1894), p. 256; cf. also J. J. Thomson, *Applications of dynamics to Physics and Chemistry* (Macmillan) 1888), p. 11.

microscopic parameters take the place of the force in the force-less mechanics of Hertz.

This ingenious hypothesis, although not developed beyond its sketchy beginnings, had nevertheless a prophetic significance. The theory of relativity, starting from entirely different consider-ations, and developing along different lines, provided an impres-sive example of the forceless mechanics of Hertz. Planetary motion around the sun was explained as caused by pure inertia, without any acting force. The planets trace out shortest lines in a Riemannian space, just as Hertz imagined for mechanical systems which are free from potential energy. The only differ-ence is that in Hertz' system the Riemannian curvature of the configuration space was caused by kinematical conditions im-posed on the hidden motions of the system, while in Einstein's theory the Riemannian structure of the physical space-time manifold is an inherent property of the geometry of the world.

Summary. If a mechanical system contains hidden microscopic motions expressible by ignorable variables these motions do not disturb the holonomic character of the macroscopic system, nor the validity of Hamil-ton's principle. They are "ignorable" because they can be eliminated. They cause an apparent potential energy which can be interpreted as the potential energy of impressed forces. This phenomenon led to the spec-ulation of Hertz that the potential energy of any me-chanical system may have its origin in hidden motions of a kinosthenic nature.

6. The time as kinosthenic variable; Jacobi's principle; the principle of least action. We consider a scleronomic or "conser-vative" mechanical system whose Lagrangian function does not contain the time explicitly. Let us imagine that we do not use the time t as the independent variable, but that all the $n + 1$ vari-ables $q_1, \ldots, q_n$ and t are given as functions of some parameter τ.

The system has now $n + 1$ degrees of freedom. Denoting derivatives with respect to τ by a prime, the action integral appears in the following form:

$$A = \int_{\tau_1}^{\tau_2} L \left(q_1, \ldots, q_n; \frac{q_1'}{t'}, \ldots \frac{q_n'}{t'} \right) t' d\tau. \qquad (56.1)$$

The time is now a kinosthenic variable since only t' appears in the integrand but not t itself.

We can make use of the theorem that the momentum connected with a kinosthenic variable is a constant of the motion. For this purpose let us form p_t, the momentum which is associated with the time t:

$$\begin{aligned} p_t &= \frac{\partial (Lt')}{\partial t'} = L - \left(\sum_{i=1}^{n} \frac{\partial L}{\partial \dot{q}_i} \frac{q_i'}{t'^2} \right) t' \\ &= L - \sum_{i=1}^{n} p_i \dot{q}_i \\ &= - \left(\sum_{i=1}^{n} p_i \dot{q}_i - L \right). \end{aligned} \qquad (56.2)$$

The expression in the last parenthesis is exactly that which we encountered before when dealing with the law of the conservation of energy (cf. section 3). We called it the "total energy" and denoted it by Λ. For the usual mechanical systems Λ is the sum of the kinetic and potential energies, $T + V$. We then have the important theorem —which holds whether the system is conservative or not—that *the momentum associated with the time t is the negative of the total energy.* If t is kinosthenic, i.e. if our system is conservative, we get at once

$$p_t = - \Lambda = \text{const.} = - E, \qquad (56.3)$$

which gives a new derivation of the energy theorem.

However, more than this can be done. We know that a kinosthenic variable can be eliminated, reducing the original variational problem by one degree of freedom. In the present case we can eliminate t from the variational problem and obtain a new variational problem which determines the motion in space but says nothing as to how the path is described with respect to the time t.

According to the general reduction process we first modify the Lagrangian function. In the present case we have to put

$$\overline{L} = Lt' - p_t t' = (L - p_t)t' = \sum_{i=1}^{n} p_i \dot{q}_i t', \qquad (56.4)$$

and thus our new action integral becomes

$$\overline{A} = \int_{\tau_1}^{\tau_2} \sum_{i=1}^{n} p_i \dot{q}_i \, t' \, d\tau, \qquad (56.5)$$

which, in view of (53.15), we can write

$$\overline{A} = 2 \int_{\tau_1}^{\tau_2} T \, t' \, d\tau. \qquad (56.6)$$

In the literature of the eighteenth century this integral frequently appeared in the form

$$\overline{A} = 2 \int T \, dt, \qquad (56.7)$$

but Jacobi pointed out that this is unsatisfactory because the time t cannot be used as an independent variable in the variational problem. Indeed, our process of reduction is not finished yet. We have to eliminate the kinosthenic variable t by means of the momentum equation (56.3) and this makes t a dependent variable. Without this elimination the variational principle (56.6) or (56.7) cannot be used.

We now make use of the symbolic C-point which represents the mechanical system in the configuration space. We know that the kinetic energy of the system can be written as the kinetic energy of a single particle of mass 1:

$$T = \tfrac{1}{2} \left(\frac{\overline{ds}}{dt} \right)^2, \qquad (56.8)$$

where $\overline{ds}$ is the line-element of the configuration space (see chap. I, section 5). Since our independent variable is no longer t but τ, we must write

$$T = \tfrac{1}{2} \left(\frac{\overline{ds}}{d\tau} \right)^2 / t'^2. \qquad (56.9)$$

Hence the elimination of t' from the energy theorem (56.3) gives

$$t' = \frac{1}{\sqrt{2(E - V)}} \frac{\overline{ds}}{d\tau}. \qquad (56.10)$$

Moreover, in the integrand of (56.6) we can put
$$T = E - V, \qquad (56.11)$$
and (56.6) finally becomes
$$\overline{A} = \int_{\tau_1}^{\tau_2} \sqrt{2\,(E - V)}\ \frac{\overline{ds}}{d\tau}\ d\tau$$
$$= \int_{\tau_1}^{\tau_2} \sqrt{2(E - V)}\ \ \overline{ds}. \qquad (56.12)$$

We have now completed the elimination process and have obtained the action integral of the reduced system. The time t does not appear in $\overline{A}$, which is moreover independent of the parameter τ. However, $\overline{ds}$ is not a complete differential and it would be quite wrong to believe that the integrand of $\overline{A}$ is $\sqrt{2(E - V)}$ and that $\overline{ds}$ corresponds to the differential of the independent variable. In order to prevent this misunderstanding, we put a dash above ds. *Some* parameter τ must be chosen as independent variable. In particular, we may take τ to be one of the q_i, for example q_n, giving all the other q_i as functions of q_n. This at once reduces the variational problem from n to $n - 1$ degrees of freedom.

The principle of minimizing the integral (56.12) in order to find the path of the mechanical system is called "Jacobi's principle." The time does not appear in its formulation. It determines the path of the C-point in the configuration space, not the motion in time. Yet this last part of the problem can easily be solved by integrating the equation (56.10) which will give t as a function of τ. This last equation is the expression of the energy theorem and no part of the variation problem. However, it completes the variation problem by determining how the motion occurs in time.

Jacobi's principle is a fundamental principle of mechanics. If we restrict ourselves to the case of a single particle, the line element $\overline{ds}$ becomes identical with the line-element of ordinary three-dimensional space in arbitrary curvilinear coordinates. Jacobi's principle then presents a remarkable analogy with Fermat's principle of least time in optics which determines the optical path by minimizing the integral

$$I = \int_{\tau_1}^{\tau_2} n \, \overline{ds}, \qquad (56.13)$$

where n is the refractive index, which may change from point to point, thus behaving in a manner similar to the quantity $\sqrt{2(E - V)}$. Maupertuis claimed that the phenomena of nature can be derived by minimizing a certain quantity which he called "action"—a quantity which, if suitably formulated, coincides with Jacobi's integral (56.12)—and he tried to show that the law of the refraction of light is deducible from the principle of least action as well as from Fermat's principle of least time. He thus pointed to that remarkable analogy between optical and mechanical phenomena which was observed much earlier by John Bernoulli and which was later fully developed in Hamilton's ingenious optico-mechanical theory. This analogy played a fundamental role in the development of modern wave-mechanics.

The link in this analogy is Jacobi's principle on the one hand and Fermat's principle on the other. The analogy does not go beyond the *path* described by a moving point in mechanics and by a light ray in optics. How the motion occurs *in time* is entirely different in the mechanical and in the optical case (cf. chap. VIII, section 7).

Jacobi criticized Lagrange's treatment of the "principle of least action," pointing out the importance of varying between definite limits, which is not possible if the time is used as the independent variable. In that case the upper limit of the action integral has to be varied in a definite manner, in order to satisfy the conservation of energy throughout the actual and the varied path. Nevertheless, the procedure of Euler and Lagrange in their formulation of the principle of least action is, when properly understood, entirely correct, and their principle is only formally different from Jacobi's principle. Jacobi's principle, as we have seen above, can be conceived as the result of the following steps.

1. In the kinetic energy replace differentiation with respect to t by differentiation with respect to the parameter τ:

$$T' = \tfrac{1}{2} \Sigma \, a_{ik} q_i' q_k' = T t'^2. \qquad (56.14)$$

2. Minimize the action integral

$$\overline{A} = 2 \int_{\tau_1}^{\tau_2} \frac{T'}{t'} \, d\tau,$$ (56.15)

after eliminating t' by means of the energy relation

$$\frac{T'}{t'^2} + V = E.$$ (56.16)

Now the "action" used by Euler and Lagrange was the same integral which serves as the basis for Jacobi's principle except for the use of the parameter τ. Moreover, Euler and Lagrange used (56.16) as an *auxiliary condition* which is equivalent to eliminating t' from that relation. We know that an auxiliary condition can be treated by elimination, but it can likewise be treated by the λ-method. The first method corresponds to Jacobi's procedure, the second to the procedure of Lagrange. The latter method leads to the following new form of the action integral:

$$\overline{\overline{A}} = \int_{\tau_1}^{\tau_2} \left[2 \frac{T'}{t'} + \lambda \left(\frac{T'}{t'^2} + V \right) \right] d\tau.$$ (56.17)

It so happens that in the present problem the undetermined multiplier λ can easily be evaluated. Since t' is one of our variables, we obtain by minimizing with respect to t'

$$-\frac{2T'}{t'^2} - \frac{2\lambda T'}{t'^3} = 0,$$ (56.18)

which gives

$$\lambda = -t'.$$ (56.19)

Hence the resulting integral becomes:

$$\overline{\overline{A}} = \int_{\tau_1}^{\tau_2} \left(\frac{T'}{t'^2} - V \right) t' d\tau = \int_{\tau_1}^{\tau_2} (T - V) t' d\tau.$$ (56.20)

Since, however, the new variational problem is a *free* problem, without any auxiliary conditions, there is now no reason why we should not use the time t as independent variable. This gives the action integral

$$\overline{\overline{A}} = \int_{t_1}^{t_2} (T - V) \, dt = A,$$ (56.21)

which leads back to *Hamilton's principle*. This deduction explains how Lagrange arrives at the Lagrangian equations of motion—which are actually direct consequences of Hamilton's principle—although starting from an entirely different principle, the "principle of least action."

Summary. In the parametric representation of motion the time is an additional position coordinate which can participate in the process of variation. The momentum associated with the time is the negative of the total energy. For scleronomic systems the time becomes a kinosthenic variable and the corresponding momentum a constant. This yields the energy theorem of conservative systems. The elimination of the time as an ignorable coordinate gives a new principle which determines only the path of the mechanical system, not the motion in time. This is Jacobi's principle which is analogous to Fermat's principle in optics. The same principle can be formulated as the "principle of least action." In the latter, the time-integral of double the kinetic energy is minimized with the auxiliary condition that both actual and varied motions shall satisfy the energy theorem during the motion. If this principle is treated by the λ-method, the resulting equations are the Lagrangian equations of motion.

7. Jacobi's principle and Riemannian geometry. As pointed out in chapter I, section 5, the geometrical structure of the configuration space is not in general Euclidean, but Riemannian. If a mechanical system consists of N free particles, then the configuration space is Euclidean of $3N$ dimensions. But if there are any constraints between these particles, then the configuration space is a curved subspace of less than $3N$ dimensions, the geometry of which can be characterized by a Riemannian line element. This line element is defined by the kinetic energy of the mechanical system, expressed in curvilinear coordinates q_k:

$$\overline{ds}^2 = 2T dt^2 = \Sigma \, a_{ik} dq_i dq_k. \tag{57.1}$$

Jacobi's principle brings out vividly the intimate relationship which exists between the motion of conservative holonomic systems and the geometry of curved spaces. We introduce, in addition to the line element $\overline{ds}$ of the configuration space, another Riemannian line element $\overline{d\sigma}$ defined by

$$\overline{d\sigma}^2 = (E - V)\overline{ds}^2. \tag{57.2}$$

According to (56.12), Jacobi's principle requires the minimizing of the definite integral

$$\overline{A} = \int_{\tau_1}^{\tau_2} \overline{d\sigma}. \tag{57.3}$$

This is the same as *finding the shortest path between two definite end-points in a certain Riemannian space.* We can associate with the motion of a mechanical system under the action of the potential energy V the motion of a point *along some geodesic of a given Riemannian space.* The problem of finding the solution of a given dynamical problem is mathematically equivalent to the problem of finding these geodesics.

In particular, let us restrict ourselves to the case where the potential energy V vanishes, i.e. where the motion occurs in the absence of any impressed forces. In that case we can dispense with the introduction of the additional line element $\overline{d\sigma}$ and can operate directly with the line element $\overline{ds}$ of the configuration space. Since $V = 0$, Jacobi's principle now requires:

$$\delta \int_{\tau_1}^{\tau_2} \overline{ds} = 0. \tag{57.4}$$

This means that *the C-point of the mechanical system moves along a shortest line or geodesic of the configuration space.* Moreover, the energy theorem gives

$$\frac{1}{2}\left(\frac{\overline{ds}}{dt}\right)^2 = E, \tag{57.5}$$

which shows that *the C-point moves with constant velocity.*

Here we have a truly beautiful generalization of the law of inertia, asserted by Leonardo da Vinci and Galileo: "A particle moves under its own inertia in a straight line, with constant velocity." This law still holds even for the most complicated mechanical systems subjected to arbitrary holonomic constraints. However, the "particle" which represents that system does not move in the ordinary three-dimensional space but in an n-dimensional Riemannian manifold.

Example. A particle is forced to stay on a curved surface; no impressed force is present. The Riemannian space is now two-dimensional and its geometry identical with the intrinsic geometry of the given surface. The moving particle traces out a geodesic line of that surface.

Problem. Show the same by using Hamilton's principle.

Notice the close relation of this geodesic principle to the dynamical principle of Einstein's theory. There, too, the problem of motion is equivalent to the finding of the geodesics of a Riemannian space. That Riemannian space has four dimensions, since space and time together form one unified manifold of four dimensions. The law of inertia gives the motion of the planets without the introduction of any specific force of gravity. Jacobi's principle holds for the relativistic mechanics of a particle. The only difference is that now the Riemannian structure of the space-time manifold is an innate property of the world and not a consequence of kinematical constraints.

Summary. Jacobi's principle links up the motion of holonomic conservative systems with Riemannian geometry. In particular, if the system moves under its own inertia without impressed forces, the representative C-point describes a geodesic or shortest path of the configuration space, which is an n-dimensional Riemannian space. The energy theorem adds that the motion occurs with constant velocity. This is a natural generalization of the ordinary law of inertia which states that a particle under the action of its own inertia moves in a straight line with uniform velocity.

8. Auxiliary conditions; the physical significance of the Lagrangian λ-factor. In the general treatment of the Calculus of Variations we have discussed in detail how mechanical systems with kinematical constraints can be handled (cf. chap. II, section 12). If these constraints are holonomic and appear in the form of algebraic relations between the variables:

$$f_i(q_1, \ldots, q_n, t) = 0, \ (i = 1, \ldots, m), \tag{58.1}$$

we can proceed in two different ways. We may eliminate m coordinates q_i with the help of the auxiliary conditions, thus reducing our system to a free system without kinematical constraints. Or we may dispense with eliminations and utilize the Lagrangian λ-method. In this method the integrand L of the given variational problem is modified to $\overline{L}$ by adding the left-hand sides of the equations (58.1), after multiplying each equation by some undetermined factor $\lambda_i(t)$.[1] Then the problem is again handled as a free problem, discarding the given auxiliary conditions.

Since the λ_i are undetermined factors, we can use $-\lambda_i$ as factors equally well and write $\overline{L}$ in the form

$$\overline{L} = L - (\lambda_1 f_1 + \ldots + \lambda_m f_m). \tag{58.2}$$

Moreover, in ordinary problems of classical mechanics L appears in the form $T - V$. We can combine the modification of L with the potential energy V by saying that V has been changed to $\overline{V}$ defined as follows:

$$\overline{V} = V + \lambda_1 f_1 + \ldots + \lambda_m f_m. \tag{58.3}$$

This elegant mathematical method has a striking and impressive physical counterpart. The fact that we make the variation problem free after modifying the Lagrangian function L means that we drop the given kinematical conditions and consider the mechanical system as without constraints. But then the modification of V to $\overline{V}$ means that *we add to the potential energy of the impressed forces the potential energy of the forces which maintain the given kinematical constraints.* These forces are given by:

[1] The λ-method remains valid even if the auxiliary conditions contain not only the q_i but the $\dot{q}_i$. We encountered such a situation in section 6 where the energy equation was an auxiliary condition of the variation.

$$K_i = - \frac{\partial}{\partial x_i} (\lambda_1 f_1 + \ldots + \lambda_m f_m)$$

$$= - \left(\lambda_1 \frac{\partial f_1}{\partial x_i} + \ldots + \lambda_m \frac{\partial f_m}{\partial x_i} \right). \qquad (58.4)$$

We see that the same conditions that we encountered before in statics (cf. chap. III, section 5) prevail in an exactly similar manner in dynamics. Once more *the Lagrangian λ-method provides the forces of reaction which maintain kinematical constraints.*

However, the following point requires special attention. Let us assume that the given kinematical conditions (58.1) and the potential energy V are independent of the time t. Then the system is conservative, as we can see at once if we use the elimination method of surplus variables in treating auxiliary conditions. But then the forces which maintain the given constraints must likewise be conservative, which means that their potential energy cannot depend on t. On the other hand, the λ_i are functions of t, which gives the impression that the potential energy of the forces of reaction

$$V_0 = \lambda_1 f_1 + \ldots + \lambda_m f_m \qquad (58.5)$$

is also a function of t.

This apparent contradiction comes from the fact that we know the potential energy of the forces of reaction *only along the C-curve.* A more detailed analysis removes the contradiction and brings about an interesting physical interpretation of the factors λ_i.

It will be sufficient to consider a single auxiliary condition

$$f(q_1, \ldots, q_n) = 0, \qquad (58.6)$$

since the results can easily be extended to an arbitrary number of such conditions. We shall assume that the potential energy V_1 of the forces which maintain the condition (58.6) has the form

$$V_1 = \phi(f). \qquad (58.7)$$

[This assumption needs revision. It is obvious that (58.6) has no invariant significance since the same condition could have been given in the more general form

$$g f = 0, \qquad (58.8)$$

where $g(q_1, \ldots, q_n)$ is any other function of the coordinates which does *not* vanish along the surface $f = 0$. Hence the argument of ϕ in (58.7) should actually be changed to gf, where g is some function of the q_i. See Problem 4 at the end of this section.]

We know that the constraint (58.6) is maintained by strong forces, so that the equation (58.6) cannot be greatly violated. Hence we need the potential V_1 only for *small* values of f. Hence, we can expand the function ϕ in a Taylor series in the neighbourhood of $f = 0$. Since a constant has no significance in the potential energy, we can write

$$V_1 = \phi'(0) f + \tfrac{1}{2} \phi''(0) f^2, \tag{58.9}$$

neglecting terms of higher order than the second.

Now we know that the forces of reaction are dormant as long as the constraint (58.6) is *exactly* fulfilled. This means that $f = 0$ must be a state of equilibrium of these forces, and we can further assert that the given constraint could not be maintained if it were not a state of *stable* equilibrium. Hence V_1 must have a *minimum* at $f = 0$, which means that

$$\phi'(0) = 0, \tag{58.10}$$

and that $\phi''(0)$ must be *positive*. We put

$$\phi''(0) = 1/\epsilon, \tag{58.11}$$

where ϵ is a small positive constant. The smallness of ϵ expresses the fact that the constraint (58.6) is maintained by *strong* forces. The permanent exact fulfillment of the condition (58.6) would require *infinitely strong* forces, i.e. $\epsilon = 0$.

Since actually ϵ is a small but *finite* constant, the constraint (58.6) will not be *exactly* fulfilled during the motion. The actual motion will occur according to the law

$$f(q_1, \ldots, q_n) = \epsilon \rho(t), \tag{58.12}$$

where $\rho(t)$ is some function of the time. In view of the smallness of ϵ we can say that the condition $f = 0$ will be *macroscopically* but not *microscopically* fulfilled. We can call the right-hand side of (58.12) the "microscopic error" of the given constraint $f = 0$. This error changes with the time t.

Now, due to the potential energy

$$V_1 = \frac{1}{2\epsilon} f^2,$$ (58.13)

the force

$$K_i = -\frac{1}{\epsilon} f \frac{\partial f}{\partial x_i},$$ (58.14)

while by the Lagrangian λ-method

$$K_i = -\lambda \frac{\partial f}{\partial x_i}.$$ (58.15)

The comparison of (58.14) and (58.15) yields

$$\lambda = \frac{f}{\epsilon},$$ (58.16)

and in view of (58.12)

$$\lambda = \rho(t).$$ (58.17)

This relation shows that *the Lagrangian λ-factor is a measure of the microscopic violation of the equation of constraint $f = 0$.* The right-hand side of that equation is actually not zero, but ϵ times λ, where ϵ is a small positive constant which tends towards zero as the force which maintains the constraint increases toward infinity.

Here we have the explanation of why the potential energy V_1 of the force of reaction depends apparently on the time t, although *actually* the force is a conservative force.

We have once more established the result—as done before for systems in equilibrium—that *holonomic auxiliary conditions are mechanically equivalent to monogenic forces.* We can add that *scleronomic holonomic conditions are equivalent to conservative monogenic forces.*

Problem 1. Consider a diatomic molecule of the following structure. The inner forces have a potential energy of the form

$$V_1 = \frac{A}{2} \left(r^2 + \frac{1}{r^2} \right),$$

where r is the mutual distance of the two atoms; A is a large constant. Show that the motion of the system under the action of impressed forces of potential energy V is macroscopically equivalent to the motion of a diatomic molecule without any V_1 but with the auxiliary condition $r = 1$.

Problem 2. Show that in the same problem the Lagrangian factor λ is equal to the microscopic violation of the condition $r = 1$, multiplied by $4A$.

Problem 3. Consider a heavy particle of mass m which is constrained to stay on the sphere $r = 1$.

Show that the Lagrangian λ is here proportional to the microscopic elastic penetration of the sphere, caused by the pressure on the surface; the factor λ, and with it the force of reaction, becomes zero at the point where the mass leaves the sphere and continues its motion by falling freely.

Problem 4. Consider the more general problem which arises owing to the fact that the equation $f = 0$ has no invariant significance, on account of which the simple relation (58.7) should be replaced by the more general assumption

$$V_1 = \phi\,(fg), \qquad (58.18)$$

where g is an unknown function of the coordinates. In order to eliminate this unknown function, we need additional information not contained in the equation of constraint $f = 0$. We know that the surface $f = 0$ will be an equipotential surface of the forces of reaction. Let us assume that we know in the configuration space a *second* neighbouring equipotential surface. This surface can be characterized by giving at each point of the surface $f = 0$ the *infinitesimal distance* v between the two equipotential surfaces $\phi = 0$ and $\phi = p$, where p is a positive infinitesimal constant. Then the following relation can be established for λ:

$$\lambda = \frac{2p}{v^2|\mathrm{grad}\,f\,|^2}\,f. \qquad (58.19)$$

Once more λ is proportional to f — i.e. to the microscopic violation of the constraint $f = 0$ — but the factor of proportionality is generally no longer constant, although it is still necessarily a *positive* quantity.

This additional information is required only for the *interpretation* of the factor λ. The force of reaction does not depend on the unknown function g and is always furnished by the Lagrangian λ-method.[1]

Summary. The Lagrangian λ-method in dealing with auxiliary conditions has the physical significance that we replace the given kinematical constraints by forces which maintain those constraints. The λ-method furnishes the forces of reaction associated with a given constraint. These forces are exerted through a microscopic violation of the constraints, and the factor λ can be interpreted as a measure of this violation. This violation changes during the motion, thus making the λ_i functions of the time t, in spite of the conservative nature of the forces which maintain the given constraints.

[1] The author is not aware of a similar discussion on the physical significance of the λ-factor in the literature.

9. Non-holonomic auxiliary conditions and polygenic forces.

If the kinematical conditions do not appear in the form of algebraic relations between the coordinates but as non-integrable differential relations of the form (26.1), we can no longer reduce the number of degrees of freedom by eliminating the surplus variables. The Lagrangian λ-method, however, continues to hold. From (26.1) we now have, in fact, for the Lagrangian equations:

$$\frac{d}{dt}\frac{\partial L}{\partial \dot{q}_k} - \frac{\partial L}{\partial q_k} = - (\lambda_1 A_{1k} + \lambda_2 A_{2k} + \ldots + \lambda_m A_{mk}). \quad (59.1)$$

The A_{ik} are given functions of the q_i. They take the place of the $\partial f_i / \partial x_k$ which occur when the conditions are holonomic.

This can be interpreted physically by observing that the work function of the forces which maintain the given non-holonomic conditions does not now exist, *but the forces themselves are once more furnished by the λ-method.* Let us put

$$K_i = - (\lambda_1 A_{1i} + \ldots + \lambda_m A_{mi}). \quad (59.2)$$

These K_i can be interpreted physically as the components of the force which acts on the mechanical system in order to maintain the given non-holonomic conditions. This force is now of a polygenic nature. Once more we see that *non-holonomic auxiliary conditions are mechanically equivalent to polygenic forces.*

In view of this equivalence, we conclude that, when properly modified, the Lagrangian equations of motion are applicable even to the case of polygenic forces. This is indeed the case. We can characterize a polygenic system of forces by means of the virtual work of these forces. Let that work be

$$\overline{\delta w} = \rho_1 \, \delta q_1 + \rho_2 \, \delta q_2 + \ldots + \rho_n \, \delta q_n. \quad (59.3)$$

The only difference compared with a monogenic force is that this work can no longer be expressed as the variation of a scalar function. Let us assume that all the impressed forces of monogenic type are absorbed in the usual way into the Lagrangian function L, while the polygenic forces are given by their virtual work (59.3). Then we obtain the equations of motion in the form

$$\frac{d}{dt}\frac{\partial L}{\partial \dot{q}_i} - \frac{\partial L}{\partial q_i} = \rho_i. \quad (59.4)$$

Here again, as in (59.1), the polygenic force produces a "right-hand side" of the Lagrangian equations.

Summary. The method of the Lagrangian multiplier remains valid even in the case of non-holonomic auxiliary conditions. The forces exerted on account of these conditions are once more furnished. These forces are of a polygenic nature. Non-holonomic auxiliary conditions and polygenic forces have the same effect in the Lagrangian equations of motion: they produce a "right-hand side" of these equations.

10. Small vibrations about a state of equilibrium. One of the most beautiful examples of the power of the analytical method is the application of the Lagrangian equations to the theory of small vibrations about a state of stable equilibrium. This theory is of basic importance for the elastic behaviour of solids, for the vibrations of structures, for the theory of specific heat and other fundamental problems. The most impressive feature of this theory is its great *generality.* No matter how simple or how complicated a mechanical system is, its motion near a state of equilibrium is always describable in the same terms. The actual calculation gets complicated if the number of degrees of freedom is large. But the theoretical aspects of the problem remain unchanged.

The simplifications which occur in this problem are due to the fact that the vibrations are *small.* We know that the geometry of the configuration space is not Euclidean but Riemannian. But we also know that the curved Riemannian space flattens out more and more if we restrict ourselves to smaller and smaller regions. This behaviour of the Riemannian space finds its analytical expression in the fact that the line element

$$\overline{ds^2} = \sum_{i,\,k=1}^{n} a_{ik}\, dq_i\, dq_k, \qquad (510.1)$$

where the a_{ik} are functions of the q_i, can be replaced by a separate line element with *constant* coefficients if we do not leave the immediate neighbourhood of the point P. The a_{ik} change so little in that neighbourhood that we can replace them by their values at the point P.

Let the point P, which is the C-point representing the state of the mechanical system in the configuration space, be a point of equilibrium. For the sake of simplicity, let us agree that the point P shall be at the origin of our reference system, which means that its coordinates are $q_i = 0$. We now consider the line element (510.1), with the a_{ik} constant, as applicable to the *whole* of space. The new space is Euclidean, but the error we thus commit tends towards zero as we approach the point P.

In the new space the q_i are no longer curvilinear, but *rectilinear*, coordinates, and we have to know a few basic facts about the analytical nature of such coordinates. In the first place, in our flat space not only the differential form (510.1), but also the finite form

$$s^2 = \sum_{i,\,k=1}^{n} a_{ik}\, q_i\, q_k \qquad (510.2)$$

has a simple geometrical significance. It gives the distance s of the point $Q = (q_1, \ldots, q_n)$ from the origin $P = (0, \ldots, 0)$.

Now let us assume that we are in an n-dimensional Euclidean space and that we choose n basic vectors

$$\mathbf{u}_1,\, \mathbf{u}_2,\, \ldots,\, \mathbf{u}_n, \qquad (510.3)$$

which may be of arbitrary length and orientation so long as they are linearly independent of each other—i.e. no $\mathbf{u}_k$ is equal to a linear combination of the other basic vectors. Then we can express the radius vector $\mathbf{R}$ in that space in the form

$$\mathbf{R} = q_1 \mathbf{u}_1 + q_2 \mathbf{u}_2 + \ldots + q_n \mathbf{u}_n. \qquad (510.4)$$

If we form the "dot-product" of $\mathbf{R}$ with itself, we obtain

$$\mathbf{R}^2 = s^2 = \sum_{i,\,k=1}^{n} (\mathbf{u}_i \cdot \mathbf{u}_k)\, q_i\, q_k. \qquad (510.5)$$

Comparison with (510.2) shows that the coefficients a_{ik} of the line element have the following significance:

$$a_{ik} = a_{ki} = \mathbf{u}_i \cdot \mathbf{u}_k. \qquad (510.6)$$

In the special case of *rectangular* coordinates the $\mathbf{u}_i$ are chosen as mutually perpendicular vectors of unit length. Then we have

$$\mathbf{u}_i \cdot \mathbf{u}_k = 0, \qquad (i \neq k)$$
$$\mathbf{u}_i^2 = 1, \tag{510.7}$$

and the square of the distance s assumes the customary Pythagorean form

$$s^2 = q_1^2 + q_2^2 + \ldots + q_n^2. \tag{510.8}$$

However, we intentionally do not specialize our unit vectors, and operate with an arbitrary rectilinear reference system. We do require, however, that our Euclidean space shall be a "real" space in which the distance between P and Q cannot be zero unless P and Q coincide. This implies that the quadratic form (510.5) is a "positive definite" form which will always give a positive s^2 no matter what values we assign to the q_i except the trivial values $q_i = 0$ when s becomes zero.

In the configuration space this condition is actually fulfilled because the kinetic energy which determines the line element (510.1), and with it the distance (510.2), can never become negative, and even the value zero is only possible if all dq_i vanish. This guarantees the positive definite character of the expression (510.2).

Let us now consider two *different* points Q and Q', belonging to the two vectors

$$\overline{PQ} = \mathbf{R} = q_1\mathbf{u}_1 + \ldots + q_n\mathbf{u}_n, \tag{510.9}$$

and

$$\overline{PQ'} = \mathbf{R}' = q_1'\mathbf{u}_1 + \ldots + q_n'\mathbf{u}_n. \tag{510.10}$$

We form the dot-product of these two vectors and obtain

$$\mathbf{R} \cdot \mathbf{R}' = \sum_{i,k=1}^{n} (\mathbf{u}_i \cdot \mathbf{u}_k) q_i q_k'$$
$$= \sum_{i,k=1}^{n} a_{ik} q_i q_k'. \tag{510.11}$$

We notice that the vanishing of the sum (510.11) means that the two vectors $\mathbf{R}$ and $\mathbf{R}'$ are *orthogonal* to each other.

After making these general observations which apply to the analytical geometry of a Euclidean space of n dimensions, we now come to the study of the potential energy $V(q_1, \ldots, q_n)$ of the mechanical system. We can expand this function in the neighbourhood of the origin, $q_i = 0$, in a Taylor series:

$$V = V_0 + \sum_{i=1}^{n} \left(\frac{\partial V}{\partial q_i}\right)_0 q_i + \frac{1}{2} \sum_{i,k=1}^{n} \left(\frac{\partial^2 V}{\partial q_i \partial q_k}\right)_0 q_i q_k + \ldots . \quad (510.12)$$

Now we have assumed that the origin P of our reference system is a point of *equilibrium*. Hence, V must have a stationary value at that point (cf. chap. II, section 2 and chap. III, section 1) and thus the linear terms of the expansion (510.12) drop out. Since an additive constant in the potential energy is always irrelevant, the expansion starts with the *second* order terms. We need not go beyond these terms because terms of the third order become negligible if the q_i remain sufficiently small. Hence we can write

$$V = \frac{1}{2} \sum_{i,k=1}^{n} b_{ik} q_i q_k, \quad (510.13)$$

where

$$b_{ik} = b_{ki} = \left(\frac{\partial^2 V}{\partial q_i \partial q_k}\right)_0 . \quad (510.14)$$

Let us now consider the equation $V = \frac{1}{2}$, or

$$\Sigma \, b_{ik} q_i q_k = 1. \quad (510.15)$$

Geometrically, this equation represents a *surface* of the n-dimensional space, and we can say more specifically that it represents a surface of the *second* order in that space. We may think of an ellipsoid or a hyperboloid of our ordinary space, only the number of dimensions is increased from 3 to n. It is remarkable that the analytical geometry of surfaces of the second order can be developed equally well in *any* number of dimensions. The theory of these surfaces is of great importance in almost all branches of mathematical physics. The entire mathematical background of elasticity, acoustics and wave mechanics can be formulated in terms of an analytical geometry of such surfaces in a space of infinitely many dimensions.

In mechanics we encounter this theory in connection with the oscillations of a mechanical system about a state of equilibrium. We know from ordinary solid analytic geometry that the study of surfaces of the second order is greatly facilitated if we let the axes of our reference system coincide with certain axes of symmetry, e.g. an ellipsoid or hyperboloid has three "principal axes" which are mutually perpendicular.

If the equation of the surface is not given with reference to these principal axes, these must be found and the reference system rotated into the new position. In problems of physics this "transformation to principal axes" is of the greatest importance because the major part of the analytical problem is usually solved when this transformation has been effected.

We shall now show how the problem of small oscillations of a mechanical system about a state of equilibrium is intimately linked up with that of determining the principal axes of the potential energy V. The principal axes of a surface of the second order have certain extremum properties. They can be found by looking for those points of the surface whose distances from the origin have stationary values. Thus our problem is to find the stationary value of

$$s^2 = \sum_{i,k=1}^{n} a_{ik} q_i q_k, \tag{510.16}$$

under the auxiliary condition that we move on the surface

$$\sum_{i,k=1}^{n} b_{ik} q_i q_k = 1. \tag{510.17}$$

This is an ordinary algebraic extremum problem with an auxiliary condition, which can be handled by the λ-method. We drop the auxiliary condition (510.17) and minimize the function

$$\sum a_{ik} q_i q_k - \frac{1}{\lambda} \sum b_{ik} q_i q_k. \tag{510.18}$$

Since λ is an undetermined factor, it is permissible to replace it by $-1/\lambda$, which will be more suitable for our present purpose.

But then we see that instead of minimizing (510.18) we can equally well minimize the function

$$F(q_1, \ldots, q_n) = \sum b_{ik} q_i q_k - \lambda \sum a_{ik} q_i q_k. \tag{510.19}$$

This gives the following interpretation of our extremum problem: find the stationary values of the function

$$2V = \sum b_{ik} q_i q_k, \tag{510.20}$$

under the auxiliary condition

$$\Sigma a_{ik}q_iq_k = 1. \tag{510.21}$$

The auxiliary condition (510.21) means that we stay on a sphere of radius 1. At each point of this sphere the potential energy V has a certain value. *We wish to find those special points Q_i of the unit sphere for which V becomes stationary.*

The conditions for the stationary value of the function F give us the following linear equations:

$$b_{11}q_1 + \ldots + b_{1n}q_n = \lambda(a_{11}q_1 + \ldots + a_{1n}q_n),$$

$$\cdot \quad \cdot \quad \cdot \quad \cdot \quad \cdot \quad \cdot \quad \cdot \quad \cdot \quad \cdot \quad \cdot \quad \cdot \quad \cdot \quad \cdot \quad \cdot$$

$$\cdot \quad \cdot \quad \cdot \quad \cdot \quad \cdot \quad \cdot \quad \cdot \quad \cdot \quad \cdot \quad \cdot \quad \cdot \quad \cdot \quad \cdot \quad \cdot \tag{510.22}$$

$$\cdot \quad \cdot \quad \cdot \quad \cdot \quad \cdot \quad \cdot \quad \cdot \quad \cdot \quad \cdot \quad \cdot \quad \cdot \quad \cdot \quad \cdot \quad \cdot$$

$$b_{n1}q_1 + \ldots + b_{nn}q_n = \lambda(a_{n1}q_1 + \ldots + a_{nn}q_n).$$

These are homogeneous linear equations for the q_i which have a non-zero solution provided that the determinant of the system is zero. We thus get the following fundamental determinant condition, called the "characteristic equation," from which the "characteristic values" λ_i can be determined:

$$\begin{vmatrix} b_{11} - \lambda a_{11}, \ldots, b_{1n} - \lambda a_{1n} \\ \cdot \quad \cdot \quad \cdot \quad \cdot \quad \cdot \quad \cdot \quad \cdot \\ \cdot \quad \cdot \quad \cdot \quad \cdot \quad \cdot \quad \cdot \quad \cdot \\ \cdot \quad \cdot \quad \cdot \quad \cdot \quad \cdot \quad \cdot \quad \cdot \\ b_{n1} - \lambda a_{n1}, \ldots, b_{nn} - \lambda a_{nn} \end{vmatrix} = 0. \tag{510.23}$$

This is an algebraic equation of the nth order for λ which must have n (possibly complex) roots. We denote these roots by

$$\lambda_1, \lambda_2, \ldots, \lambda_n. \tag{510.24}$$

It is possible that some of the roots are multiple roots. We consider such a case a "degenerate" case, and remove the degeneracy by modifying the coefficients a_{ik} or b_{ik} by arbitrarily small amounts. This separates the multiple roots, and thereafter we perform a limiting process. Hence it is justifiable to exclude multiple roots and assume that *all the λ_i are different from one another.*

To each λ_i we can find a corresponding solution of the linear equations (510.22). This solution leaves a common factor of all the q_i arbitrary, but this factor is uniquely determined (except for the sign $\pm$) by satisfying the auxiliary condition (510.21).

The values $q_1, \ldots, q_n$ which belong to a certain λ_i, can be considered as the components of a vector p_i of unit length. This vector will be called the ith "principal axis" of the surface (510.17) We have n such principal axes

$$p_1, p_2, \ldots, p_n, \qquad (510.25)$$

corresponding to the n characteristic values (510.24).

These principal axes have a number of remarkable properties that we shall mention briefly:

1. *The λ-roots of the characteristic equation are invariants with respect to arbitrary linear transformations of the coordinates q_i.* If we multiply the equation (510.22) in succession by $q_1, \ldots, q_n$ and form the sum, we obtain on the right-hand side $\lambda \cdot 1$ in view of the auxiliary conditions (510.21), while the left-hand side becomes $2V$. This gives the relation

$$\lambda = 2V. \qquad (510.26)$$

Hence any linear transformation of the coordinates q_i which leaves the value of V unchanged also leaves the λ_i unchanged. While the n principal axes p_i give the *directions* in which the potential energy V assumes its stationary values, the roots λ_i give those stationary values V_i *themselves*, according to the relation

$$V_i = \lambda_i/2. \qquad (510.27)$$

2. *The λ-roots are all real, and thus the principal axes are n real vectors of the n-dimensional Euclidean space.* The roots λ_i of an algebraic equation of the nth degree are in general *complex;* that they are real in the case of the characteristic equation (510.23) is due to the *symmetry* of the elements a_{ik} and b_{ik} of the determinant:

$$a_{ik} = a_{ki}, \quad b_{ik} = b_{ki}. \qquad (510.28)$$

In order to prove this important theorem, we proceed as follows. Let us assume that a certain λ_i is complex and let us solve the linear equations (510.22) associated with this complex λ. The q_i are now complex numbers and we know that any algebraic relation between complex numbers remains true if i is changed to $-i$. Hence we can write down the equations (510.22) once more, changing q_i to $q_i{}^*$ and λ to λ^*, where the "star" denotes the complex conjugate. Now we multiply the first set of equations in succession by $q_1{}^*, \ldots, q_n{}^*$, and the second set of equations by $q_1, \ldots, q_n$ and form the sum in each case. On the left-hand sides we get the *same* sum in both cases and thus comparison of the right-hand sides gives the relation

$$(\lambda - \lambda^*) \sum_{i,k=1}^{n} a_{ik}q_iq_k{}^* = 0. \qquad (510.29)$$

If we replace q_k by $\alpha_k + i\beta_k$, where α_k and β_k are real, (510.29) becomes, in view of (510.28),

$$(\lambda - \lambda^*)\,(s_1{}^2 + s_2{}^2) = 0, \qquad (510.30)$$

where s_1 and s_2 are real distances of the n-dimensional Euclidean space. Hence

$$\lambda = \lambda^*, \qquad (510.31)$$

which means that λ must be real.

3. *The principal axes $\mathbf{p}_i$ are mutually perpendicular and thus form an orthogonal reference system for the n-dimensional space.* This fundamental property of the principal axes of a surface of the second order can be proved with the help of the same relation (510.29) that we used before in proving the reality of the λ_i. We merely *interpret* this equation in a new way. Let us assume now that λ and λ^* are two different characteristic roots and that q_i and $q_i{}^*$, respectively define the two principal axes associated with these two λ_i. In the present case the first factor of (510.29) cannot vanish because λ and λ^* are different. Hence the *second* factor must vanish, which gives

$$\sum_{i,k=1}^{n} a_{ik}q_iq_k{}^* = 0. \qquad (510.32)$$

According to (510.11), the significance of this equation, if written in vector form, is

$$\mathbf{p} \cdot \mathbf{p},^* = 0. \qquad (510.33)$$

which means that the two principal axes $\mathbf{p}$ and $\mathbf{p}^*$—which can be identified with *any* pair of axes $\mathbf{p}_i$ and $\mathbf{p}_k$—are orthogonal.

With the help of these fundamental properties, we can now take these n principal axes as the axes of a new reference system which is *rectangular*. Analytically, this transformation is performed according to the following scheme. Let us denote the solution $q_i, \ldots, q_n$ obtained by solving the principal axis problem for a certain λ_i, by

$$a_{1i}, a_{2i}, \ldots, a_{ni}. \tag{510.34}$$

We arrange these solutions in successive *columns*, according to the following scheme:

$$
\begin{array}{cccc}
p_1, & p_2, & \ldots, & p_n \\
\hline
a_{11}, & a_{12}, & \ldots, & a_{1n} \\
\cdot & \cdot & & \cdot \\
\cdot & \cdot & & \cdot \\
\cdot & \cdot & & \cdot \\
a_{n1}, & a_{n2}, & \ldots, & a_{nn}
\end{array}
$$

Then we apply the linear transformation:

$$
\begin{aligned}
q_1 &= a_{11} u_1 + \ldots + a_{1n} u_n, \\
&\quad \cdot \quad \cdot \quad \cdot \quad \cdot \quad \cdot \quad \cdot \quad \cdot \\
q_n &= a_{n1} u_1 + \ldots + a_{nn} u_n.
\end{aligned}
\tag{510.35}
$$

The $u_1, \ldots, u_n$ are the *rectangular* coordinates of a certain point Q which in the old system had the *rectilinear* coordinates $q_1, \ldots, q_n$.

The simplification that this transformation accomplishes is quite remarkable. In the first place, the distance s^2 of a point Q from the origin appears now in the Pythagorean form

$$s^2 = u_1^2 + u_2^2 + \ldots + u_n^2, \tag{510.36}$$

because we operate with ordinary Cartesian coordinates. Hence the general quadratic form (510.2) is transformed to a purely *diagonal* form.

Now let us see what happens to the potential energy V in consequence of this transformation. At first we know only that V must be some quadratic form of the new coordinates u_k with new coefficients b'_{ik}:

$$V = \frac{1}{2} \sum_{i,k=1}^{n} b'_{ik} u_i u_k. \tag{510.37}$$

But now we can set up the problem of the principal axes once more in our new reference system. We now get the equations

$$\sum_{a=1}^{n} b'_{ia} u_a = \lambda u_i, \qquad (510.38)$$

together with the auxiliary condition

$$u_1^2 + \ldots + u_n^2 = 1. \qquad (510.39)$$

We know that the λ of the equation (510.38) can be identified successively with the n *previous* values $\lambda_1, \ldots, \lambda_n$, since the λ_i are invariants of transformation. Furthermore, in the new reference system the solution to the problem of the principal axes is known *explicitly*, seeing that the principal axes coincide with the coordinate axes. Hence the solution which belongs to a given λ_i must be

$$u_i = 1, \text{ all other } u_k = 0. \qquad (510.40)$$

If we put these values into the equations (510.38), we have

$$\begin{aligned} b'_{ii} &= \lambda_i, \\ b'_{ik} &= 0, \quad (i \neq k). \end{aligned} \qquad (510.41)$$

This shows that the potential energy V assumes in the new reference system the form

$$V = \tfrac{1}{2} (\lambda_1 u_1^2 + \lambda_2 u_2^2 + \ldots + \lambda_n u_n^2), \qquad (510.42)$$

and we obtain the remarkable result that not only the square of the distance, s^2, but also the potential energy V, assumes in the new reference system a diagonal form. *By one linear transformation of the coordinates two general quadratic forms can be reduced simultaneously to diagonal form.*

For our mechanical problem this transformation accomplishes a great simplification of both kinetic and potential energies. The kinetic energy in the new reference system becomes

$$T = \tfrac{1}{2} \left(\overline{\frac{ds}{dt}} \right)^2 = \tfrac{1}{2} (\dot{u}_1^2 + \ldots + \dot{u}_n^2), \qquad (510.43)$$

while the potential energy becomes

$$V = \tfrac{1}{2} (\lambda_1 u_1^2 + \ldots + \lambda_n u_n^2). \qquad (510.44)$$

The Lagrangian equations of motion are now:

$$\ddot{u}_1 + \lambda_1\, u_1 = 0,$$

$$\cdot \quad \cdot \quad \cdot \quad \cdot \quad \cdot \quad \cdot \qquad (510.45)$$

$$\ddot{u}_n + \lambda_n\, u_n = 0.$$

These differential equations are completely *separated* and easily integrable. They are the differential equations of simple harmonic motion, giving

$$u_i = A_i \cos\sqrt{\lambda_i}\, t + B_i \sin\sqrt{\lambda_i}\, t, \qquad (510.46)$$

where A_i and B_i are arbitrary constants of integration.

Problem: Consider a double pendulum, composed of two single pendulums of the same length and mass, the second being suspended from the bob of the first. Use as position coordinates the two infinitesimal angles θ_1 and θ_2 between thread and plumb line.

1. Form the kinetic energy and the potential energy of the system.

2. Show from the form of the line element that the two axes Q_1 and Q_2 of the oblique reference system are at an angle of $45°$ to each other.

3. Show that the two principal axes p_1 and p_2 bisect the angle between Q_1 and Q_2 and its supplementary angle.

4. Show that the frequencies connected with the two principal axes are

$$\nu_1 = \sqrt{2 - \sqrt{2}}\ \nu,$$
$$\nu_2 = \sqrt{2 + \sqrt{2}}\ \nu,$$

where ν is the natural frequency of each of the simple pendulums.

We now have to distinguish between two fundamentally different possibilities. We know that the λ_i are always *real* numbers. But we cannot guarantee that they are always *positive*. In the general case some of the λ_i may be positive and some negative. Let us assume that *all λ_i are positive*. In this case the solution (510.46) shows that the C-point *oscillates around the point of equilibrium P*. The motion consists of a superposition of monochromatic vibrations of frequency

$$\nu_i = \sqrt{\lambda_i}, \qquad (510.47)$$

with small but constant amplitudes and arbitrary phases. These vibrations occur in the direction of the n mutually orthogonal principal axes p_i. The superposition of all these vibrations gives

"Lissajous figures" in the n-dimensional configuration space. These vibrations are called "normal vibrations" on account of their mutual orthogonality. Their number is always equal to the number of degrees of freedom of the mechanical system. The frequencies ν_i change from one principal axis to the other, and thus we speak of the "spectrum" of the normal vibrations.

A beautiful example is the case of a solid body, the molecules of which are near a state of equilibrium but in permanent irregular vibrations on account of the thermal motion. All these vibrations can be analytically represented by one single C-point which is placed in a Euclidean space of $3N$ dimensions, if N is the number of molecules constituting the solid body. The motion of the C-point can be pictured as harmonic vibrations of definite frequencies along definite mutually perpendicular axes. Each degree of freedom is connected with one axis. The spectrum of these vibrations extends from the very low elastic and acoustic vibrations up to the very high infra-red vibrations. The distribution of the amplitudes and phases is governed by statistical laws and is a function of the absolute temperature T.

Let us now assume that at least *one* of the characteristic roots λ_i — for example λ_n — is *negative*. In this case we put

$$\sqrt{-\lambda_n} = \nu_n, \tag{510.48}$$

and obtain as the solution of the last Lagrangian equation of motion

$$u_n = A_n e^{\nu_n t} + B_n e^{-\nu_n t}. \tag{510.49}$$

This solution is no longer periodic, but *exponential*. The smallest impulse along the p_n axis is sufficient to bring the C-point into a motion which is no longer oscillatory but which recedes from the point P of equilibrium more and more.

We see the study of the small vibrations about a state of equilibrium leads to the conclusion that we have to distinguish between two types of equilibrium which are widely different in their physical behaviour. The first case is realized if all the roots of the characteristic equation are positive. In this case we call the equilibrium "stable," because small disturbances will not throw the system out of its state of equilibrium, but merely produce vibrations about that state. The second state is realized if at least one of the characteristic roots is negative. In that case the smallest disturbance suffices to bring the system away from the original state of equilibrium, which means that the equilibrium, even if it exists at a particular instant, cannot be maintained. An equilibrium of this nature is called "unstable." In the limiting case where none of the λ_i are negative but one or more are *zero*, we speak of a "neutral" equilibrium.

We have to bring the discussions of this section, which explore the mechanical events in the neighbourhood of a state of equilibrium, into relation with the *macroscopic* picture of the configuration space. Our coordinates with which we operated have the significance of *local* coordinates; they represent small local variations δq_i of the coordinates q_i^0 from the point of equilibrium P. The potential energy V that we have considered represents likewise the *local variation* of the potential energy V in the neighbourhood of the point P, obtained by expanding V in a Taylor series. The linear terms drop out since we expand about a point of equilibrium. The expansion starts with the second order terms and gives what we called in our general discussions (cf. chap. II, section 3) the "second variation" of V. We now see that the same second variation which was decisive for the minimum properties of a stationary value is decisive for the stability or instability of a state of equilibrium. If all λ_i are positive, then the second variation is a positive definite form which means that the potential energy increases in every direction from P. Hence V has a *local minimum* at P. Stability of an equilibrium position and the minimum of the potential energy are equivalent statements. If at least one of the λ_i is negative, the second variation changes its sign and the stationary value of the potential energy no longer leads to a true extremum. At the same time the corresponding equilibrium is unstable.

Stability questions are of the greatest importance for certain problems of elasticity. A thin elastic structure suddenly collapses if the load gets too large. This collapse does not occur by a breakdown of the elastic forces but by "buckling." The phenomenon of buckling has the significance that at a certain load the elastic structure changes its state of stable equilibrium to one of unstable equilibrium. The analytical problem of buckling leads automatically to an investigation of the second variation $\delta^2 V$ of the potential energy at a point where δV is zero. If the characteristic values of $\delta^2 V$ can be found, then the signs of the λ_i decide at once whether the equilibrium is stable or not. As long as the smallest λ_i is positive, the structure is stable; if the smallest λ_i is negative, the structure is unstable. The critical load is reached at the moment when the smallest λ_i vanishes.

It is frequently very difficult to find the characteristic spectrum of the second variation and another approach to the problem is more accessible. We minimize the second variation and investigate whether the minimum is positive or negative. In the first case, the equilibrium is stable; in the second case, unstable. The condition for buckling is obtained by setting the minimum value of $\delta^2 V$ equal to zero.

Summary. The motion of any mechanical system near a state of stable equilibrium can conveniently be studied in the configuration space. This space is now Euclidean and the q_i-variables are rectilinear coordinates of that space. The principal axes of the potential energy designate n mutually perpendicular directions in the configuration space which can be chosen as the axes of a natural reference system. The C-point performs harmonic vibrations in these directions, with definite frequencies which change from axis to axis. The amplitudes and phases of these vibrations, called "normal vibrations," are arbitrary, depending on the initial conditions. The general motion of the system is a superposition of the normal vibrations, resulting in Lissajous figures described by the C-point in the configuration space. The stability of the equilibrium requires that all the roots of the characteristic equation be positive, since otherwise the oscillatory character of the motion is destroyed.

CHAPTER VI

THE CANONICAL EQUATIONS OF MOTION

"Was it a God who wrote these signs
Which soothe the inner tumult's raging,
Which fill the lonely heart with joy
And, with mysteriously hidden might,
Unriddle Nature's forces all around?"

GOETHE

(Faust gazing at the sign of
the Macrocosmos, *Faust*,
Part I, Scene 1.)

Introduction. The principle of least action and its generalization by Hamilton transfer the problems of mechanics into the realm of the calculus of variations. The Lagrangian equations of motion, which follow from the fact that a definite integral is stationary, are the basic differential equations of theoretical mechanics. Yet we are not at the end of the road. The Lagrangian function is quadratic in the velocities. Hamilton discovered a remarkable transformation which renders the Lagrangian function linear in the velocities, at the same time doubling the number of mechanical variables. This transformation is not limited to the particular Lagrangian function which occurs in mechanics. Hamilton's transformation reduces all Lagrangian problems to a specially simple form, called by Jacobi the "canonical" form. The dynamical equations associated with this new Lagrangian problem, the "canonical equations," replace the n original Lagrangian differential equations of the second order by $2n$ new differential equations of the first order, remarkable for their simple and symmetrical structure. The discovery of these differential equations started a new era in the development of theoretical mechanics.

1. Legendre's dual transformation. The French mathematician Legendre (1752-1833) discovered an important transformation in his studies connected with the solution of differential equations. This transformation has remarkable properties and is well adapted to many problems of analysis. In mechanics it brings about a new form of the Lagrangian equations which is well suited to further mathematical research. Before applying this transformation to the Lagrangian equations, we discuss its general mathematical properties.

Let us start with a given function of n variables $u_1, \ldots, u_n$,

$$F = F(u_1, \ldots, u_n). \tag{61.1}$$

We introduce a new set of variables $v_1, \ldots, v_n$, by means of the following transformation:

$$v_i = \frac{\partial F}{\partial u_i}. \tag{61.2}$$

The so-called "Hessian"—i.e., the determinant formed by the second partial derivatives of F—we assume to be different from zero, guaranteeing the independence of the n variables v_i. In that case the equations (61.2) are solvable for the u_i as functions of the v_i.

We now define a new function G as follows:

$$G = \sum_{i=1}^{n} u_i v_i - F. \tag{61.3}$$

We express the u_i in terms of the v_i, and substitute in (61.3). The function G can then be expressed in terms of the new variables v_i alone:

$$G = G(v_1, \ldots, v_n). \tag{61.4}$$

We now consider the infinitesimal variation of G produced by arbitrary infinitesimal variations of the v_i. The combination of (61.4) and (61.3) gives

$$\begin{aligned}
\delta G &= \sum_{i=1}^{n} \frac{\partial G}{\partial v_i} \, \delta v_i \\
&= \sum_{i=1}^{n} (u_i \delta v_i + v_i \delta u_i) - \delta F \\
&= \sum_{i=1}^{n} \left[u_i \delta v_i + \left(v_i - \frac{\partial F}{\partial u_i} \right) \delta u_i \right].
\end{aligned} \tag{61.5}$$

Since G is a function of the v_i alone, we should express the u_i as functions of the v_i. This expresses the variations of the u_i in terms of the variations of the v_i. However, examination of (61.5) shows that this elimination is rendered unnecessary by the fact that *the coefficient of δu_i is automatically zero*, since we have defined the v_i according to (61.2). But then (61.5) gives at once

$$u_i = \frac{\partial G}{\partial v_i} \, . \tag{61.6}$$

This result expresses a remarkable duality of Legendre's transformation from which it has received the name "dual transformation." The following scheme brings out this duality:

	Old system	*New system*
Variables:	$u_1, \ldots, u_n$;	$v_1, \ldots, v_n$.
Function:	$F = F(u_1, \ldots, u_n)$;	$G = G(v_1, \ldots, v_n)$.

Transformation

$v_i = \dfrac{\partial F}{\partial u_i}$,	(α)	$u_i = \dfrac{\partial G}{\partial v_i}$,
$G = \Sigma u_i v_i - F,$	(β)	$F = \Sigma u_i v_i - G,$
$G = G(v_1, \ldots, v_n).$	(γ)	$F = F(u_1, \ldots, u_n).$

$$\tag{61.7}$$

Just as the new variables are the partial derivatives of the old function with respect to the old variables, so the old variables are the partial derivatives of the new function with respect to the new variables.

The transformation given by (61.7) is entirely symmetrical. "Old system" and "new system" refer to the fact that we start on the left and go to the right. But actually we can equally well start from the right and go to the left. *"Old" and "new" systems are entirely equivalent for this transformation.*

We enlarge our transformation in one further respect. Let us assume that F is actually a function of two sets of variables: $u_1, \ldots, u_n$, and $w_1, \ldots, w_m$:

$$F = F(w_1, \ldots, w_m; \, u_1, \ldots, u_n). \tag{61.8}$$

The w_i are independent of the u_i. They occur in F as mere parameters, but *do not participate in the transformation*, which is performed exactly as before. The new function G will likewise contain them. We call the u_i the *active*, the w_i the *passive* variables of the transformation.

We now go back to the equation (61.5), and find the complete variation of G, letting all the v_i and w_i vary arbitrarily. We then get an additional relation which we did not have before. The relation (61.2) remains unchanged; but we now have in addition

$$\frac{\partial F}{\partial w_i} = -\frac{\partial G}{\partial w_i}. \tag{61.9}$$

Summary. Legendre's transformation changes a given function of a given set of variables into a new function of a new set of variables. The old and the new variables are related to each other by a point transformation. This transformation has the remarkable property that it is entirely symmetrical in both systems, and the same transformation that leads from the old to the new system leads back from the new to the old system.

---

2. Legendre's transformation applied to the Lagrangian function. The Lagrangian function L of a variation problem is generally some function of the n position coordinates q_i, the n velocities $\dot{q}_i$, and the time t:

$$L = L(q_1, \ldots, q_n; \dot{q}_1, \ldots, \dot{q}_n; t). \tag{62.1}$$

For the present we do not think of the $\dot{q}_i$ as velocities in the actual motion. For our discussion they are merely n variables, entirely independent of the q_i.

We now apply Legendre's transformation, considering the $\dot{q}_1, \ldots, \dot{q}_n$ as active variables and all the others as passive variables. Hence the $\dot{q}_i$ correspond to the u_i, and the $q_1, \ldots, q_n$ and t to the w_i of the general scheme. Thus the number m of passive variables is $n + 1$ in the present problem.

Following the general scheme of Legendre's transformation, we proceed in three steps.

1. We introduce the "new variables," which are now called "momenta," and are denoted by p_i:

$$p_i = \frac{\partial L}{\partial \dot{q}_i}. \tag{62.2}$$

2. We introduce the new function, which is now denoted by H and called the "total energy":

$$H = \sum_{i=1}^{n} p_i \dot{q}_i - L. \tag{62.3}$$

3. We express the new function H in terms of the new variables p_i by solving the equations (62.2) for the $\dot{q}_i$, and substituting in (62.3). We thus obtain

$$H = H(q_1, \ldots, q_n; p_1, \ldots, p_n; t), \tag{62.4}$$

H being now called the "Hamiltonian function." The basic features of the transformation are as follows:

Old system	*New system*
Function: Lagrangian function L,	Hamiltonian function H.
Variables: velocities,	momenta.

Passive variables: position coordinates, time.

The dual nature of the transformation is expressed in the following scheme:

$$p_i = \frac{\partial L}{\partial \dot{q}_i}, \qquad \qquad \dot{q}_i = \frac{\partial H}{\partial p_i},$$

$$H = \Sigma p_i \dot{q}_i - L, \qquad \qquad L = \Sigma p_i \dot{q}_i - H, \tag{62.5}$$

$$H = H(q_1, \ldots, q_n; p_1, \ldots, p_n; t), \quad L = L(q_1, \ldots, q_n; \dot{q}_1, \ldots, \dot{q}_n; t).$$

Just as we started from the Lagrangian function L and constructed in three steps the Hamiltonian function H, so we can start from the Hamiltonian function H and construct in three steps the Lagrangian function L.

The equation (61.9) of our general scheme now takes the form

$$\frac{\partial L}{\partial q_i} = -\frac{\partial H}{\partial q_i},$$

$$\frac{\partial L}{\partial t} = -\frac{\partial H}{\partial t}. \tag{62.6}$$

Summary. Legendre's transformation can be applied to the Lagrangian function L, considering the velocities $\dot{q}_i$ as active variables of the transformation and the position coordinates q_i and the time t as passive variables. The velocities are transformed into the momenta; the Lagrangian function is transformed into the Hamiltonian function.

3. Transformation of the Lagrangian equations of motion. The Langrangian equations of motion are differential equations of the second order in the position coordinates q_i. However, they can be written as differential equations of the first order if we introduce as intermediate quantities the momenta p_i, defined by

$$p_i = \frac{\partial L}{\partial \dot{q}_i} \; . \tag{63.1}$$

The Lagrangian equations can then be written in the form

$$\dot{p}_i = \frac{\partial L}{\partial q_i} \; . \tag{63.2}$$

We have merely called a certain set of quantities "momenta" in order to simplify the writing of the Lagrangian equations. Yet the introduction of the p_i has the effect of replacing the original system of n differential equations of the second order by a system of $2n$ differential equations of the first order, namely the equations (63.1) and (63.2). The introduction of the p_i has the effect that no derivatives of an order higher than the first are required in forming the equations of motion. The procedure is analogous to expressing "mass times acceleration" as "rate of change of momentum," after defining momentum as "mass times velocity" in vectorial mechanics.

We now apply Legendre's transformation as in the preceding section. Equation (63.1) can be written

$$\dot{q}_i = \frac{\partial H}{\partial p_i} \; , \tag{63.3}$$

because of the duality of the transformation. The equations

(63.1) and (63.3) are equivalent to each other. Notice that these equations do not express any physical law. The equations (63.1) merely *define* the momenta, and the equations (63.3) rewrite these equations by expressing the velocities in terms of the momenta, instead of the momenta in terms of the velocities.

The Lagrangian equations of motion are contained in (63.2). Applying Legendre's transformation, these can be written [cf. (62.6)]

$$\dot{p}_i = - \frac{\partial H}{\partial q_i} . \tag{63.4}$$

Thus, finally, the Lagrangian equations of motion have been replaced by a new set of differential equations, called the "canonical equations of Hamilton":

$$\dot{q}_i = \frac{\partial H}{\partial p_i} , \qquad \dot{p}_i = - \frac{\partial H}{\partial q_i} \tag{63.5}$$

These equations are entirely equivalent to the original Lagrangian equations and are merely a mathematically new form. Yet the new equations are vastly superior to the originals. For derivatives with respect to t appear only on the left-hand sides of the equations, since the Hamiltonian function does not contain any derivatives of q_i or p_i with respect to t.

This outstanding system of equations appears for the first time in one of Lagrange's papers (1809) which deals with the perturbation theory of mechanical systems. Lagrange did not recognize the basic connection of these equations with the equations of motion. It was Cauchy who (in an unpublished memoir of 1831) first gave these equations their true significance. Hamilton made the same equations the foundation of his admirable mechanical investigations. The reference to "Hamilton's canonical equations" is thus fully justified, although Hamilton's paper appeared in 1835.

Summary. The Lagrangian equations of motion are differential equations of the second order. The application of Legendre's transformation produces a remarkable separation between differentiation with respect to time and algebraic processes. The new equations form a simultaneous system of $2n$ differential equations of the first order. They are called the "canonical equations."

4. The canonical integral. The equations (63.5) have a double origin. The first set of equations holds on account of Legendre's transformation and can be considered as an implicit definition of the momenta p_i. The second set of equations is a consequence of the variational principle. Yet the conspicuous symmetry of the complete set of equations suggests that they must be deducible from one single principle. This is actually the case.

Because of the duality of Legendre's transformation, we can start with the Hamiltonian function H and construct the Lagrangian function L. We obtain L by means of the relation

$$L = \sum_{i=1}^{n} p_i \dot{q}_i - H. \tag{64.1}$$

We then have to eliminate the p_i by expressing them as functions of the q_i and $\dot{q}_i$. Closer inspection shows, however, that this elimination *need not be performed*.

Let us investigate how the variation of the p_i influences the variation of the action integral. The variation of (64.1) with respect to the p_i gives

$$\delta L = \sum_{i=1}^{n} \left(\dot{q}_i - \frac{\partial H}{\partial p_i} \right) \delta p_i, \tag{64.2}$$

but the coefficient of δp_i is zero on account of Legendre's transformation. This shows that *an arbitrary variation of the p_i has no influence on the variation of L.* But then it has no influence either on the integral of δL with respect to the time. This gives the following important result. Originally we required the vanishing of the first variation of the action integral A, with the restriction that the p_i are not free variables but certain given functions of the q_i and $\dot{q}_i$. The variation of p_i is thus determined by the variation of q_i. However, since the variation of p_i has no influence on the variation of the action integral, we can enlarge the validity of the original variational principle and state that the action integral assumes a stationary value *even if the p_i are varied arbitrarily*, which means *even if the p_i are considered as a second set of independent variables.*

Hence it is not necessary to change anything in the Lagrangian function (64.1). We can form the action integral

$$A = \int_{t_1}^{t_2} [\, \Sigma \; p_i \dot{q}_i - H(q_1, \ldots, q_n; p_1, \ldots, p_n; t) \,] \; dt, \quad (64.3)$$

and require that it assume a stationary value for arbitrary variations of the q_i *and* the p_i. This new variation problem has $2n$ variables and the Euler-Lagrange differential equations can be formed with respect to all the q_i and the p_i. If we do that, we get the following $2n$ differential equations:

$$\frac{d}{dt} \frac{\partial L}{\partial \dot{q}_i} - \frac{\partial L}{\partial q_i} \equiv \frac{dp_i}{dt} + \frac{\partial H}{\partial q_i} = 0,$$

$$\frac{d}{dt} \frac{\partial L}{\partial \dot{p}_i} - \frac{\partial L}{\partial p_i} \equiv 0 - \dot{q}_i + \frac{\partial H}{\partial p_i} = 0. \quad (64.4)$$

These are exactly the canonical equations (63.5), *now conceived as a unified system of 2n differential equations, derived from the action integral* (64.3). We no longer need the original Lagrangian function and the Legendre transformation by which H was obtained. We have a new variational principle which is equivalent to the original principle, yet superior on account of the simpler structure of the resulting differential equations. These equations are no longer of the second order, but of the first. The derivatives are all separated and not intermingled with algebraic operations.

This remarkable simplification is achieved on account of the simple form of the new action integral (64.3). We call this form the "canonical integral." The integrand has again the form, "kinetic energy minus potential energy," since the second term of the integrand is a mere function of the position coordinates—which are now the q_i *and* the p_i—while the first term depends on the velocities. It is the *kinetic part* of the canonical integral which accounts for its remarkable properties. The "kinetic energy" is now a *simple linear function of the velocities* $\dot{q}_i$,[1] namely:

$$p_1 \dot{q}_1 + p_2 \dot{q}_2 + \ldots + p_n \dot{q}_n. \quad (64.5)$$

[1] In classical mechanics the kinetic energy happens to be a quadratic function of the velocities. This, however, is by no means necessary. The transformation to the canonical form can be performed for arbitrary Lagrangian functions, however complicated

We notice that each q_k is associated with its own p_k. For this reason the p_k are called the "conjugate momenta" and the variables of the variational problem can appropriately be arranged in pairs:

$$\begin{pmatrix} q_1 \\ p_1 \end{pmatrix}, \begin{pmatrix} q_2 \\ p_2 \end{pmatrix}, \ldots, \begin{pmatrix} q_n \\ p_n \end{pmatrix}. \tag{64.6}$$

Hamilton's ingenious transformation of the Lagrangian equations amounts to the realization that an arbitrary variational problem, however complicated, can be changed into an equivalent problem with twice the number of variables, which in its kinetic part is normalized to a simple form. This transformation does not require any integration, but only differentiations and eliminations.

The ordinary problems of mechanics lead to Lagrangian functions which do not contain derivatives higher than the first. In the general case of a variational problem derivatives up to any order m may occur in the integrand. Even these problems can be normalized by the canonical integral, so that the canonical equations of Hamilton can be considered as *the normal form into which any set of differential equations arising from a variational problem can be transformed*; the transformation once more required nothing but differentiations and eliminations.[1]

Problem. Consider the problem of the loaded elastic bar, treated in chapter II, section 15, particularly the integral (215.3). Introduce the preliminary "momentum" u by the definition

$$u = \frac{\partial L}{\partial \ddot{y}},$$

and transform the given integral according to the Hamiltonian method into

$$\int_0^l [u\ddot{y} - \phi(y, \dot{y}, u)]dx.$$

An integration by parts changes the first term of the integrand into $-\dot{u}\dot{y}$, and now we have a regular Lagrangian problem in y and u which can be transformed

---

[1]This was first recognized by the Russian mathematician M. Ostrogradsky: "Mémoire sur les equations différentielles relative au problème des Isopérimètres" (*Mém. de l'Acad. de St. Pétersb.*, 1850, vol. VI, p. 385). Quoted after Cayley's first *Report*, p. 186; see Bibliography.

into the Hamiltonian form, giving two pairs of canonical equations for the four variables y, u, p_1, p_2; these take the place of the original single differential equation of the fourth order for y. Show the equivalence of the canonical system with the original differential equation. This method is obviously applicable to any number of variables, reducing the second derivatives to first derivatives. In the general case of mth derivatives we start with the highest derivatives, reducing them to derivatives of the $(m - 1)$th order; then we repeat the process, until finally only first derivatives remain in the integrand. Then the Hamiltonian transformation brings the integrand into the canonical form.

Notice that an arbitrary system of differential equations can be transformed into the form

$$\dot{q}_i = f_i(q_1, \ldots, q_n, t), \qquad (64.7)$$

by introducing suitable independent variables $q_1, \ldots, q_n$. However, differential equations arising from a variational principle have the outstanding property that the right-hand sides are *derivable from a single function H* by differentiation. *The function H characterizes the entire set of equations.*

A peculiar feature of the canonical equations is their appearance in pairs. If, however, the conjugate variables q_k, p_k are replaced by the complex variables

$$\frac{q_k + ip_k}{\sqrt{2}} = u_k,$$

$$\frac{q_k - ip_k}{\sqrt{2}} = u_k^* \qquad (64.8)$$

then the double set of canonical equations (63.5) can be replaced by the following *single set of complex equations*:

$$\frac{du_k}{idt} = -\frac{\partial H}{\partial u_k^*} . \qquad (64.9)$$

Because of the duality of Legendre's transformation, *every Hamiltonian problem can be associated with a corresponding Lagrangian problem.* For this purpose we express the p_i in terms of q_i and $\dot{q}_i$, making use of the implicit equations

$$\dot{q}_i = \frac{\partial H}{\partial p_i}, \qquad (64.10)$$

and substitute these p_i in the integrand. This gives the Lagrangian function L, which depends only on the q_i and $\dot{q}_i$.

Summary. Hamilton's canonical equations can be considered as the solution of a Lagrangian problem with an integrand of a particularly simple structure. The variables of this variational problem are the q_i and the p_i which are varied independently of each other. The integrand of this variational problem is normalized to the form

$$\Sigma p_i \dot{q}_i - H(q_1, \ldots, q_n; \ p_1, \ldots, p_n; \ t).*$$

5. The phase space and the phase fluid. The variables of the canonical integral are the q_i and the p_i. Hence the new variational problem has $2n$ degrees of freedom. Accordingly, if we wish to picture the new situation geometrically, we have to use a space of $2n$ dimensions. The "position" of the mechanical system now includes the original position coordinates of Lagrangian mechanics along with the momenta. The great American physicist Gibbs called this q-p space, in which a single point C represents the extended "position" of the mechanical system, the "phase space." In Lagrangian mechanics we spoke of the "configuration space" which included the variables q_i alone. In Hamiltonian mechanics we speak of the "phase space" which includes the q_i and the p_i as a set of $2n$ variables.

While we found it advantageous to establish a definite kind of geometry in the configuration space, and that geometry turned out to be of the Riemannian type, the situation is different in the new phase space. The phase space has no definite metrical structure, and merely for convenience we shall assume that the q_i and p_i are plotted as rectangular coordinates of a $2n$-dimensional Euclidean space. The Euclidean geometry is as good as any other geometry since there is no inherent reason for introducing a metric in the phase space.

*Cf. Appendix I, pp. 397–400.

The doubling of dimensions which characterizes the phase space seems at first sight an unnecessary complication. Yet for the theoretical investigation of problems of motion the new procedure has great advantages. One of the most conspicuous advantages comes into evidence if we consider the geometrical representation of the *totality of paths*, pictured at first in the Lagrangian configuration space and then in the Hamiltonian phase space. As long as it is a question of a single path, the moving C-point traces out a curve in both cases. However, many theoretical investigations would be hampered if we were to single out one particular path from the totality of all possible paths. Many questions in mechanics cannot be answered satisfactorily if we single out one particular solution of the equations of motion corresponding to a given special choice of initial conditions. The *complete* solution has to be obtained, which is adjusted to arbitrary initial conditions.

Now in the Lagrangian configuration space we get a hopeless criss-cross of lines if we try to picture the totality of paths. Motion can start from every point of the configuration space *in every direction* and with arbitrary initial velocity. It is impossible to get any well-ordered representation of all these lines. But let us consider the phase space of the Hamiltonian equations. These equations are not of the *second* but of the *first* order. They determine the velocity of the moving C-point if its position is given. The motion may start from any point of the phase space, *but after giving one point P the motion is uniquely determined*. In analytical language we can say that a complete solution of the canonical equations (63.5) is obtained if the q_i and p_i are given as functions of the time t and $2n$ constants of integration which may be identified as the $2n$ position coordinates at time $t = 0$:

$$q_i = f_i(q_1^0, \ldots, q_n^0; p_1^0, \ldots, p_n^0; t),$$
$$p_i = g_i(q_1^0, \ldots, q_n^0; p_1^0, \ldots, p_n^0; t). \qquad (65.1)$$

This complete solution of the canonical equations can be pictured in a well-ordered way, without any overlapping, if the $2n$ coordinates q_i, p_i are considered as different dimensions of the phase space. In fact, the geometrical picture is even more complete if we add one more dimension by introducing the time t as a $(2n + 1)$th coordinate. Cartan[1] called this $(2n + 1)$-dimensional space the "state space" (*espace des états*). In the state space the problem of motion is completely geometrized, and the complete solution of the canonical equations is pictured as an infinite manifold of curves which fill a $(2n + 1)$-dimensional space. These curves never cross each other. Indeed, such crossing would mean that two tangents are possible at the same point of the state space, but that is excluded because of the canonical equations which give a unique tangent at any point of the space.

The geometrical and analytical picture we get here is in complete analogy with the motion of a *fluid*. Let us imagine the motion of an ordinary three-dimensional fluid as we encounter it in hydrodynamics. The motion of that fluid can be described in two ways: we can use the "particle description" or we can use the "field description."

In the first case we follow the individual fluid particle and give its position as it changes with the time t, starting from a certain initial position x_0, y_0, z_0:

$$x = f(x_0, y_0, z_0, t),$$
$$y = g(x_0, y_0, z_0, t), \qquad (65.2)$$
$$z = h(x_0, y_0, z_0, t).$$

On the other hand, we can consider the "velocity field" as it exists at a certain time t, given by

$$\dot{x} = \frac{\partial f}{\partial t}, \qquad \dot{y} = \frac{\partial g}{\partial t}, \qquad \dot{z} = \frac{\partial h}{\partial t}. \qquad (65.3)$$

We now solve the equations (65.2) for the x_0, y_0, z_0, expressing them as functions of x, y, z, t. If we substitute these expressions in (65.3), we obtain $\dot{x}$, $\dot{y}$, $\dot{z}$, expressed as functions of x, y, z and t:

$$\dot{x} = u(x, y, z, t),$$
$$\dot{y} = v(x, y, z, t), \qquad (65.4)$$
$$\dot{z} = w(x, y, z, t).$$

This is the *field description* of the fluid motion which determines the velocity vector at any point of space, at any time. If the par-

[1]Cartan, *Invariants intégraux*, p. 4; see bibliography.

ticle description is given, we can obtain the field description by differentiations and eliminations. On the other hand, if the field description is given, we can obtain the particle description by *integrating* the equations (65.4). The two descriptions are equivalent.

This hydrodynamical picture carries over completely into the phase-space. The only difference is that instead of the 3 coordinates x, y, z we have the $2n$ coordinates $q_1, \ldots, q_n; p_1, \ldots, p_n$. The particle description corresponds to the integrated form of these equations. The behaviour of the $2n$-dimensional phase fluid is similar to that of an ordinary fluid.

Summary.   The configuration space of the canonical equations comprises $2n$ dimensions, viz. the n position coordinates q_i and the n momenta p_i, which are the independent variables of the variational problem. This $2n$-dimensional space is called the "phase space." If the time is considered as an additional variable the $(2n + 1)$-dimensional space is called the "state space." Geometrically, the motion can be pictured as that of a $2n$-dimensional fluid, called the "phase fluid." Each individual stream-line of the moving fluid represents the motion of the mechanical system under specified initial conditions; the fluid as a whole represents the complete solution, for arbitrary initial conditions.

6. The energy theorem as a consequence of the canonical equations.

The canonical equations,

$$\dot{q}_k = \frac{\partial H}{\partial p_k},$$

$$\dot{p}_k = -\frac{\partial H}{\partial q_k},$$

(66.1)

assume a new significance, if interpreted hydrodynamically, in connection with the motion of the phase fluid. They determine the velocity of a fluid particle at a certain point of the phase space at a certain time. Now it so happens that certain types of fluid motion which are of special interest in ordinary hydrodynamics are also of interest in the motion of the phase fluid.

One such type is the *steady* motion of a fluid. In this the velocity field is independent of the time t. Although the fluid is in motion and the particles constantly change position, yet the velocity *at a definite point of space* is constant. This means, for the field description (65.4), that the right-hand sides of the equations are independent of t.

The same situation arises for the phase fluid in the case of a conservative (scleronomic) system. Here the Lagrangian function L does not depend on t, and consequently the Hamiltonian function H is likewise independent of t:

$$H = H(q_1, \ldots, q_n; p_1, \ldots, p_n). \tag{66.2}$$

But then the right-hand sides of the equations (66.1) are independent of the time, which shows at once that *the phase fluid associated with a conservative system is in a state of steady motion.*

A second fundamental theorem can be derived for such systems, and that is the *energy theorem.* Differentiation of (66.2) gives

$$\frac{dH}{dt} = \sum_{i=1}^{n} \left(\frac{\partial H}{\partial q_i} \dot{q}_i + \frac{\partial H}{\partial p_i} \dot{p}_i \right). \tag{66.3}$$

The canonical equations (66.1) show at once that

$$\frac{dH}{dt} = 0, \tag{66.4}$$

and thus

$$H = \text{const.} = E. \tag{66.5}$$

This theorem gives H the physical significance of "total energy." If L is equal to $T - V$ and T is a quadratic form in the velocities, while V is independent of the velocities, then

$$H = T + V. \tag{66.6}$$

In relativistic mechanics both conditions are violated. Yet the energy theorem (66.5) remains true, but for an H which is defined according to the general law (62.3):

$$H = \sum_{i=1}^{n} \frac{\partial L}{\partial \dot{q}_i} \; \dot{q}_i - L. \tag{66.7}$$

The energy theorem (66.5) has an interesting geometrical interpretation in connection with the moving phase fluid. The equation

$$H\,(q_1, \ldots, q_n; p_1, \ldots, p_n) = E \tag{66.8}$$

represents a surface of the $2n$-dimensional phase space. If the constant E assumes arbitrary values we obtain an infinite family of surfaces which fill the phase space. The meaning of the energy theorem is that *a fluid particle which starts its motion on a definite energy surface remains constantly on that surface, no matter how long we follow its motion.*

Summary. If the Hamiltonian function H is independent of the time t, we have a conservative mechanical system. Such systems are distinguished by two special properties of the associated phase fluid:

1. The motion of the phase fluid is *steady.*

2. Each fluid particle remains permanently on the "energy surface" $H = E$.

---

7. Liouville's theorem. A physical fluid sometimes resists changes of volume by strong forces. We speak of an "incompressible fluid" if an arbitrary volume of the fluid is carried along unchanged during the motion. Analytically the incompressibility of a fluid shows up in two ways. In the particle description (65.2), the condition of incompressibility is that the "functional determinant" of the x, y, z with respect to the x_0, y_0, z_0 is everywhere equal to 1. In the field description (65.4), the condition of incompressibility is that

$$\text{div } \mathbf{v} \equiv \frac{\partial u}{\partial x} + \frac{\partial v}{\partial y} + \frac{\partial w}{\partial z} = 0. \tag{67.1}$$

The phase fluid associated with the canonical equations has the interesting property that it *imitates the behaviour of an incompressible fluid.* This fluid is now $2n$-dimensional and thus the natural generalization of equation (67.1) is

$$\text{div } \mathbf{v} \equiv \sum_{i=1}^{n} \left(\frac{\partial \dot{q}_i}{\partial q_i} + \frac{\partial \dot{p}_i}{\partial p_i} \right) = 0. \tag{67.2}$$

The canonical equations (66.1) show directly that this relation is fulfilled, and not only for conservative systems but for any systems. Now Green's theorem which transforms the volume integral of the divergence into a surface integral of the flux, holds in n dimensions no less than in three dimensions. In view of this transformation, the divergence equation (67.2) can be changed into the statement that *the total flux of the phase fluid, taken for any closed surface of the phase space, is always zero.* But this means that *the phase fluid moves like an incompressible fluid.* This theorem was first discovered by Liouville (1838) and is thus called "Liouville's theorem."

Liouville's theorem permits us to add a new conservation law to the previous energy theorem. We may cut out an arbitrary $2n$-dimensional region of the phase space of volume

$$\sigma = \int dq_1, \dots, dq_n; \, dp_1, \dots, dp_n, \tag{67.3}$$

and investigate what happens as the points of this region move with the phase fluid. Although the region may become distorted, yet *its volume remains unchanged during the motion:*

$$\sigma = \text{const.} \tag{67.4}$$

Example: If the mechanical system has but one degree of freedom, the phase space becomes two-dimensional and the state space three-dimensional. Hence for this simple case the behaviour of the phase fluid can be completely visualized in our ordinary space. The "energy surfaces" are here reduced to curves and these curves give directly the streamlines of the two-dimensional phase fluid. Moreover, although the picture of the streamlines is static and does not contain the velocity with which the fluid particles travel along the streamlines, this velocity is deducible from the distance between neighbouring streamlines, in view of the incompressibility of the phase fluid.

Problem. The two figures below belong to the following two problems:
1. Linear oscillator:

$$L = \tfrac{1}{2} m\dot{q}^2 - \tfrac{1}{2} k^2 q^2. \tag{67.5}$$

2. Elastic reflection of a particle between two rigid walls.

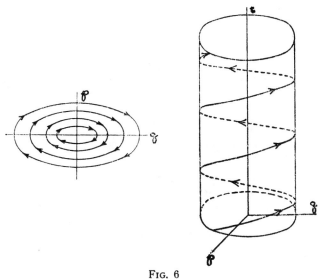

FIG. 6
Linear oscillator.
(A) Phase space.　　　　(B) State space.

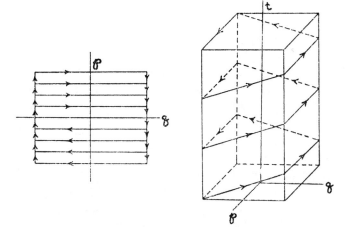

FIG. 7
Elastic reflection.
(A) Phase space.　　　　(B) State space

Give an analysis of Figs. 6 and 7. The streamlines of the first problem are concentric ellipses, of the second problem rectangles.

Assume the law

$$V = \tfrac{1}{2} k^2 q^{2m} \tag{67.6}$$

where m varies from 1 to infinity. Show that the ellipses of the oscillator bulge out more and more, as m increases, and come nearer and nearer to rectangles. For $m = \infty$ we get the streamline distribution of the elastic reflection. The helix of Fig. 6 (B) is then deformed to the broken line of Fig. 7 (B). The "walls" are at the points $q = \pm 1$.

Summary. The phase fluid of the $2n$-dimensional phase space behaves like an incompressible fluid. An arbitrary region, cut out of the fluid and carried along by the fluid particles, changes its shape during the motion but remains of constant volume.

8. Integral invariants, Helmholtz' circulation theorem. The French mathematician Poincaré (1859-1912) called any integral associated with the phase fluid which has the property of remaining unchanged during the motion an "integral invariant." The volume σ of the phase fluid, considered in the last paragraph, is one example of such an integral invariant. Another important example is a quantity introduced by Helmholtz and called the "circulation."

We return to an earlier theorem, considered in Chapter v, section 3 [cf. (53.3)] which gives the variation of the action integral for arbitrary variations of the q_i, even though they may not vanish at the two end-points:

$$\delta A = \left[\sum_{i=1}^{n} p_i \, \delta q_i \right]_{t_1}^{t_2}. \tag{68.1}$$

This important theorem holds for the Hamiltonian form of mechanics without any modification, since the variation of the p_i has no influence on δA. Hence (68.1) holds for arbitrary variations of the q_i *and* the p_i.

We make the following use of this theorem. Let us draw an arbitrary closed curve L in the phase space at a certain time t_1.

This is to be a "material line," i.e. its points are rigidly attached to fluid particles which carry the line with them. Hence we shall find the line L at another time t_2 somewhere else in the phase space. It will again be a closed curve. Let us assume that the curve at time t_1 is given in the parameter form

$$q_i = f_i(\tau),$$
$$p_i = g_i(\tau). \tag{68.2}$$

We can now form the following line integral, extended along the closed curve L:

$$\Gamma = \oint \sum_{i=1}^{n} p_i dq_i = \oint \sum_{i=1}^{n} p_i q_i' d\tau. \tag{68.3}$$

This quantity is an invariant of the motion:

$$\Gamma = \text{const.} \tag{68.4}$$

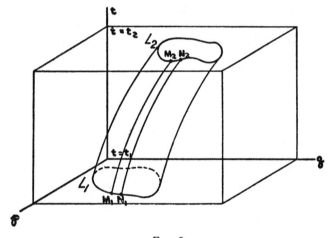

FIG. 8

Fig. 8 illustrates the situation. The phase space at the times t_1 and t_2 is pictured as two cross-sections of the $(2n + 1)$-dimensional state space. The point M_1 is carried over into the point M_2 by the moving fluid, and the neighbouring point N_1 into N_2. The line M_1M_2 is the world line of a fluid particle, and so is the line N_1N_2. The action integral has the value A between M_1 and M_2, and the value $A + dA$ between N_1 and N_2. Since the line

N_1N_2 can be considered as a variation of the line M_1M_2, we can apply the theorem (68.1) and obtain

$$dA = \left[\sum_{i=1}^{n} p_i dq_i \right]_{t_1}^{t_2}. \tag{68.5}$$

If we integrate this equation between two points τ_1 and τ_2 of the curve L, we obtain

$$\Delta A = \left[\int_{\tau_1}^{\tau_2} \sum_{i=1}^{n} p_i dq_i \right]_{t_1}^{t_2}. \tag{68.6}$$

However, if we go completely round L, so that the end-point and the initial point coincide, then ΔA vanishes. This gives

$$\left[\oint \sum_{i=1}^{n} p_i dq_i \right]_{t_1}^{t_2} = 0, \tag{68.7}$$

and, since t_1 and t_2 are arbitrary, the theorem (68.4) follows.

Helmholtz made an interesting application of this "circulation theorem." Let us think of the motion of a particle in a field of force of potential energy V. The associated phase space here is six-dimensional. Yet we do not need this six-dimensional space in order to visualize the theorem (68.4). Instead of considering p_1, p_2, p_3, as three additional coordinates, we can draw a vector with components p_1, p_2, p_3 at points q_1, q_2, q_3 of the configuration space. Let us choose as q-coordinates the ordinary rectangular coordinates x, y, z. We then have a closed curve in ordinary three-dimensional space, and along that curve a continuous vector field with components p_1, p_2, p_3. But the momentum vector $\mathbf{P}$ is merely $m\mathbf{v}$, where $\mathbf{v}$ is the velocity. Hence the circulation becomes

$$\Gamma = m \oint \mathbf{v} \cdot d\mathbf{s}, \tag{68.8}$$

where $d\mathbf{s}$ is the directed element of length along the closed curve. This picture dispenses with the six-dimensional phase space and replaces it by the three-dimensional configuration space.

The configuration space generally has no relation to the actual physical space. However, the configuration space of a single particle coincides with the physical space. The various paths drawn in this space represent the paths of the particle, described under various initial conditions. Now these paths can also be considered as the streamlines of a so-called "ideal fluid," i.e. a physical fluid (not necessarily incompressible) which is frictionless and the temperature of which is constant. It is true that the fluid particles are acted upon by forces which originate from the surrounding particles, but Euler's hydrodynamical equations of motion show that these forces have a potential and are thus equivalent to an external monogenic force. Hence the conditions for the validity of Hamilton's principle are satisfied and the streamlines of the moving fluid are the same as the streamlines of the configuration space, to which the circulation theorem can be applied. We thus get the circulation theorem of Helmholtz which states that

$$\oint \mathbf{v} \cdot d\mathbf{s} = \text{const.,} \tag{68.9}$$

the integral being taken round any closed curve in the ideal fluid and carried along by the fluid particles. Hence, if the circulation round every closed curve is zero at $t = 0$, then the same property holds permanently. This means that a fluid which is initially free from vortices remains so permanently—i.e. vortices cannot be created or destroyed.

Summary. The "circulation" is an invariant of the motion of the phase fluid. This is the quantity $\Sigma p_i dq_i$, integrated along an arbitrary closed curve of the phase space. The invariance of the circulation has the same significance for the phase fluid that the theorem of Helmholtz has for an ideal physical fluid, asserting the indestructible character of vortices.

---

9. The elimination of ignorable variables. Although the canonical equations have a much simpler structure than the original Lagrangian equations, we do not possess any general method for the integration of these equations. The ignorable variables again play a particularly important role in the integration of the equations of motion. Whenever a variable is ignorable, a partial integration and reduction of the given mechanical problem becomes possible. The process of reduction, however, is much simpler in the Hamiltonian than in the Lagrangian form of mechanics.

If the Lagrangian function L does not contain a certain q_i—let us say q_n—the Hamiltonian function H will not contain it either. The last of the canonical equations then gives

$$\dot{p}_n = 0, \tag{69.1}$$

from which

$$p_n = \text{const.} = C_n. \tag{69.2}$$

The constancy of a kinosthenic momentum is thus once more established.

Since the p_i are now independent variables, there can be no objection to using the condition (69.1) during the process of variation. The principle of varying between definite limits is not violated since p_n is not varied at all.

If p_n is kept constant, it can be replaced in the Hamiltonian function by C_n. Moreover, in the kinetic part of the action integral we get

$$\int_{t_1}^{t_2} p_n \dot{q}_n dt = C_n \int_{t_1}^{t_2} \dot{q}_n dt = C_n \left[q_n \right]_{t_1}^{t_2}. \tag{69.3}$$

This is a pure boundary term which is a constant as far as variation goes and can be dropped. Hence, the action integral is reduced to

$$\overline{A} = \int_{t_1}^{t_2} \left(\sum_{i=1}^{n-1} p_i \dot{q}_i - H \right) dt. \tag{69.4}$$

The same procedure holds for any number of ignorable variables.

Problem. Since every Hamiltonian problem is associated with a Lagrangian problem, show that the dropping of ignorable variables in the Hamiltonian treatment amounts in the Lagrangian treatment to the Routhian reduction process, studied before in chapter v, section 4.

After solving the reduced mechanical problem, the ignorable variables are finally obtained as functions of t by straight integrations, since the right-hand sides of the corresponding canonical equations

$$\dot{q}_m = \frac{\partial H}{\partial p_m} \tag{69.5}$$

are then explicit functions of t.

Summary. The elimination of ignorable variables in the Hamiltonian form of mechanics is a very simple process. We drop the contribution of the ignorable variables to the kinetic part of the canonical integral, while in the Hamiltonian function the ignorable momenta are replaced by constants.

10. The parametric form of the canonical equations. We have seen how the concept of the "state space" completely geometrizes the problem of motion associated with the canonical equations. The totality of solutions of the canonical equations can be pictured as an infinite family of non-intersecting curves which fill the state space.

The strange feature about the state space is that it comprises an *odd* number of dimensions. While all the position coordinates q_i are associated with their momenta p_i, the time t is singled out as a variable which is not associated with its momentum. The reason is that the time is not considered as a mechanical variable but it plays the role of the independent variable.

Great advantage sometimes accrues through letting the time t become one of the mechanical variables. Instead of considering the position coordinates q_i as functions of the time t, we consider the position coordinates q_i *and the time t* as mechanical variables, giving them as functions of some unspecified parameter τ. Lagrange observed that in this case the configuration space of a single particle becomes a space of four rather than three dimensions. In relativistic mechanics this procedure is an absolute necessity since space and time are united into the four-dimensional world of Einstein and Minkowski.

The next problem is how to formulate the Hamiltonian canonical equations of motion. The original Lagrangian problem is now one of finding $q_1, \ldots, q_n$ *and* t as functions of an independent variable τ. Hence the Lagrangian configuration space has $n + 1$ dimensions, and we can add t to the position coordinates q_i by putting

$$t = q_{n+1} \tag{610.1}$$

But then the corresponding phase space must comprise $2n + 2$ dimensions, corresponding to the $n + 1$ pairs of canonical variables,

$$\begin{pmatrix} q_1 \\ p_1 \end{pmatrix}, \quad \begin{pmatrix} q_2 \\ p_2 \end{pmatrix}, \quad \cdots \cdots, \begin{pmatrix} q_n \\ p_n \end{pmatrix}, \quad \begin{pmatrix} t \\ p_t \end{pmatrix},$$

where t and p_t are merely alternative notations for q_{n+1} and p_{n+1}. The momentum p_t associated with the time t—which has as its physical interpretation the negative of the total energy Λ [cf. chapter v, section 6, (56.2)]—has joined the other variables as an additional independent variable and thus adds one more dimension to the previously conceived "state space." We shall call this new space the "extended phase space." The original Lagrangian integral, in parametric form, becomes [cf. chap. v, section 6, (56.1)],

$$A = \int_{\tau_1}^{\tau_2} L\left[q_1, \ldots, q_n, q_{n+1}, \frac{q_1'}{q_{n+1}'}, \ldots, \frac{q_n'}{q_{n+1}'} \right] q_{n+1}' d\tau. \tag{610.2}$$

We can see in advance that the ordinary routine procedure will not cover the parametric case, because the canonical equations of motion determine the variables q_i and p_i as *definite* functions of the independent variable—which in our case is τ—while actually τ is an unspecified variable, and the freedom of identifying τ with any suitable variable must show up in the solution.

Indeed, the Hamiltonian function H_1 of the extended problem *vanishes identically*. We can see this by examining the Lagrangian function

$$L_1 = L\left(q_1, \ldots, q_n, q_{n+1}; \frac{q_1'}{q_{n+1}'}, \ldots, \frac{q_n'}{q_{n+1}'} \right) q_{n+1}'. \tag{610.3}$$

The function L_1 is a homogeneous form of the first order in the $n + 1$ variables $q_1', \ldots, q'_{n+1}$. Hence, by Euler's theorem on homogeneous functions,

$$\sum_{i=1}^{n+1} \frac{\partial L_1}{\partial q_i'}\, q_i' = L_1, \tag{610.4}$$

which shows that

$$H_1 = \sum_{i=1}^{n+1} p_i q_i' - L_1 = 0. \tag{610.5}$$

Hence, *the canonical integral is reduced to the completely symmetrical form*

$$A = \int_n^n \sum_{i=1}^{n+1} p_i q_i' d\tau, \tag{610.6}$$

without any Hamiltonian function. Yet obviously something has to take the place of the function H_1, since in the absence of a Hamiltonian function all the q_i and p_i would become constants, which is obviously impossible.

Actually there *is* something which takes the place of the Hamiltonian function, namely, an *auxiliary condition* which must hold between the q_i and the p_i. The equations

$$p_i = \frac{\partial L_1}{\partial q_i'} \tag{610.7}$$

can obviously not be solvable for the q_i', otherwise we would get a definite set of q_i' associated with a definite set of p_i, but actually the p_i remain unchanged if all the q_i' are multiplied by the same factor a. Now, if the equations (610.7) are not solvable, this must mean that they are not independent of each other, which again means that there must exist an *identity* between the p_i (this identity contains the q_i as passive variables, and is thus actually a relation between the q_i and p_i).

Indeed, if we single out the last variable $q_{n+1} = t$, we obtain for the momentum associated with this variable

$$p_{n+1} = - H (q_1, \ldots, q_{n+1}; p_1, \ldots, p_n), \tag{610.8}$$

and thus our variation problem assumes the following form—the integral

$$A = \int_n^n (p_1 q_1' + \ldots + p_{n+1} q'_{n+1})\, d\tau \tag{610.9}$$

is to be stationary under the auxiliary condition:

$$p_{n+1} + H = 0. \qquad (610.10)$$

If, in the integral (610.9), we substitute for p_{n+1} its value $-H$, we obtain the standard form of the canonical integral

$$A = \int_{t_1}^{t_2} (p_1 \dot{q}_1 + \ldots + p_n \dot{q}_n - H)\, dt, \qquad (610.11)$$

provided that we return from the parameter τ to $q_{n+1} = t$ as independent variable.

However, we need not single out the variable q_{n+1}, and it is preferable to write down the identity which must exist between the q_i and p_i in the more general form

$$K(q_1, \ldots, q_{n+1};\, p_1, \ldots, p_{n+1}) = 0. \qquad (610.12)$$

This form preserves the symmetry of the $2n + 2$ canonical variables, without giving preference to any of them.

Problem 1. Consider the action integral of Jacobi's principle

$$A = \int_{\tau_1}^{\tau_2} \sqrt{2} \sqrt{E - V} \sqrt{\Sigma\, a_{ik}\, q_i'\, q_k'}\; d\tau.$$

Show that the p_i associated with this integrand satisfy the following identity:

$$\frac{\Sigma\, b_{ik}\, p_i\, p_k}{2\,(E - V)} = 1,$$

where the b_{ik} are the coefficients of the matrix which is the reciprocal of the original matrix a_{ik}.

Hence the Hamiltonian problem associated with a parametric Lagrangian problem assumes finally the following form. The canonical integral (610.9) is to be made stationary under the auxiliary condition (610.12). We can treat the auxiliary condition (610.12) by means of the Langrangian λ-method, modifying the given integral as follows:

$$\overline{A} = \int_{\tau_1}^{\tau_2} \left(\sum_{i=1}^{n+1} p_i q_i' - \lambda K \right) d\tau. \qquad (610.13)$$

Here λ is some arbitrary function of τ, in complete harmony with its unspecified nature. In the present problem the "undetermined Lagrangian factor" actually remains an undetermined factor. If we wish, we can put λ equal to 1 by proper choice of the variable τ. Then we obtain the variational integral

$$\overline{A} = \int_{\tau_1}^{\tau_2} \left(\sum_{i=1}^{n+1} p_i q_i' - K \right) d\tau, \tag{610.14}$$

which is precisely of the canonical form, except that our "extended" problem has $2n + 2$ instead of $2n$ canonical variables. The function K—the left-hand side of the auxiliary condition (610.12)—takes the place of the Hamiltonian function H; hence we shall call K the "extended Hamiltonian function."

The parametric formulation of the canonical equations thus becomes

$$\left. \begin{aligned} q_k' &= \frac{\partial K}{\partial p_k}, \\[2mm] p_k' &= -\frac{\partial K}{\partial q_k}. \end{aligned} \right\} \quad (k = 1, \dots, n+1). \tag{610.15}$$

For the special choice

$$K = p_{n+1} + H, \tag{610.16}$$

we have

$$\left. \begin{aligned} q_k' &= \frac{\partial H}{\partial p_k}, \\[2mm] p_k' &= -\frac{\partial H}{\partial q_k}, \end{aligned} \right\} \quad (k = 1, \dots, n), \tag{610.17}$$

$$q_{n+1}' = 1, \tag{610.18}$$

$$p_{n+1}' = -\frac{\partial H}{\partial q_{n+1}}. \tag{610.19}$$

Equation (610.18) shows that our normalized parameter τ now becomes equal to q_{n+1}, which is the time t. The equations (610.17) are the regular canonical equations. The last equation of the set, (610.19), gives the law according to which the negative of the total energy, p_{n+1}, changes with the time t (cf. chap. v, section 3). This equation is now included in the parametric set of equations as an independent equation, since p_{n+1} is one of the independent variables.

The general parametric formulation of the canonical equations in the form (610.15) has great theoretical advantages and can be considered as the most advanced form of the canonical equations. It shows the role of conservative systems in a new light. We notice that after adding the time t to the mechanical variables, *every system becomes conservative.* The "extended Hamiltonian function" K does not depend on the independent variable τ, and thus our system is a conservative system of the extended phase space. The motion of the phase fluid is steady and every fluid particle remains permanently on a definite surface:

$$K = \text{const.} \tag{610.20}$$

Thus the auxiliary condition $K = 0$ is permanently satisfied if only the initial values of the q_i and p_i at the τ-value τ_1 are chosen in agreement with the condition $K = 0$.

The formulation of the principle of least action by Euler and Lagrange for the case of conservative systems assumes new significance in the light of parametric systems. We remember that that principle required the minimizing of the integral of $2T$ with respect to the time, provided that the moving point satisfied the energy equation $T + V = E$. If we change from configuration space to phase space, the principle of Euler-Lagrange appears in the following form. The integral

$$A = \int_{\tau_1}^{\tau_2} (p_1 q_1' + \ldots + p_n q_n') \, d\tau \tag{610.21}$$

is to be made stationary, under the auxiliary condition

$$H - E = 0. \tag{610.22}$$

The principle expressed by the parametric canonical equations requires the integral

$$A = \int_{\tau_1}^{\tau_2} (p_1 q_1' + \ldots + p_n q_n' + p_{n+1} q_{n+1}') \, d\tau \tag{610.23}$$

to be stationary, under the auxiliary condition

$$K = 0. \tag{610.24}$$

Notice that this principle changes over immediately into the principle of Euler-Lagrange if our system is conservative and if K is given in the form $H + p_{n+1}$. Since the last variable $t = q_{n+1}$ is now ignorable, the momentum p_{n+1} can be replaced by the constant $-E$ and the last term of the integrand in (610.23) can be dropped. This is exactly the principle enunciated in (610.21) and (610.22).

The parametric form of the canonical equations gives us also a deeper insight into the mutual relations of the various minimum principles of mechanics. If the canonical integral is normalized to the form

$$A = \int_{\tau_1}^{\tau_2} \Sigma p_i q_i' d\tau, \tag{610.25}$$

the difference between the various minimum principles corresponds to different interpretations of the auxiliary condition

$$K = 0. \tag{610.26}$$

As was stated in Problem 1 of this section, the auxiliary condition of Jacobi's principle assumes the form

$$\frac{1}{2} \frac{\sum\limits_{i,\,k=1}^{n} b_{ik} p_i p_k}{E - V} = 1. \tag{610.27}$$

This same auxiliary condition can obviously be given also in the form

$$\tfrac{1}{2} \Sigma b_{ik} p_i p_k + V = E, \tag{610.28}$$

which is the auxiliary condition of the Euler-Lagrange principle. The equivalence of these two principles is thus directly established.   Moreover, if we treat the auxiliary condition (610.28) by the λ-method and go back from the phase space to the configuration space, we obtain Hamilton's principle. This shows the equivalence of all three principles for conservative systems.   The time t which plays simply the role of a parameter, is interpreted as that particular variable τ for which the undetermined factor λ becomes 1.

It is of interest to see what happens if we leave the auxiliary condition in the form (610.27) and now apply the λ-method. If we change back from phase space to configuration space, we obtain the principle that the integral

$$A = \int_{\tau_1}^{\tau_2} (E - V) \sum_{i,\,k=1}^{n} a_{ik} q_i' q_k' d\tau \tag{610.29}$$

is stationary. This is Jacobi's principle, but *without the customary square root*. The mechanical paths following from this principle are nevertheless the same as those which follow from Jacobi's principle. The difference lies only in the normalization of the independent variable τ. The τ of the customary form of Jacobi's principle is an unspecified parameter. The τ of the principle (610.29), however, is normalized in a definite manner.

Problem 2. The principle (610.29) leads to the following interesting equivalence theorem. Consider the motion of a particle whose mass m is not constant, but changes according to the law

$$m = m_0(1 - aV)$$

where a is a constant. This variable mass describes, under its own inertia, without any external field of force, exactly the same path as the constant mass m_0 under the action of the potential energy V, provided that the constant mass m_0 starts its motion in the initial direction of the mass m and moves with the total energy $E = 1/a$. Show the truth of this theorem by writing down the Lagrangian equations of motion. The equivalence holds for the *paths* only, not for the motion with respect to time.

Summary. If the time t is added to the other mechanical variables as an $(n + 1)$th position coordinate, a remarkable symmetrization of the canonical integral takes place. The canonical integral is now

$$\int_{\tau_1}^{\tau_2} \sum_{i=1}^{n+1} p_i q_i' d\tau,$$

which must be made stationary under the auxiliary condition

$$K(q_i, p_i) = 0.$$

The phase space has now $2n + 2$ dimensions and the motion of the phase fluid is always steady, while the mechanical system is always conservative. The particular properties of conservative systems are thus generalized to arbitrary systems. This parametric formulation of the canonical equations has many theoretical advantages.

CHAPTER VII

CANONICAL TRANSFORMATIONS

Introduction. We have brought the differential equations of motion into a particularly desirable form, the "canonical form." But our ultimate goal will be reached only if we can solve these equations. No direct integration method is known, and we have to follow a more indirect course. One such is the method of coordinate transformations. We try to find coordinates in which the Hamiltonian function of the canonical equations is so highly simplified that in the new coordinates the equations of motion are directly integrable. From this point of view it is obviously desirable to investigate the entire group of coordinate transformations which are associated with the canonical equations. The study of these "canonical transformations" is thus a valuable aid in the integration problem of mechanics. This theory is essentially due to Jacobi. Although he may not have possessed the imaginative capacities of Hamilton and his efforts were too much bent on the integration problem, yet the discovery of the canonical transformations is a great achievement. The resulting integration theory played an important part in the modern development of atomic physics. In Hamilton's far-reaching investigations, the integration problem assumed a more incidental importance.

1. Coordinate transformations as a method of solving mechanical problems. We have already found in the Lagrangian form of mechanics that a proper choice of coordinates can greatly facilitate the problem of solving the differential equations of motion. We may find a first integral of the Lagrangian equations whenever one of our variables happens to be kinosthenic or ignorable. Hence we try to produce ignorable variables by transforming the original set of coordinates.

In the Hamiltonian form of dynamics the situation is quite similar. Once more we have no direct method for the integration of the canonical equations and the most effective tool we possess is the method of coordinate transformations. In this procedure the Hamiltonian equations show a number of advantages compared with the Lagrangian equations. In Lagrangian mechanics the significant function is the Lagrangian function L, and this function is the difference between the kinetic energy and the potential energy. If we try to simplify the potential energy the kinetic energy may get too complicated, and vice versa. It is rather difficult to simplify simultaneously the potential and the kinetic energy. In Hamiltonian mechanics the situation is much more favourable because the significant function, the Hamiltonian H, does not contain any derivatives, but only the variables themselves. Hence, it is comparable with the potential energy of the Lagrangian problem. The kinetic energy is normalized to the form $\Sigma p_i \dot{q}_i$ and does not participate in the transformation problem. It does determine the general class of transformations which can be employed, but staying within that class we can focus our attention entirely on the Hamiltonian function H.

A further reason why the Hamiltonian equations are superior to the Lagrangian equations in their transformation properties is that the number of variables is doubled. If at first sight this increase seems more of a loss than a gain, the procedure of coordinate transformations turns the liability into an asset. It is of great advantage that we can widen the realm of possible transformations by having a larger number of variables at our disposal.

Finally, in Lagrangian mechanics we do not possess any systematic method for the simplification of the Lagrangian function. We may hit on ignorable variables by lucky guesses, but there is no systematic way of producing them. In Hamiltonian mechanics a definite method can be devised for the systematic production of ignorable variables and the simplification of the Hamiltonian function. This method, which reduces the entire integration problem to the finding of one fundamental function, the generating function of a certain transformation, plays a central role in the theory of canonical equations and opens wide perspectives, as the next chapter will show.

The method of coordinate transformations employs a point of view entirely different from that of the direct integration problem. We do not think of the q_i and p_i as functions of the time t. We forget completely about the equations of motion and consider the q_i, p_i merely as *variables*. They are the coordinates of a point of the phase space and nothing else. The specific problem of motion is entirely eliminated. However, it is important that our transformation shall preserve the canonical equations. This involves the differential form which appears in the canonical integral. If this differential form is preserved, the whole set of canonical equations is preserved. And hence we have a transformation problem which is characterized by the invariance of a certain differential form.

We call transformations which preserve the canonical equations "canonical transformations." The general theory of these transformations is the achievement of Jacobi.

Summary. The most effective tool in the investigation and solution of the canonical equations is the transformation of coordinates. Instead of trying to integrate the equations directly, we try to introduce a new coordinate system which is better adapted to the solution of the problem than the original one. There is an extended class of transformations at our disposal in this procedure. They are called "canonical transformations."

2. The Lagrangian point transformations. The position coordinates of Lagrangian mechanics are the quantities q_i. The Lagrangian equations of motion remain invariant with respect to arbitrary point transformations of these coordinates. In Hamiltonian mechanics we again have a Lagrangian problem, but now in the $2n$ variables q_i and p_i. The configuration space of Hamiltonian mechanics is the $2n$-dimensional phase space. Hence at first sight we might think that arbitrary point transformations of the phase space are now at our disposal. This would mean that the $2n$ coordinates q_i and p_i can be transformed into some new $\bar{q}_i$ and $\bar{p}_i$ by any functional relations we please. This, however, is not the case. The canonical equations

$$\dot{q}_i = \frac{\partial H}{\partial p_i},$$

$$\dot{p}_i = -\frac{\partial H}{\partial q_i}$$

(72.1)

are the result of a very *special* Lagrangian problem, namely a Lagrangian problem whose Lagrangian function is normalized to the canonical form

$$\sum_{i=1}^{n} p_i \dot{q}_i - H.$$

(72.2)

An arbitrary point transformation of the q_i and p_i into $\bar{q}_i$ and $\bar{p}_i$ would destroy the normal form of the canonical integral, and with it the canonical equations. We thus restrict ourselves to transformations which preserve the canonical form of these equations, which is guaranteed if the variational integrand has the form (72.2). *Any transformation which leaves the canonical integrand* (72.2) *invariant, leaves also the canonical equations* (72.1) *invariant.*

Before we investigate the most general group of canonical transformations, we study some special types. The Lagrangian point transformations form a very special case of the much more extended group of canonical transformations.

In all our further discussions concerning transformations, we shall adhere to a convenient custom inaugurated by Whittaker in his *Analytical Dynamics* (see bibliography). The transformed variables will be denoted by corresponding *capital letters* Q_i, P_i.[1]

Since the variables of Lagrangian mechanics are the q_i alone, a point transformation of Lagrangian mechanics is of the form:

$$q_1 = f_1(Q_1, \ldots, Q_n),$$

$$\cdot \quad \cdot \quad \cdot \quad \cdot \quad \cdot \quad \cdot$$

$$\cdot \quad \cdot \quad \cdot \quad \cdot \quad \cdot \quad \cdot$$

(72.3)

$$\cdot \quad \cdot \quad \cdot \quad \cdot \quad \cdot \quad \cdot$$

$$q_n = f_n(Q_1, \ldots, Q_n).$$

[1] If the time t enters into the transformation, however, we shall denote its transform by $\bar{t}$ rather than T, so as to avoid confusion with the kinetic energy T.

We see from the form (72.2) of the canonical integrand that the canonical equations will certainly be preserved if we transform the p_i by requiring the invariance of the differential form $\Sigma p_i \delta q_i$:

$$\sum_{i=1}^{n} p_i \delta q_i = \sum_{i=1}^{n} P_i \delta Q_i. \tag{72.4}$$

If this principle holds for arbitrary infinitesimal changes of the q_i, we can replace the canonical integral

$$A = \int_{t_1}^{t_2} \left(\sum_{i=1}^{n} p_i \, dq_i - H dt \right) \tag{72.5}$$

by the corresponding integral

$$A = \int_{t_1}^{t_2} \left(\sum_{i=1}^{n} P_i \, dQ_i - H dt \right), \tag{72.6}$$

which shows at once that the canonical equations are preserved and that the Hamiltonian function H' of the new coordinate system is the same as the H we had before, after expressing the old variables q_i, p_i in terms of the new variables Q_i, P_i. We can say that *the Hamiltonian function H is an invariant of the point transformation* (72.3).

We shall see later that the invariance of the differential form (72.4) is no inherent attribute of an arbitrary canonical transformation. The transformations which satisfy this condition are merely a *subgroup* of the general group of canonical transformations. Even within this subgroup the transformation formulae (72.3) characterize a very restricted group of transformations, distinguished by the property that the relations between the q_i and the Q_i do not involve the p_i, P_i at all. The correlated transformation of the p_i follows from the invariance principle (72.4). From the formulae (72.3) we have

$$\delta q_i = \frac{\partial f_i}{\partial Q_1} \delta Q_1 + \ldots + \frac{\partial f_i}{\partial Q_n} \delta Q_n \tag{72.7}$$

and, introducing these expressions in (72.4), we obtain the transformation of the p_i implicitly—in the form of n linear equations—by equating the coefficients of δQ_i on both sides of the equation. These equations can be solved for the p_i, provided that the functional determinant of the transformation (72.3) is different from zero.

We have to show that this transformation of the q_i and p_i is actually equivalent to a Lagrangian point transformation. The formulae (72.3) establish such a transformation as far as the q-variables are concerned, but we have to show that the transformation of the momenta, based on the invariance principle (72.4), is likewise in agreement with the corresponding transformation of Lagrangian mechanics.

We know that the momenta p_i of Hamiltonian mechanics are introduced by the definition

$$p_i = \frac{\partial L}{\partial \dot{q}_i}. \tag{72.8}$$

Correspondingly, if we introduce the new coordinates Q_i in the Lagrangian function, we have to define the new momenta P_i by

$$P_i = \frac{\partial L}{\partial \dot{Q}_i}. \tag{72.9}$$

The Lagrangian function L is an invariant of this transformation, but the p_i and P_i are "covariant" quantities, connected with each other by a definite rule. Let us consider L as a function of the $\dot{q}_i$ alone, forgetting for the moment the presence of the variables q_i and t, which will be kept constant. Then the principle of invariance gives, for arbitrary variations of the $\dot{q}_i$ and corresponding variations of the $\dot{Q}_i$

$$\delta L = \sum_{i=1}^{n} \frac{\partial L}{\partial \dot{q}_i} \delta \dot{q}^i = \sum_{i=1}^{n} \frac{\partial L}{\partial \dot{Q}_i} \delta \dot{Q}_i. \tag{72.10}$$

This yields
$$\sum_{i=1}^{n} p_i \, \delta \dot{q}_i = \sum_{i=1}^{n} P_i \, \delta \dot{Q}_i. \tag{72.11}$$

But now, differentiation of the equations (72.3) with respect to t,

$$\dot{q}_i = \frac{\partial f_i}{\partial Q_1} \dot{Q}_1 + \ldots + \frac{\partial f_i}{\partial Q_n} \dot{Q}_n, \tag{72.12}$$

shows that the transformation from the $\dot{q}_i$ to the $\dot{Q}_i$ is a *linear* transformation, and exactly the same as (72.7) which connects the δq_i and the δQ_i. For this reason the invariance principle (72.11) remains true if we replace the $\delta \dot{q}_i$ and $\delta \dot{Q}_i$ by the δq_i and δQ_i:

$$\sum_{i=1}^{n} p_i \delta q_i = \sum_{i=1}^{n} P_i \delta Q_i. \tag{72.13}$$

This, however, is the invariance principle (72.4) from which we started. We have thus established that the transformation of the momenta, associated with a Lagrangian point transformation, can be obtained from the invariance principle (72.4).

Problem 1. Express the kinetic energy of a particle in rectangular coordinates x, y, z and again in polar coordinates r, θ, ϕ. Introduce the conjugate momenta p_x, p_y, p_z and p_r, p_θ, p_ϕ on the Lagrangian basis and determine their mutual relations. Then show that the same relations are obtainable from the invariance principle

$$p_x \delta_x + p_y \delta_y + p_z \delta_z = p_r \delta r + p_\theta \delta \theta + p_\phi \delta \phi.$$

The point transformation (72.3) was "scleronomic," i.e. it did not involve the time t. In order to extend our discussion to the rheonomic case, the most natural procedure is to add the time t to the mechanical variables and consider our problem in the $(2n + 2)$-dimensional "extended phase space" which belongs to the parametric form of the canonical equations (cf. chap. VI, section 10). In this space the point transformation (72.3) automatically includes the time t, since we now have a point transformation from the old variables $q_1, \ldots, q_n, t$ to the new ones $Q_1, \ldots, Q_n, \bar{t}$. The invariance principle (72.4) now takes the form

$$\Sigma\, p_i \delta q_i + p_t \delta t = \Sigma\, P_i \delta Q_i + p_{\bar t}\, \delta_{\bar t}. \tag{72.14}$$

Moreover, it is no longer the Hamiltonian function H, but the *extended Hamiltonian function K*, which is an invariant of the transformation. This function—as we have seen in chapter VI, section 10—stands in the following relation to the ordinary Hamiltonian function H:

$$K = p_t + H, \tag{72.15}$$

and thus the invariance of K leads to the relation,

$$p_t + H = p_{\bar t} + H', \tag{72.16}$$

if H' denotes the Hamiltonian function of the transformed system. Hence we have

$$H' = H + p_t - p_{\bar{t}},\qquad(72.17)$$

which shows that in the case of a general rheonomic transformation the Hamiltonian function is no longer an invariant of the transformation.

Usually we are interested in rheonomic transformations which introduce a moving frame of reference, without, however, transforming the time itself. A transformation of this type is characterized by the equations

$$\begin{aligned}
q_1 &= f_1(Q_1, \ldots, Q_n, t),\\
&\cdots \cdots \cdots \cdots \cdots\\
&\cdots \cdots \cdots \cdots \cdots\\
&\cdots \cdots \cdots \cdots \cdots\\
q_n &= f_n(Q_1, \ldots, Q_n, t)\\
t &= \bar{t}.
\end{aligned}\qquad(72.18)$$

Let us introduce these relations in the invariance principle (72.14), varying the time alone, without the Q_i; we find

$$p_t - p_{\bar{t}} = - \sum_{i=1}^{n} p_i \frac{\partial f_i}{\partial t},\qquad(72.19)$$

and thus

$$H' = H - \sum_{i=1}^{n} p_i \frac{\partial f_i}{\partial t}.\qquad(72.20)$$

Since H' has to be expressed in terms of the new variables Q_i and P_i, we have to eliminate the p_i from the second term on the right-hand side of (72.20), expressing them as functions of the P_i. Notice that this term is *linear* in the momenta, and thus gives rise to "gyroscopic terms" in H'.

Problem 2. Consider the motion of a single particle, transforming the rectangular coordinates x, y, z to a rotating reference system:

$$\begin{aligned}
x &= x' \cos \omega t - y' \sin \omega t,\\
y &= x' \sin \omega t + y' \cos \omega t,\\
z &= z'.
\end{aligned}$$

Show that the modification of the Hamiltonian function gives rise to the apparent Coriolis force and centrifugal force in the rotating reference system.

Summary. A Lagrangian point transformation leaves the canonical equations invariant. The transformation of the momenta takes place on the basis of the invariance of the differential form $\Sigma p_i \delta q_i$. The Hamiltonian function is an invariant of the transformation if the new coordinate system is at rest relative to the old system. Otherwise the Hamiltonian function has to be modified by gyroscopic terms.

3. Mathieu's and Lie's transformations. The previous section has shown that the invariance of the differential form (72.4) leads to the invariance of the canonical equations. We can, however, achieve the invariance of the differential form (72.4) *without* restricting the transformation so that (72.3) holds. Such more general transformations—known already to Jacobi— were studied by the French mathematician Mathieu (in 1874) and are sometimes referred to as "Mathieu transformations."[1] Sophus Lie made extended use of the same transformations in his geometrical methods of solving differential equations. He called them "contact transformations."

Lie defined a contact transformation by requiring the differential form

$$dz - \sum_{i=1}^{n} p_i dq_i \qquad (73.1)$$

to be proportional to the transformed differential form

$$dZ - \sum_{i=1}^{n} P_i dQ_i. \qquad (73.2)$$

If z is introduced as an $(n + 1)th$ variable q_{n+1}, the condition may be written in the homogeneous form

$$\sum_{i=1}^{n+1} p_i dq_i = \sum_{i=1}^{n+1} P_i dQ_i, \qquad (73.3)$$

and becomes equivalent to a Mathieu transformation of $n + 1$ dimensions.

Instead of a point transformation which connects the q_i and Q_i without involving the p_i, we widen the functional relation be-

[1] Cf. E. T. Whittaker, *Analytical Dynamics* (see bibliography), p. 301.

tween the coordinates by letting the momenta participate in the transformation of the position coordinates. Our only condition is the invariance principle

$$\sum_{i=1}^{n} p_i \delta q_i = \sum_{i=1}^{n} P_i \delta Q_i. \tag{73.4}$$

Closer inspection of this principle reveals that the q_i and the Q_i cannot be totally independent of each other. Otherwise the absurd result would follow that all p_i and P_i must vanish. There must exist at least one functional relation

$$f(q_1, \ldots, q_n, Q_1, \ldots, Q_n) = 0 \tag{73.5}$$

between the q_i and Q_i *alone*, without involving the p_i and P_i. In the case of a point transformation n such relations exist. We can classify the general Mathieu transformations according to the number m of independent relations which exist between the q-coordinates alone. The smallest number is 1, the largest is n. These relations can be prescribed at will, and thus the Mathieu transformations are characterized by the following set of auxiliary conditions which have to be added to the principle (73.4):

$$\Omega_1(q_1, \ldots, q_n; Q_1, \ldots, Q_n) = 0,$$
$$\cdots\cdots\cdots\cdots\cdots\cdots\cdots\cdots\cdots\cdots$$
$$\cdots\cdots\cdots\cdots\cdots\cdots\cdots\cdots\cdots\cdots \tag{73 6}$$
$$\cdots\cdots\cdots\cdots\cdots\cdots\cdots\cdots\cdots\cdots$$
$$\Omega_m(q_1, \ldots, q_n; Q_1, \ldots, Q_n) = 0.$$

The number m is $\leqq n$. If the maximal number n is reached, we get the most restricted form of Mathieu transformations. We then return to the previously considered case of point transformations, because now we can eliminate all the q_i in terms of the Q_i, and obtain the relations (72.3).

In order to get the actual transformation between the old and new coordinates, we might eliminate m of the q_i as functions of the remaining q_k and the Q_i, and then substitute the variations of these q_i in the invariance principle (73.4). It is preferable, however, not to disturb the symmetry of the relations (73.6) but to apply the Lagrangian λ-method, considering the equations (73.6) as given auxiliary conditions of the variation. This yields the principle

$$\sum_{i=1}^{n} (p_i \delta q_i - P_i \delta Q_i) = \lambda_1 \delta \Omega_1 + \ldots + \lambda_m \delta \Omega_m, \quad (73.7)$$

and now we can vary all the q_i and Q_i arbitrarily. We thus obtain the generating formulae of an arbitrary Mathieu transformation in the following form:

$$p_i = \lambda_1 \frac{\partial \Omega_1}{\partial q_i} + \ldots + \lambda_m \frac{\partial \Omega_m}{\partial q_i},$$

$$P_i = -\left(\lambda_1 \frac{\partial \Omega_1}{\partial Q_1} + \ldots + \lambda_m \frac{\partial \Omega_m}{\partial Q_i} \right), \quad (73.8)$$

for $i = 1, 2, \ldots n$. These equations, together with the given auxiliary conditions (73.6), determine the transformation completely, although in an implicit form. We may eliminate the λ_i from the equations (73.8), obtaining $2n - m$ relations between the old and the new variables. These, together with the m auxiliary conditions (73.6), complete the transformation. By proper eliminations the old q_i, p_i can finally be expressed in terms of the new Q_i, P_i, or vice versa.

From here on, our conclusions run parallel to those of the previous section. The invariance of the differential form $\Sigma p_i \delta q_i$ guarantees the invariance of the canonical equations, and once more the Hamiltonian function H is an invariant of the transformation. Moreover, we can once more include the time amongst the variables by changing over to the parametric form of the canonical equations in which the time t joins the position co-ordinates $q_1, \ldots, q_n$ as a surplus variable. We then get the rheonomic form of the Mathieu transformations, characterized by the invariance of the differential form

$$\sum_{i=1}^{n} p_i \delta q_i + p_t \delta t = \sum_{i=1}^{n} P_i \delta Q_i + p_{\bar{t}} \delta \bar{t}, \quad (73.9)$$

together with the auxiliary conditions

$$\Omega_k(q_1, \ldots, q_n, Q_1, \ldots, Q_n, t) = 0, \quad (k = 1, \ldots, m) \quad (73.10)$$

$$t = \bar{t}.$$

The equations (73.8) are now completed by one further equation,

$$p_t - p_{\bar{t}} = \lambda_1 \frac{\partial \Omega_1}{\partial t} + \ldots + \lambda_m \frac{\partial \Omega_m}{\partial t}. \quad (73.11)$$

We are once more in the extended phase space in which the extended Hamiltonian function K is an invariant of the transformation. This involves the relation

$$p_t + H = \bar{p_t} + H', \qquad (73.12)$$

from which

$$H' = H + p_t - \bar{p_t},$$

$$= H + \lambda_1 \frac{\partial \Omega_1}{\partial t} + \ldots + \lambda_m \frac{\partial \Omega_m}{\partial t}. \qquad (73.13)$$

We see once more that the invariance of the Hamiltonian function holds only for scleronomic transformations, while the rheonomic case requires additional correction terms.

Problem. Assume $m = n$, and let the auxiliary conditions (73.6) be given in the explicit form (72.18). Show that the transformation of the Hamiltonian function according to (73.13) agrees with the result (72.20) of the previous section.

Summary. Mathieu's transformations are characterized by the requirement that the differential form $\Sigma p_i \delta q_i$ shall be an invariant of the transformation. This requires that there shall exist at least one condition between the q_i and Q_i alone not involving the p_i. Mathieu's transformations can be classified according to the number of conditions which exist between the q_i and Q_i alone, and which can be given *a priori*. The minimum is 1, the maximum is n. The latter case corresponds to the point transformations previously considered, which are thus a special sub-group of Mathieu transformations.

---

4. The general canonical transformation. In order to preserve the canonical equations, the invariance of the differential form (72.13) is not absolutely necessary. There is a more extended group of transformations which leaves the canonical equations invariant. Let us assume that the differential form

(72.13) transforms according to the following law:

$$\Sigma p_i \delta q_i = \Sigma P_i\, \delta Q_i + \delta S, \qquad (74.1)$$

where δS is the *complete differential* of a certain function S which is given as a function of the q_i and Q_i. The canonical integral now becomes

$$A = \int_{t_1}^{t_2}\left(\sum_{i=1}^{n} p_i\, dq_i - H dt\right)$$
$$= \int_{t_1}^{t_2}\left(\sum_{i=1}^{n} P_i\, dQ_i - H dt\right) + \int_{t_1}^{t_2} dS. \qquad (74.2)$$

The last term gives a pure *boundary term* with no influence on the variation, which is between definite limits. Hence, although we have modified the canonical integral, the modification amounts to the addition of a constant. And thus the vanishing of the variation of the canonical integral in the original variables guarantees the vanishing of the variation of the canonical integral in the new variables. This means that the *canonical equations of motion remain invariant under the transformation* (74.1).

Here we have the most general condition for a canonical transformation. Instead of requiring the invariance of the differential form $\Sigma p_i \delta q_i$, we stipulate that

$$\sum_{i=1}^{n} (p_i \delta q_i - P_i \delta Q_i) = \delta S, \qquad (74.3)$$

where S is given as a function of the q_i and Q_i

$$S = S(q_1, \ldots, q_n; Q_1, \ldots, Q_n). \qquad (74.4)$$

The function S is called the "generating function" of the canonical transformation. The variation of S becomes

$$\delta S = \sum_{i=1}^{n}\left(\frac{\partial S}{\partial q_i}\delta q_i + \frac{\partial S}{\partial Q_i}\delta Q_i\right). \qquad (74.5)$$

Let us substitute this expression in (74.3), assuming that no auxiliary conditions connect the q_i and Q_i, so that all the δq_i and δQ_i are free variations. This gives

$$p_i = \frac{\partial S}{\partial q_i}, \qquad (74.6)$$
$$P_i = -\frac{\partial S}{\partial Q_i}.$$

The equations (74.6) are the fundamental equations of a general canonical transformation which is free from conditions given *a priori*. It is in the nature of a canonical transformation that we cannot obtain explicitly the old variables in terms of the new variables or vice versa without eliminations. We have a *mixed* representation in which the old and the new momenta are expressed as functions of the old and new position coordinates. For an explicit representation we have to solve either the first or the second set of equations for the position coordinates.

The most general form of canonical transformation arises if the transformation possesses a generating function S and in addition there are auxiliary relations given between the q_i and Q_i as in the case of the Mathieu transformations. The application of the Lagrangian λ-method shows that the transformation equations now have the form

$$p_i = \frac{\partial \bar{S}}{\partial q_i},$$

$$P_i = -\frac{\partial \bar{S}}{\partial Q_i}, \tag{74.7}$$

where the "modified generating function" $\bar{S}$ is defined as follows:

$$\bar{S} = S + \lambda_1 \Omega_1 + \ldots + \lambda_m \Omega_m. \tag{74.8}$$

So far we have considered only scleronomic transformations. But the rheonomic case can be included at once if we treat the problem again in the extended phase space in which the time is an additional position coordinate. The generating function S has now the form

$$S = S(q_1, \ldots, q_n; Q_1, \ldots, Q_n; t), \tag{74.9}$$

with the auxiliary condition

$$t = \bar{t}. \tag{74.10}$$

The definition of the function S by means of the extended differential form

$$\sum_{i=1}^{n} p_i \delta q_i + p_t \delta t - \left(\sum_{i=1}^{n} P_i \delta Q_i + p_{\bar{t}} \delta \bar{t} \right) = \delta S, \tag{74.11}$$

shows that we obtain once more the equations (74.6) which are characteristic of a canonical transformation; but they are augmented by one further equation:

$$p_t - \bar{p_t} = \frac{\partial S}{\partial t}. \tag{74.12}$$

Finally, the invariance of the extended Hamiltonian function $K = p_t + H$ leads to the following transformation law for the ordinary Hamiltonian function H

$$H' = H + p_t - \bar{p_t}$$

$$= H + \frac{\partial S}{\partial t}. \tag{74.13}$$

Problem. Give the proof of the equation (74.13) by proceeding in a different way. Leave the time t as independent variable and consider a canonical transformation in which the time appears as a parameter. Then obtain the relation (74.13) by distinguishing between dS in the canonical integral and the δS in the definition of a canonical transformation. In the first case the time is varied, in the second case not.

Summary. The general form of a canonical transformation involves a generating function which controls the transformation. Any function of the q_i and Q_i may be chosen as the generating function of a canonical transformation. In addition to this function certain relations may be given *a priori* between the q_i and the Q_i. We then have a "conditioned" canonical transformation. The number of prescribed conditions may vary from one to n. The formulae of a canonical transformation have the peculiarity that they do not give the transformation in explicit form. Instead of determining the new variables in terms of the old ones or *vice versa*, we get a mixed representation. The old and new momenta are expressed in terms of the old and new position coordinates.

---

5. The bilinear differential form. In every transformation theory we have certain basic quantities which do *not* change with the transformation. They are the basic invariants which determine the nature of the transformation. When we began the study of canonical transformations, we established the invariance of the differential form $\Sigma p_i \delta q_i$ to insure the invariance of the canonical equations. But later we saw that the invariance of the canonical equations holds under even more general conditions. The necessary and sufficient condition for a canonical transformation was given in the form that the difference of two differential forms has to be a complete differential δS of a certain function S. Under these circumstances the question arises as to what is actually the basic invariant of the canonical transformations.

This problem is closely related to the problem of the *geometrical structure* of the phase space. We have seen how much the theory of dynamics benefited by assigning a definite geometrical structure to the Lagrangian configuration space. There we were given a Riemannian line element $\overline{ds}$, whose square was prescribed as a quadratic differential form in the variables q_i. This $\overline{ds}^2$ was the basic invariant of the Lagrangian point transformation and it was this infinitesimal distance which—if augmented by the proper boundary conditions—determined the geometrical structure of the configuration space.

Do we have something similar in Hamiltonian phase space? Is there any invariant differential form which could take over the role of the $\overline{ds}^2$ of configuration space? Actually there exists a fundamental differential form associated with the canonical transformations which is an invariant of these transformations, although it is widely different from the Riemannian $\overline{ds}^2$. It is also quadratic in the differentials but it is associated with *two* displacements and thus does not resemble a distance. The geometry of the phase space is thus not of the ordinary metrical kind. It resembles more a geometry in which *areas* can be measured but not distances. Since the fundamental differential invariant of canonical transformations is linear in both infinitesimal displacements, we call it the "bilinear differential form." The entire theory of canonical transformations can be based on this invariant differential form.

The differential quantity

$$\sum_{i=1}^{n} p_i\,dq_i - \sum_{i=1}^{n} P_i\,dQ_i = dS \tag{75.1}$$

reminds us of the work of a monogenic force. That work was the complete differential of a definite function from which the force itself could be derived. If we seek to test whether a given force is monogenic or not, we let that force act on a particle, carrying it round an arbitrary path back to the origin. If the total work of the force is zero for any closed path, the force is monogenic, otherwise not.

A similar criterion can be applied to the differential form (75.1). Let us integrate (75.1) round any closed curve L of the phase space. Then we obtain two line integrals on the left-hand

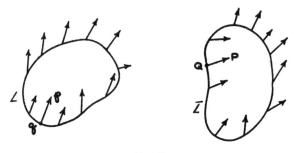

FIG. 9

side, since every p, q point is connected with a corresponding P, Q point by the transformation. The integral on the right-hand side vanishes. Hence we get an invariance principle in which the undetermined function S is no longer present:

$$\Gamma = \oint \sum_{i=1}^{n} p_i\,dq_i = \oint \sum_{i=1}^{n} P_i\,dQ_i. \tag{75.2}$$

For any closed curve of the phase space, the quantity Γ, the "circulation," can be formed and is invariant with respect to an arbitrary canonical transformation.

Now, instead of thinking of the $2n$-dimensional phase space, we can visualize our procedure by staying in the space of the q-variables alone, plotting at every point q_i the quantities p_i as vector components. An arbitrary closed curve L of the phase space means in this picture a closed curve L of the q-space, associated with an arbitrary continuous pencil of vectors along that curve (see Fig. 9).

The equation (75.2) characterizes a canonical transformation with the help of an integral invariant. We now wish to show how this integral invariant can be changed into a differential invariant. For this purpose we consider the given closed curve L as the contour of a certain two-dimensional region K. We assume the vector p_i to be given at every point of this region, which we

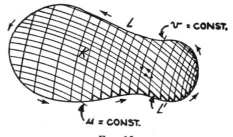

Fig. 10

characterize analytically by two "Gaussian curvilinear coordinates" u and v:

$$\begin{aligned} q_i &= q_i(u, v), \\ p_i &= p_i(u, v). \end{aligned} \qquad (75.3)$$

The parametric lines $u = $ const. and $v = $ const. form a dense network of lines. Instead of going round the smooth contour L, we follow the zig-zag line L' which is formed of infinitesimal portions of parametric lines. In the limit, as the parametric lines become arbitrarily dense, the circulation Γ taken along the zig-zag line L' approaches indefinitely the circulation taken along the smooth contour L.

We notice that we can evaluate the circulation Γ in two ways. We can go along the contour, as we have done before. But we can also go round all the numerous little parallelograms, formed by the two families of parametric lines $u = $ const. and $v = $ const. The final result is the same since all contributions which come from the interior cancel out, due to the fact that each interior line is described once in each direction. The contribution from the border, on the other hand, remains unaltered. This leads to a transformation of a line integral into a surface integral.

Let us assume that $f(u, v)$ is an arbitrary function of the region K and let $d'f$ denote the change of f due to the change of v alone, keeping u constant:

$$d'f = \frac{\partial f}{\partial v} \, dv. \tag{75.4}$$

Similarly let $d''f$ denote the change of f due to the change of u alone, keeping v constant:

$$d''f = \frac{\partial f}{\partial u} \, du. \tag{75.5}$$

Now the circulation round an infinitesimal parallelogram formed by parametric lines can be written as follows:

$$\sum_{i=1}^{n} [d'(p_i d''q_i) - d''(p_i d'q_i)]$$

$$= \sum_{i=1}^{n} (d'p_i d''q_i - d''p_i d'q_i) \tag{75.6}$$

$$= \sum_{i=1}^{n} \left(\frac{\partial q_i}{\partial u} \frac{\partial p_i}{\partial v} - \frac{\partial p_i}{\partial u} \frac{\partial q_i}{\partial v} \right) du \, dv,$$

and we obtain the following fundamental transformation of a line integral into a surface integral:

$$\oint_L \sum_{i=1}^{n} p_i dq_i = \int_K \sum_{i=1}^{n} \left(\frac{\partial q_i}{\partial u} \frac{\partial p_i}{\partial v} - \frac{\partial p_i}{\partial u} \frac{\partial q_i}{\partial v} \right) du \, dv. \tag{75.7}$$

Problem. Investigate the theorem (75.7) for the **three-dimensional case** $n = 3$ and show that it reduces to the integral transformation known as "Stokes' theorem." The integral transformation (75.7) generalizes Stokes' theorem to any number of dimensions. Notice that this theorem does not depend on any specific metrical properties of space.

With the help of this transformation the invariance of the circulation Γ can be stated in a new form. That circulation can now be written as a *surface integral*, extended over the region K. Since this region is entirely arbitrary, the invariance of Γ involves the invariance of the integrand

$$\sum_{i=1}^{n} \left(\frac{\partial q_i}{\partial u} \frac{\partial p_i}{\partial v} - \frac{\partial p_i}{\partial u} \frac{\partial q_i}{\partial v} \right). \tag{75.8}$$

This again is equivalent to the invariance of the differential form

$$\sum_{i=1}^{n} (d'p_i d''q_i - d''p_i d'q_i), \tag{75.9}$$

which can be associated with any two independent infinitesimal displacements $d'q_i, d'p_i$, and $d''q_i, d''p_i$ of the phase space. *This is the "bilinear differential form," the invariance of which is a necessary and sufficient condition for a canonical transformation.* The generating function S is eliminated from this condition. In fact, the definition of the canonical transformation with the help of the S-function can be considered as the *integrated form* of the statement that the bilinear differential form is the determining invariant of a canonical transformation.

Summary. The condition that a transformation be canonical can be formulated without the use of the generating function S. The invariance of the circulation round any closed curve of the phase space is a characteristic property of canonical transformations. The same property can be expressed in differential form. We obtain a definite differential expression, the "bilinear differential form," which is an invariant of canonical transformations. The bilinear differential form parallels the $\overline{ds}^2$ of metrical geometry. However, while the distance element is assigned to *one* infinitesimal displacement, the bilinear differential is assigned to *two* infinitesimal displacements. Hence it resembles an area-element rather than a distance-element.

---

6. The bracket expressions of Lagrange and Poisson. Lagrange anticipated many of the results which are more systematically deducible from the theory of canonical transformations.[1] He noticed that a certain expression is of prime im-

[1] Lagrange, *Oeuvres*, vol. vi, p. 713. The original paper came out in 1808.

portance for the theory of perturbations. Let us consider two sets of variables, q_i and p_i, which are given as functions of the two parameters u and v. Then we form the "Lagrange bracket"

$$[u, v] = \sum_{i=1}^{n} \left(\frac{\partial q_i}{\partial u} \frac{\partial p_i}{\partial v} - \frac{\partial q_i}{\partial v} \frac{\partial p_i}{\partial u} \right). \tag{76.1}$$

It is obviously anti-symmetric with respect to a change of the variables u, v:

$$[u, v] = - [v, u]. \tag{76.2}$$

This bracket expression is closely connected with the theory of canonical transformations.

We notice that this is exactly the quantity encountered in the last section when we transformed the circulation expressed as a line integral to a surface integral. In the new notation we can write

$$\oint_L \sum_{i=1}^{n} p_i \, dq_i = \int_K [u, v] \, du \, dv. \tag{76.3}$$

Since the invariance of the circulation, taken round any closed curve L, is a characteristic property of a canonical transformation, we see that the same property can also be expressed as the invariance of the Lagrange bracket $[u, v]$: *The canonical transformations are those transformations of the variables q_i, p_i to Q_i, P_i which leave the Lagrange bracket invariant, no matter how the q_i, p_i depend on u and v.* The meaning of this invariance is that if we change our coordinates q_i, p_i to new coordinates Q_i, P_i by a canonical transformation, and then form the Lagrange bracket in the new system of coordinates, we get the same value for it that we had before.

We have seen that the characterization of a canonical transformation by the generating S-function does not lead to explicit transformation equations. We get the transformation in implicit form and obtain the final equations by elimination. The question is obviously of interest: how can we decide whether a given explicit transformation is canonical or not? We have no way of *producing* explicit canonical transformations but we can actually *test* a given transformation as to its canonical nature. We can characterize a canonical transformation by a system of differential relations. If these relations are satisfied, our transformation is canonical, and conversely, if our transformation is canonical, these relations must be satisfied.

These differential relations follow at once from the fact that an arbitrary Lagrange bracket does not change with the canonical transformation. Let us assume that we are given the transformation from the old to the new coordinates in explicit form:

$$q_i = f_i(Q_1, \ldots, Q_n; P_1, \ldots, P_n),$$
$$p_i = g_i(Q_1, \ldots, Q_n; P_1, \ldots, P_n). \tag{76.4}$$

We can now pick out any pair of variables Q_i and Q_k or Q_i and P_k or P_i and P_k, and consider them as the two parameters u, v for which the Lagrange bracket is to be formed, considering the other variables as constants. We then do the same in the new coordinate system. In the new system, all the Q_i and P_i are independent of each other, and we obtain directly from the definition of the Lagrange bracket:

$$[Q_i, Q_k] = 0, \quad [P_i, P_k] = 0, \quad [Q_i, P_k] = \delta_{ik}, \tag{76.5}$$

where the "Kronecker delta," $\delta_{ik} = 1$ or 0 according as $i = k$ or $i \neq k$.

Problem 1. Making use of the law of implicit differentiation show that, for any transformation of the coordinates q_i, p_i to Q_i, P_i which satisfies the conditions (76.5), an arbitrary Lagrange bracket $[u, v]$ is invariant. This shows that the conditions (76.5) are not only necessary, but also sufficient, to guarantee the canonical nature of a transformation.

Problem 2. Show that the bilinear differential form

$$\sum_{i=1}^{n} (d'p_i d''q_i - d''p_i d'q_i)$$

is likewise an invariant of the transformation if the conditions (76.5) hold.

Soon after Lagrange's investigation, Poisson (in 1809) introduced another type of bracket expression which is the natural counterpart of the Lagrange bracket. Instead of considering the q_i and p_i as functions of u and v, we consider a pair of variables u and v as given functions of the coordinates q_i, p_i:

$$u = u(q_1, \ldots, q_n; p_1, \ldots, p_n),$$
$$v = v(q_1, \ldots, q_n; p_1, \ldots, p_n). \tag{76.6}$$

Then the following quantity is known as a "Poisson bracket":

$$(u,v) = \sum_{i=1}^{n} \left(\frac{\partial u}{\partial q_i} \frac{\partial v}{\partial p_i} - \frac{\partial v}{\partial q_i} \frac{\partial u}{\partial p_i} \right). \tag{76.7}$$

These brackets have likewise the property of anti-symmetry:

$$(u, v) = -(v, u). \tag{76.8}$$

They are closely related to the Lagrange brackets. Let us assume that we have $2n$ variables $u_1, \ldots, u_{2n}$ which are given functions of the q_i, p_i. Alternatively, we may regard the q_i, p_i as functions of the $u_1, \ldots, u_{2n}$. For the first type of relations the Poisson brackets can be formed, for the second type the Lagrange brackets. One kind of bracket determines the other. Hence, if the Lagrange brackets are invariants of a canonical transformation, so are the Poisson brackets. This gives an alternative formulation of the conditions for a canonical transformation. *The canonical transformations are those transformations which leave the Poisson bracket (u, v) invariant, no matter how the functions u and v depend on the coordinates q_i, p_i.*

We now apply this invariance principle to the inverse form of the canonical transformation (76.5), expressing the new coordinates in terms of the old:

$$\begin{aligned} Q_i &= Q_i(q_1, \ldots, q_n; p_1, \ldots, p_n), \\ P_i &= P_i(q_1, \ldots, q_n; p_1, \ldots, p_n). \end{aligned} \tag{76.9}$$

We identify u and v with any two of the Q_i, P_i, and form the Poisson brackets in the new and likewise in the old coordinate system. For the latter system they can be directly evaluated, as before. Hence the property of invariance gives

$$(Q_i, Q_k) = 0, \quad (P_i, P_k) = 0, \quad (Q_i, P_k) = \delta_{ik}. \tag{76.10}$$

These conditions are equivalent to the original conditions (76.5).

Problem. Consider $u_1, \ldots, u_{2n}$ as given functions of the q_i, p_i, and solve these relations for the q_i, p_i; establish the following reciprocity relations between the Lagrange and the Poisson brackets:

$$\sum_{a=1}^{2n} [u_i, u_a] (u_k, u_a) = \delta_{ik}. \tag{76.11}$$

Because of these relations, the conditions (76.10) are immediate consequences of the conditions (76.5).

Summary. A given generating function defines a canonical transformation in implicit form. Although explicit canonical transformations cannot be given by any formula, a particular transformation can be tested as to its canonical or non-canonical character. For this purpose the Lagrange or the Poisson bracket expressions can be employed. These brackets are closely related to canonical transformations, which can be defined as those transformations of the conjugate variables which leave an arbitrary Lagrange or Poisson bracket invariant.

7. Infinitesimal canonical transformations. Let us perform an arbitrary canonical transformation, defined by the condition

$$\sum_{i=1}^{n} (p_i \delta q_i - P_i \delta Q_i) = \delta S. \tag{77.1}$$

Now let the variables Q_i, P_i be transformed into other variables $\bar{q}_i, \bar{p}_i$, by a second canonical transformation:

$$\sum_{i=1}^{n} (P_i \delta Q_i - \bar{p}_i \delta \bar{q}_i) = \delta S'. \tag{77.2}$$

Then the sum of these two equations gives:

$$\sum_{i=1}^{n} (p_i \delta q_i - \bar{p}_i \delta \bar{q}_i) = \delta(S + S'), \tag{77.3}$$

which shows that the direct transition from the q_i, p_i to the final $\bar{q}_i$, $\bar{p}_i$ is in itself a canonical transformation. This means that *the product of two canonical transformations is once more a canonical transformation.* We say that the canonical transformations "have the group property," which plays a fundamental role in Lie's theory of differential equations. There the complete solution of a set of differential equations is reduced to the study of a continuous group of transformations. The canonical equations of motion are in the closest relation to such continuous transformations, as the following developments will show.

We consider a generating function S which contains a parameter t:

$$S = S(q_1, \ldots, q_n; Q_1, \ldots, Q_n; t). \tag{77.4}$$

This t will later be identified with the time; at present, however, it is merely a variable parameter.

A generating function of the form (77.4) gives rise to an infinity of canonical transformations since to each value of t there corresponds a definite canonical transformation. We now consider the transformation which belongs to a definite value t, and another neighbouring transformation which belongs to the value $t + \Delta t$. We can think of the transformation as a mapping of the space on itself. The transformation I maps a point P of the phase space into some point Q, while the transformation II maps the same point P into some other point Q'. Since S is a *continuous* function of t, the two points Q and Q' are near to each other. Because of the group property of canonical transformations, the transition from Q to Q' is in itself a canonical transformation. We have here a special type of transformation in which *each point of the phase space is transformed into a neighbouring point.* Thus we can put

$$\begin{aligned} Q_k &= q_k + \Delta q_k, \\ P_k &= p_k + \Delta p_k, \end{aligned} \tag{77.5}$$

where Δq_k and Δp_k are small quantities, whose product and squares are negligible.

Such a transformation is called an "infinitesimal canonical transformation." It has the outstanding property that, while an arbitrary canonical transformation does not permit an explicit representation, *an infinitesimal canonical transformation can be obtained in explicit form.*

For our present purpose it will be more convenient to denote the original variables from which we start by capital letters Q_i, P_i. The new variables will be denoted by q_i, p_i if the parameter assumes the value t, and by $q_i + \Delta q_i$, $p_i + \Delta p_i$, if the parameter assumes the value $t + \Delta t$. We now have:

$$\sum_{i=1}^{n} (p_i \delta q_i - P_i \delta Q_i) = \delta S,$$

$$\sum_{i=1}^{n} [(p_i + \Delta p_i)\delta(q_i + \Delta q_i) - P_i \delta Q_i] = \delta S',$$

(77.6)

where

$$S' = S(q_1 + \Delta q_1, \ldots, q_n + \Delta q_n; Q_1, \ldots, Q_n; t + \Delta t)$$

$$= S + \sum_{i=1}^{n} \frac{\partial S}{\partial q_i} \Delta q_i + \frac{\partial S}{\partial t} \Delta t$$

$$= S + \sum_{i=1}^{n} p_i \Delta q_i + \frac{\partial S}{\partial t} \Delta t. \tag{77.7}$$

Forming the difference of the two equations (77.6), we obtain the following remarkable relation:

$$\sum_{i=1}^{n} (\Delta p_i \delta q_i - \Delta q_i \delta p_i) = \delta \frac{\partial S}{\partial t} \Delta t. \tag{77.8}$$

In order to draw further conclusions, we construct a new function. The original function S has the form (77.4), and its partial derivative with respect to t has a similar form. We now introduce the p_i by means of the determining equations of a canonical transformation:

$$p_i = \frac{\partial S}{\partial q_i}. \tag{77.9}$$

From these equations we obtain the Q_i as functions of the q_i and p_i. Then we substitute for Q_i in $\frac{\partial S}{\partial t}$ and call the resulting function $-B$:

$$\frac{\partial S}{\partial t} = -B(q_1, \ldots, q_n; p_1, \ldots, p_n; t). \tag{77.10}$$

This is the function to be constructed.

If we make use of this new function on the right-hand side of (77.8), we obtain

$$\sum_{i=1}^{n} (\Delta p_i \delta q_i - \Delta q_i \delta p_i) = - \sum_{i=1}^{n} \left(\frac{\partial B}{\partial q_i} \delta q_i + \frac{\partial B}{\partial p_i} \delta p_i \right) \Delta t. \tag{77.11}$$

Comparison of the two sides of this equation yields

$$\Delta q_i = \frac{\partial B}{\partial p_i} \Delta t,$$

$$\Delta p_i = -\frac{\partial B}{\partial q_i} \Delta t. \tag{77.12}$$

These are the explicit equations of an infinitesimal canonical transformation. Instead of the "absolute coordinates" $q_i + \Delta q_i$, $p_i + \Delta p_i$ of the new reference system, the "relative coordinates" Δq_i, Δp_i can be used. These coordinates are explicitly expressed in terms of a single function B which characterizes the transformation, and can be chosen as an arbitrary function of the variables q_i and p_i.

Summary. In an infinitesimal canonical transformation the points of the phase space are displaced by infinitesimal amounts. The formulae of such a transformation can be given in explicit form. The transformation is characterized by an arbitrarily chosen function of the variables q_i, p_i.

---

8. The motion of the phase fluid as a continuous succession of canonical transformations. The result of the last section throws new light on the nature of the dynamical equations. If we let Δt tend towards zero after dividing both sides of the equations (77.12) by Δt, we obtain in the limit the differential equations

$$\frac{dq_i}{dt} = \frac{\partial B}{\partial p_i},$$

$$\frac{dp_i}{dt} = -\frac{\partial B}{\partial q_i}. \tag{78.1}$$

These are nothing but the canonical equations of motion, if the variable parameter t is identified with the time and the function B is identified with the Hamiltonian function H. For the full understanding of this result, let us assume that we watch the motion of the phase fluid during a certain time interval Δt. Let us assume that the fluid particles are marked so that the position of each particle is recognizable. At a certain instant t a snapshot is taken of the moving fluid; then at the instant $t + \Delta t$ another snapshot. All the fluid particles have moved away from their original positions but their displacements are infinitesimal. *These infinitesimal displacements represent a canonical transformation of the phase fluid.* The process can be repeated any number of times. *The whole motion of the phase fluid is nothing but a continuous succession of canonical transformations.*

This amazing result—first conceived by Hamilton, although with a slightly different interpretation—reveals the role of canonical transformations in the study of the phenomena of motion from a new angle. *The entire motion of the mechanical system can be considered as a transformation problem.* The successive states of the phase fluid represent a continuously changing mapping of the space on itself. *This mapping is constantly canonical.*

But even more can be said. The successive transformations of the phase fluid are connected with each other. At the end of the previous paragraph we constructed the Hamiltonian function $B = H$ by starting from an arbitrary generating function S which contains a parameter t. But now we can proceed in the opposite direction. A given problem of motion presents us with the Hamiltonian H as a function of the q_i, p_i and (possibly) t. We replace the p_i by $\dfrac{\partial S}{\partial q_i}$ and try to find the original function S from which the equation

$$\frac{\partial S}{\partial t} = - H \tag{78.2}$$

arose. This means an *integration problem.* The first procedure was simpler because it required nothing but differentiation and elimination. The new problem is more difficult and will be studied in greater detail in the next chapter. One can show that for any given H a corresponding S-function can be found; in fact, even an infinity of possible S-functions. After constructing this S-function we have arrived at a function

$$S = S(q_1, \ldots, q_n; Q_1, \ldots, Q_n; t) \tag{78.3}$$

which generates an infinite family of canonical transformations. A point Q_i, P_i is transformed into a point q_i, p_i, the position of which changes continuously with the time t, and this motion represents exactly the motion of the mechanical system if we let t change while the Q_i, P_i are kept constant. The motion of the entire phase fluid is nothing but *the successive evolution of a time-dependent canonical transformation.*

This new viewpoint shows also the invariants of the motion in a new light. These invariants are actually *invariants of any canonical transformation.* The invariance of the circulation, discussed in chapter VI, section 8, is a characteristic feature of canonical transformations: in fact it defines these transformations. Liouville's theorem [cf. chapter VI, section 7] brought the invariance of the volume into evidence, based on the incompressibility of the phase fluid. This theorem may be expressed in the form that the value of the functional determinant of the transformation which connects any two states of motion with each other is always equal to 1. This, too, is a general property of canonical transformations. *The value of the functional determinant of any canonical transformation is equal to 1.*

Problem. Write down the functional determinant Δ of a canonical transformation and multiply it by itself. Show that $\Delta^2 = 1$. The exclusion of the possibility $\Delta = -1$ requires further consideration, but for the motion of the phase fluid the choice $+1$ follows from the continuity of motion.

Summary. A time-dependent function generates an infinite family of canonical transformations. The successive phases of this transformation can be conceived as a continuously changing mapping of the space on itself, and that again means the motion of a fluid. This motion satisfies the canonical equations. We thus get an entirely new conception of the canonical equations. The motion of the phase fluid can be conceived as the successive stages of an infinite family of continuous canonical transformations.

9. Hamilton's principal function and the motion of the phase fluid. The results of our discussion have obviously a bearing on the integration problem of the dynamical equations. We know that the relation between the generating function S and the given Hamiltonian function H [cf. equation (78.2)] is

$$\frac{\partial S}{\partial t} + H = 0, \tag{79.1}$$

always understanding that the p_i of the Hamiltonian function H are replaced by $\partial S/\partial q_i$. Let us assume that we are able to find a generating function S which satisfies this partial differential equation. Then we have succeeded in obtaining the motion of the phase fluid as the successive phases of a time-dependent canonical transformation, with a given generating function S. After appropriate differentiations and eliminations this transformation can be found in explicit form. The transformation equations take the form

$$q_i = f_i(Q_1, \ldots, Q_n; P_1, \ldots, P_n; t),$$
$$p_i = g_i(Q_1, \ldots, Q_n; P_1, \ldots, P_n; t). \tag{79.2}$$

This, however, amounts to a *complete integration* of the dynamica problem because all the mechanical variables are expressed as explicit functions of the time t and $2n$ constants $Q_1, \ldots, Q_n$; $P_1, \ldots, P_n$ which can be adjusted to arbitrary initial conditions. Actually these constants are the coordinates of that fixed point Q_i, P_i which is transformed into the variable point q_i, p_i. This latter point changes its position with the time t in view of the fact that our transformation depends on t. But the fact remains that the entire motion of the phase fluid is described in explicit form. In this description the coordinates Q_i, P_i play the role of arbitrary constants of integration.

Hence the problem of integration is shifted to the problem of finding the generating function of a given continuous sequence of infinitesimal canonical transformations. If this task is accomplished, the rest is plain differentiation and elimination.

It was Hamilton who first hit on the idea of finding a fundamental function which can yield all the equations of motion by simple differentiations and eliminations, without any integration.

He first proved the existence of such a function in geometrical optics, where he called it the "characteristic function"; this function was of great advantage for many involved problems. Later, in his dynamical researches, Hamilton encountered the same function again and this time he called it the ' principal function." In view of the common variational background of optics and mechanics the two concepts are equivalent, since Hamilton's discovery belongs actually to the calculus of variations and the special form of the variational integral—which is the time in Fermat's optical principle and the action in Lagrange's mechanical principle—has no bearing on the problem.

Hamilton did not have available the theory of canonical transformations and made his discovery from an entirely different starting point. The principal function of Hamilton is not a speculative mathematical device for producing a transformation of a definite kind, but has a very definite physical significance. In order to explain Hamilton's procedure, let us start with a conservative system whose Lagrangian function L and Hamiltonian function H do not contain the time explicitly. This is the situation in optics and this was Hamilton's starting point in both optics and mechanics. The generalization to nonconservative systems can be made by a very simple device: we reduce the problem to the conservative case by adding the time t to the mechanical variables.

We consider the principle of least action in Lagrange's formulation, but change over from configuration space to phase space (cf. chapter v, section 6). Find the stationary value of the "action"

$$A = 2 \int_{\tau_1}^{\tau_2} T dt = \int_{\tau_1}^{\tau_2} \sum_{i=1}^{n} p_i \dot{q}_i dt = \int_{\tau_1}^{\tau_2} \sum_{i=1}^{n} p_i dq_i, \qquad (79.3)$$

under the auxiliary condition that the C-point of the phase space stays on the energy surface

$$H(q_1, \ldots, q_n, p_1, \ldots, p_n) - E = 0. \qquad (79.4)$$

We know that the same principle is expressible in Jacobi's form. Minimize the integral

$$A = \sqrt{2} \int_{\tau_1}^{\tau_2} \sqrt{E - V} \, d\bar{s} \qquad (79.5)$$

without any auxiliary condition. As we have seen, we can introduce the Riemannian line-element

$$\overline{d\sigma} = \sqrt{2(E - V)} \; \overline{ds}, \tag{79.6}$$

in which case Jacobi's principle assumes a purely geometrical significance: determine the shortest lines or geodesics of a given Riemannian manifold. Let us assume that we have found these geodesics and that we evaluate the integral (79.5) along these geodesics. What we get is the arc length of a geodesic between two points M and N or, in simpler language, the "distance" between the two points M and N. This distance will obviously be a function of the coordinates q_i of the two end-points M and N. *This distance, expressed in the coordinates of the two end-points, is Hamilton's principal function.* Obviously any two points $\bar{q}_1, \ldots, \bar{q}_n,$ and $q_1, \ldots, q_n$ of the configuration space (at least in a sufficiently limited region) can be connected by a geodesic, and thus Hamilton's principal function can be formed for any pair of points $\bar{q}_i$ and q_i. If we denote this principal function by W, then

$$W = W(q_1, \ldots, q_n; \bar{q}_1, \ldots, \bar{q}_n). \tag{79.7}$$

Now we know from our previous discussions that the variation of any integral A is zero whenever the variation occurs between definite limits, while if we vary the limits the variation of A can be expressed in the form of a boundary term (cf. chapter VI, section 8):

$$\delta A = \left[\Sigma \; p_i \delta q_i \right]_{\tau_1}^{\tau_2} = \sum_{i=1}^{n} p_i \delta q_i - \sum_{i=1}^{n} \bar{p}_i \; \delta \bar{q}_i. \tag{79.8}$$

On the other hand, the function W is by definition nothing but the definite integral A, expressed in terms of the q_i and $\bar{q}_i$ between which the integral has been taken. This gives

$$\delta A = \delta W = \sum_{i=1}^{n} \left(\frac{\partial W}{\partial q_i} \delta q_i + \frac{\partial W}{\partial \bar{q}_i} \delta \bar{q}_i \right). \tag{79.9}$$

Comparison of the last two equations yields the following relations:

$$p_i = \frac{\partial W}{\partial q_i},$$
$$\bar{p}_i = - \frac{\partial W}{\partial \bar{q}_i}. \tag{79.10}$$

These equations show once more that *two positions of the moving phase fluid are connected with each other by means of a canonical transformation*. But now we can say more; the role of W in the equations (79.10) shows that Hamilton's principal function is the generating function of that canonical transformation which converts one state of the moving phase fluid into a later state.

Notice that the generating function S of a time-dependent canonical transformation is more general than Hamilton's W-function, because a general time-dependent canonical transformation transforms an arbitrary point Q_i, P_i of the phase space into the moving point q_i, p_i while Hamilton's transformation transforms the *initial position* $\bar{q}_i$, $\bar{p}_i$ of the moving fluid particle into some later position q_i, p_i.

In our whole construction the coordinates q_i, p_i have never left the energy surface $H = E$. Our variables form actually a $(2n - 1)$-dimensional manifold for which the phase space outside the energy surface does not exist. (Remember that the equation (610.27), associated with Jacobi's principle, came out *as an identity* satisfied by the variables q_i, p_i.) For this reason the function W satisfies automatically the partial differential equation

$$H\left(q_1, \ldots, q_n; \frac{\partial W}{\partial q_1}, \ldots, \frac{\partial W}{\partial q_n}\right) - E = 0. \qquad (79.11)$$

But the same can be said of the coordinates $\bar{q}_i$, $\bar{p}_i$ of the initial point which belongs to the same manifold. Hence the function W satisfies also a *second* partial differential equation, obtained by replacing p_i by $-\dfrac{\partial W}{\partial \bar{q}_i}$:

$$H\left(\bar{q}_1, \ldots, \bar{q}_n; -\frac{\partial W}{\partial \bar{q}_1}, \ldots, -\frac{\partial W}{\partial \bar{q}_n}\right) - E = 0. \qquad (79.12)$$

Hamilton's principal function is thus restricted by a *pair* of partial differential equations.

Since the explicit equations of a canonical transformation can be obtained by simple differentiations and eliminations if the generating function is given, we see that we can express explicitly the coordinates q_i, p_i in terms of the $\bar{q}_i$, $\bar{p}_i$, which means that we obtain explicitly the path of the C-point which starts from some given position in the configuration space. *This is Hamilton's outstanding discovery.* If the principal function W is given, the entire dynamical problem is reduced to differentiations and eliminations.

We now proceed to the time-dependent case, which holds for conservative and non-conservative systems alike. Our procedure is exactly the same as that for conservative systems: we merely add the time t to the position coordinates and thus consider the "extended action"

$$A = \int_{\tau_1}^{\tau_2} \left(\sum_{i=1}^{n} p_i q_i' + p_t t' \right) d\tau = \int_{\tau_1}^{\tau_2} \left(\sum_{i=1}^{n} p_i dq_i + p_t dt \right), \quad (79.13)$$

under the auxiliary condition

$$K(q_1, \ldots, q_n, t; p_1, \ldots, p_n, p_t) = 0. \quad (79.14)$$

Once more we can define a line element $\overline{d\sigma}$ for the extended configuration space $q_1, \ldots, q_n, t$ by putting

$$\overline{d\sigma} = L dt = L t' d\tau, \quad (79.15)$$

although the geometry established by this line element will no longer be of the Riemannian type. Again we can connect two points $\bar{q}_1, \ldots, \bar{q}_n, \bar{t}$ and $q_1, \ldots, q_n, t$ of the $(n + 1)$-dimensional space by a shortest line and measure the arc length

$$A = \int_{\tau_1}^{\tau_2} \overline{d\sigma} = \int_{\tau_1}^{\tau_2} L t' d\tau \quad (79.16)$$

along this line. Furthermore, we get a definite "distance" associated with the position of two points of space and that distance will again be a function of the coordinates of the two end-points:

$$W = W(q_1, \ldots, q_n, t; \bar{q}_1, \ldots, \bar{q}_n, \bar{t}). \quad (79.17)$$

Hence we have constructed Hamilton's principal function. Varying again between arbitrary limits we establish the relations

$$\begin{aligned} p_i &= \frac{\partial W}{\partial q_i}, & \bar{p}_i &= - \frac{\partial W}{\partial \bar{q}_i}, \\ p_t &= \frac{\partial W}{\partial t}, & \bar{p}_t &= - \frac{\partial W}{\partial \bar{t}}, \end{aligned} \quad (79.18)$$

which show the canonical nature of the transformation. The partial differential equation now takes the form

$$K\left(q_1, \ldots, q_n, t; \frac{\partial W}{\partial q_1}, \ldots, \frac{\partial W}{\partial q_n}, \frac{\partial W}{\partial t}\right) = 0, \quad (79.19)$$

or, expressing K in terms of the Hamiltonian function $K = p_t + H$,

$$\frac{\partial W}{\partial t} + H\left(q_1, \ldots, q_n; \frac{\partial W}{\partial q_1}, \ldots, \frac{\partial W}{\partial q_n}; t\right) = 0,$$

$$-\frac{\partial W}{\partial \bar{t}} + H\left(\bar{q}_1, \ldots, \bar{q}_n; -\frac{\partial W}{\partial \bar{q}_1}, \ldots, -\frac{\partial W}{\partial \bar{q}_n}; \bar{t}\right) = 0. \quad (79.20)$$

These are the two partial differential equations which hold for W.

We notice that in view of these two conditions for W the last two equations of the transformation (79.18) follow from the previous equations and can be dropped. In the remaining equations we obtain q_i from the second set and substitute for them in the first set. This gives the transformation equations:

$$q_i = f_i(\bar{q}_1, \ldots, \bar{q}_n; \bar{t}, t),$$
$$p_i = g_i(\bar{q}_1, \ldots, \bar{q}_n; \bar{t}, t), \quad (79.21)$$

which solve the problem of motion in explicit form by giving the coordinates q_i, p_i of the moving point at any time t if the initial position at the time $\bar{t}$ is given.

This integration scheme of Hamilton was simplified and improved by Jacobi. The principal function of Hamilton is very restricted and has to satisfy two simultaneous partial differential equations. The solution of this problem would be practically impossible without the broader integration scheme of Jacobi. The generating function S of a time-dependent canonical transformation generates the entire motion of the phase fluid, subject only to the partial differential equation

$$\frac{\partial S}{\partial t} + H\left(q_1, \ldots, q_n; \frac{\partial S}{\partial q_1}, \ldots, \frac{\partial S}{\partial q_n}; t\right) = 0. \quad (79.22)$$

A second differential equation is no longer necessary, since the point $Q_1, \ldots, Q_n$ need *not* lie on the extended energy surface $K = 0$. Moreover, S is a function of $q_1, , \ldots, q_n; Q_1, \ldots, Q_n; t$ only, while Hamilton's principal function contains in addition to these variables the surplus variable $\bar{t}$.

The literature frequently refers to the differential equation (79.22) as "the partial differential equation of Hamilton-Jacobi." The name is well justified because, in spite of the fundamental importance of Hamilton's distance function, his original scheme was impractical for actual integration purposes. Hamilton's great discovery gave Jacobi the clue to the canonical transformations which put Hamilton's own method in a broader framework. With the help of Jacobi's much less restricted S-function, even Hamilton's W-function can be obtained. But it would be practically impossible to find the W-function directly by solving two simultaneous partial differential equations. The relation between the two theories will be discussed in greater detail in the next chapter.

Summary. While the generating function of a canonical transformation is a purely mathematical quantity, Hamilton introduced a principal function which is closely related to the variational integral. In its geometrical interpretation this function has a striking significance. It gives the distance between two points in a properly defined metrical geometry, expressed as a function of the coordinates of the two endpoints. Hamilton's principal function is the generating function of that particular canonical transformation which connects two states of the phase fluid, belonging to two different times, directly with each other, without the help of an intermediary outside point.

CHAPTER VIII

THE PARTIAL DIFFERENTIAL EQUATION OF
HAMILTON-JACOBI

*Put off thy shoes from off thy feet, for the place whereon
thou standest is holy ground.* EXODUS III, 5

Introduction. We have done considerable mountain climbing. Now we
are in the rarefied atmosphere of theories of excessive beauty and we are
nearing a high plateau on which geometry, optics, mechanics, and wave mechanics meet on common ground. Only concentrated thinking, and a considerable amount of re-creation, will reveal the full beauty of our subject in which
the last word has not yet been spoken. We start with the integration theory
of Jacobi and continue with Hamilton's own investigations in the realm of
geometrical optics and mechanics. The combination of these two approaches
leads to de Broglie's and Schroedinger's great discoveries, and we come to the
end of our journey.

**1. The importance of the generating function for the prob-
lem of motion.** In the theory of canonical transformations
no other theorem is of such importance as that a canonical transformation is completely characterized by knowing one single
function S, the generating function of the transformation. This
parallels the fact that the canonical equations are likewise characterized by one single function, the Hamiltonian function H.
These two fundamental functions can be linked together by a
definite relation. In order to solve the problem of motion it
suffices to consider the Hamiltonian function and try to simplify
it to a form in which the canonical equations become directly
integrable. For this purpose a suitable canonical transforma-

tion can be employed. But this transformation depends on one single function. And thus the problem of solving the entire canonical set can be replaced by the problem of solving a single equation. This equation happens to be a partial differential equation.

From the practical viewpoint not much is gained. The solution of a partial differential equation—even *one* equation—is no easy task, and in most cases is not simpler than the original integration problem. Those problems which can be solved explicitly by means of the partial differential equation are, in the majority of cases, the same problems which can be solved also by other means. For this reason the Hamiltonian methods were long considered as of purely mathematical interest and of little practical importance. The philosophical value of these methods, the entirely new understanding they furnished for the deeper problems of mechanics, remained unnoticed except by a few scientists who were impressed by the extraordinary beauty of the Hamiltonian developments. Among these we may mention especially Jacobi; while later there were Helmholtz, Lie, Poincaré, and, in our day, de Broglie and Schroedinger. In contemporary physics, the Hamiltonian methods gained recognition because of the optico-mechanical analogy which was made clear by Hamilton's partial differential equation. Since the advent of Schroedinger's wave mechanics, which is based essentially on Hamilton's researches, the leading ideas of Hamiltonian mechanics have found their way into the textbooks of theoretical physics. Yet even so, the *technical* side of the theory is primarily stressed, at the cost of the philosophical side.

From the point of view adopted here, the purely technical side of the subject is of minor importance. The principal interest is focused on the basic significance of the theory and the interrelation of its various aspects. We shall thus discuss in succession the Jacobian and the Hamiltonian theories and exhibit the central role which the partial differential equation of Hamilton-Jacobi plays in these developments.

Summary. Canonical transformations are characterized by one single function, the generating function. The problem of finding a proper canonical transformation which simplifies the Hamiltonian function to a form in which the equations are directly integrable is thus equivalent to the problem of constructing one single function. This function is determined by a single partial differential equation. The problem of solving the entire system of canonical equations can be replaced by the problem of solving this one equation.

2. Jacobi's transformation theory. Let us consider a conservative mechanical system with a given Hamiltonian function H which does not depend on the time t. We wish to transform the mechanical variables $q_1, \ldots, q_n; p_1, \ldots, p_n$ into a new set of variables $Q_1, \ldots, Q_n; P_1, \ldots, P_n$ by a canonical transformation. We do not specify this canonical transformation, except for a single condition, namely, that the Hamiltonian function H shall be one of the new variables, say Q_n:

$$Q_n = H(q_1, \ldots, q_n; p_1, \ldots, p_n). \tag{82.1}$$

The other $2n - 1$ equations of the transformation are arbitrary except for the requirement that the transformation shall be canonical.

Let us assume that we have succeeded in finding such a transformation. Then it is easy to integrate the canonical equations in the new system of coordinates. Since the Hamiltonian function H is an invariant of a canonical transformation, the Hamiltonian function H' of the new system is equal to Q_n. But this means that in the new system *all variables are ignorable* and we can obtain a complete integration of the equations of motion.

The first set of equations

$$\dot{Q}_i = \frac{\partial H'}{\partial P_i},$$

gives at once

$$Q_i = \text{const.} = a_i, \quad (i = 1, 2, \ldots, n - 1). \tag{82.2}$$

We wish to separate the last subscript n because the constant associated with Q_n has a special significance; it is nothing but the *energy constant E*:

$$Q_n = \text{const.} = E. \qquad (82.3)$$

Now the second set of equations

$$\dot{P}_i = -\frac{\partial H'}{\partial Q_i},$$

yields

$$P_i = \text{const.} = -\beta_i, \ (i = 1, \ldots, n-1), \qquad (82.4)$$

while the last equation gives

$$P_n = \tau - t. \qquad (82.5)$$

We can interpret this solution of the canonical equations in geometrical language as follows. The original world lines of the moving phase fluid fill the state space with an infinite family of curves. By a canonical transformation a mapping of the space on itself is produced which *straightens out these world lines to an infinite bundle of parallel straight lines*, inclined at an angle of 45° to the time axis.

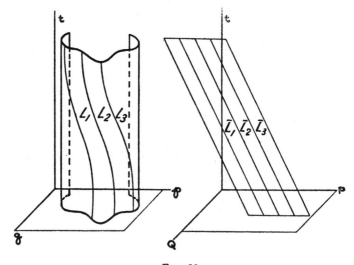

FIG. 11

The remarkable feature of this transformation is the fact that by merely flattening out the cylindrical surfaces $H = E$ into the parallel planes $Q_n = E$ we ensure automatically that all the previously curved world lines of the phase fluid are now straightened out to parallel straight lines.

We see that the original integration problem has been reduced to that of finding a canonical transformation which satisfies the single condition (82.1). This condition requires that if we express the given Hamiltonian function in the new variables Q_i, P_i, it becomes Q_n.

Now we cannot solve this problem explicitly since we have no means of obtaining a canonical transformation expressing the q_i, p_i in terms of the Q_i, P_i. We interject an *intermediate step*. Instead of introducing the new variables Q_i and P_i at once, we first introduce the Q_i only, keeping the old q_i but eliminating the p_i. After that we eliminate also the q_i.

The first step can be accomplished explicitly since we have the formula

$$p_i = \frac{\partial S}{\partial q_i} , \qquad (82.6)$$

which expresses the p_i in terms of the q_i and Q_i. If we put these expressions in H, the Hamiltonian function becomes a function of the q_i and Q_i.

Now let us assume that as a result of this elimination the Hamiltonian function becomes simply Q_n:

$$H\left(q_1, \ldots, q_n; \frac{\partial S}{\partial q_i}, \ldots, \frac{\partial S}{\partial q_n}\right) = Q_n. \qquad (82.7)$$

Here the q_i are not present at all. *Hence the second step, the elimination of the q_i, can be omitted and our task is completed.*

The equation (82.7) is a partial differential equation for the function S. It is not enough to find *some* solution of this differential equation. In order that S shall serve as a generating function, it must have the form

$$S = S(q_1, \ldots, q_n; Q_1, \ldots, Q_n). \qquad (82.8)$$

Now the Q_i, except for Q_n, do not appear explicitly in the differential equation; they occur merely as parameters. From the point of view of integrating (82.7), the first $n - 1$ of the Q_i play the role of *integration constants*.

In the general theory, any solution of a partial differential equation of the first order which contains as many constants of integration as there are variables is called a "complete solution." Although in the present case the number of variables is n, and thus a complete solution should contain n constants of integration, actually *one of the constants is irrelevant*.

Since S itself is not present in the differential equation, but only the partial derivatives of S, a solution is determined only to an *additive constant*. This constant, however, is irrelevant for the canonical transformation and can be omitted. The remaining $n - 1$ constants of integration in the complete solution are identified with $Q_1, \ldots, Q_{n-1}$.

Hence we can describe the process of constructing the function S as follows. Find a complete solution of the differential equation (82.7) with $n - 1$ essential constants of integration $a_1, \ldots, a_{n-1}$:

$$S = S(q_1, \ldots, q_n; a_1, \ldots, a_{n-1}; Q_n). \qquad (82.9)$$

Then replace these constants by the variables $Q_1, \ldots, Q_{n-1}$.

When this is done, we have found the complete solution of the canonical equations of motion. The equations

$$p_i = \frac{\partial S}{\partial q_i},$$
$$P_i = -\frac{\partial S}{\partial Q_i}, \qquad (82.10)$$

contain implicitly the transformation of the q_i, p_i into the Q_i, P_i. After the proper eliminations the q_i, p_i can be expressed as functions of the Q_i, P_i:

$$q_i = f_i(Q_1, \ldots, Q_n; -P_1, \ldots, -P_n),$$
$$p_i = g_i(Q_1, \ldots, Q_n; -P_1, \ldots, -P_n). \qquad (82.11)$$

But the values of the new coordinates are all known, as the solution (82.2-5) shows. If we substitute these expressions in (82.11), the q_i and p_i are determined *as explicit functions of the time t* and $2n$ constants of integration $a_1, \ldots, a_{n-1}, E, \beta_1, \ldots, \beta_{n-1}, \tau$:

$$q_i = f_i(a_1, \ldots, a_{n-1}, E, \beta_1, \ldots, \beta_{n-1}, t - \tau),$$
$$p_i = g_i(a_1, \ldots, a_{n-1}, E, \beta_1, \ldots, \beta_{n-1}, t - \tau). \tag{82.12}$$

Problem. In the case of one single degree of freedom the partial differential equation (82.7) reduces to an ordinary differential equation and is solvable by quadratures. Consider the problem of the linear oscillator

$$V = \tfrac{1}{2} mk^2 x^2. \tag{82.13}$$

Solve the Hamilton-Jacobi differential equation, obtaining the following canonical transformation

$$x = \frac{1}{k} \sqrt{\frac{2Q}{m}} \cos kP,$$
$$p = \sqrt{2mQ} \sin kP.$$

The ellipses $H = E$ of the original coordinate system are transformed into the straight lines $Q = E$ of the new coordinate system. Now a canonical transformation of one pair of variables preserves the area. How can an ellipse

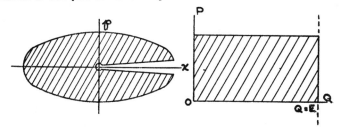

FIG. 12

which encloses a definite area be transformed into a straight line? The solution of this apparent contradiction is provided by the *non-single-valued* nature of the transformation. In order to make the transformation single-valued, we restrict P to the range 0 to $2\pi/k$ and we "cut" the x, p plane along the x-axis. But then the ellipse cannot be closed except by the method shown in Fig. 12. The corresponding figure in the transformed reference system is the shaded rectangle which has actually the same area as the original ellipse.

If we review the procedure from the standpoint of solving the canonical equations, we notice that it is not necessary to change the constants of integration to the variables Q_i and then the Q_i back to constants again. Similarly, the variable Q_n can be identified in advance with the energy constant E. Hence the following "recipe" contains the entire procedure in the technical sense:

1. Write down the energy equation

$$H(q_1, \ldots, q_n, p_1, \ldots, p_n) = E. \qquad (\alpha)$$

2. Replace the p_i by the partial derivatives of some function S with respect to q_i, as

$$H\left(q_1, \ldots, q_n; \frac{\partial S}{\partial q_1}, \ldots, \frac{\partial S}{\partial q_n}\right) = E. \qquad (\beta)$$

3. Find a complete solution of this equation with $n - 1$ nontrivial constants of integration:

$$S = S(q_1, \ldots, q_n; a_1, \ldots, a_{n-1}; E). \qquad (\gamma)$$

4. Form the equations

$$
\begin{aligned}
\frac{\partial S}{\partial a_i} &= \beta_i, \\
\frac{\partial S}{\partial E} &= t - \tau.
\end{aligned}
\qquad (\delta)
$$

5. Solve these equations for the q_i, obtaining

$$q_i = f_i(a_1, \ldots, a_{n-1}, E, \beta_1, \ldots, \beta_{n-1}, t - \tau). \qquad (\epsilon)$$

This completes the solution.

We now come to the general case of a rheonomic system which does not satisfy the law of the conservation of energy. According to the method followed in earlier sections we can always generalize the conclusions drawn for conservative systems by merely adding the time t to the position coordinates q_i and formulating our problem as a conservative problem in the extended phase space. We have the canonical integral

$$A = \int_{\tau_1}^{\tau_2}\left(\sum_{i=1}^{n} p_i q_i' + p_t t'\right) d\tau, \qquad (82.14)$$

which must be made stationary under the auxiliary condition

$$K(q_1, \ldots, q_n, t; p_1, \ldots, p_n, p_t) = 0. \qquad (82.15)$$

Let us assume that we apply a canonical transformation which is arbitrary except for the condition that one of the equations of the transformation assumes the form

$$\bar{t} = K(q_1, \ldots, q_n, t; p_1, \ldots, p_n, p_t). \qquad (82.16)$$

Then in the new coordinate system the auxiliary condition becomes

$$\bar{i} = 0, \tag{82.17}$$

and the canonical integral is reduced to the form

$$A = \int_{\tau_1}^{\tau_2} \sum_{i=1}^{n} P_i Q_i' \, d\tau \tag{82.18}$$

without any further auxiliary condition. But this implies that *all Q_i and P_i are constants*:

$$Q_i = \alpha_i, \tag{82.19}$$
$$P_i = \beta_i.$$

Once more the solution of the canonical equations in the new system of coordinates is very easily obtained, and the result is even more symmetrical than before, because all variables are now on an equal footing.

The condition (82.16) is equivalent to the solution of the partial differential equation

$$K\left(q_1, \ldots, q_n, t; \frac{\partial S}{\partial q_1}, \ldots, \frac{\partial S}{\partial q_n}, \frac{\partial S}{\partial t}\right) = \bar{i}, \tag{82.20}$$

and once more a "complete solution" is required which contains n constants of integration $\alpha_1, \ldots, \alpha_n$, to be identified with the n variables $Q_1, \ldots, Q_n$. Moreover, in view of the auxiliary condition (82.17), the variable $\bar{i}$ remains permanently zero for the actual motion and we can introduce this value for $\bar{i}$ in advance. If we remember that the connection between K and the ordinary Hamiltonian function H was $K = p_t + H$, we obtain the partial differential equation

$$\frac{\partial S}{\partial t} + H\left(q_1, \ldots, q_n; t; \frac{\partial S}{\partial q_1}, \ldots, \frac{\partial S}{\partial q_n}\right) = 0, \tag{82.21}$$

which is the partial differential equation of Hamilton-Jacobi, now formulated for arbitrary rheonomic systems.

The remarkable feature of this transformation method is that merely finding a canonical transformation which flattens out the "energy surface" $K = 0$ into the plane $\bar{i} = 0$ is sufficient to transform *all the stream lines of the moving phase fluid into parallel straight lines.*

We have followed a line of reasoning which has carried over the properties of conservative systems to the general rheonomic case, by merely adding two more dimensions—t and p_t—to the phase space. But we can follow an alternative method which retains the t as the independent variable while remaining within the ordinary phase space. We can consider a canonical transformation of the q_i, p_i into the Q_i, P_i, *without* adding the time t to the active variables of the transformation. The time is present in the transformation merely *as a parameter*, i.e. the equations of the transformation between the old and new variables are constantly changing. Under such a time-dependent canonical transformation the Hamiltonian function H is not an invariant. As the equation (74.13) has shown, the Hamiltonian function H' of the new system of coordinates has the value

$$H' = \frac{\partial S}{\partial t} + H. \tag{82.22}$$

We can now require that

$$H' = 0. \tag{82.23}$$

This gives exactly the condition (82.21), i.e. the partial differential equation of Hamilton-Jacobi, now interpreted as the condition that the Hamiltonian function be transformed into zero by a time-dependent canonical transformation. If the Hamiltonian function of the new coordinate system is zero, then the canonical equations give at once that all the Q_i and P_i are constants during the motion. We thus come back to our previous integration method, now derived from a somewhat different point of view.

Here again we can formulate the results of the transformation theory in the form of a "recipe" which contains the procedure of integration in the technical sense (and there are happy souls who think that they understand the entire Hamilton-Jacobi theory if they understand this recipe).

1. Obtain an arbitrary complete solution of the partial differential equation (82.21), i.e. a solution which contains n essential constants of integration $a_1, \ldots, a_n$:

$$S = S(q_1, \ldots, q_n, t; a_1, \ldots, a_n). \tag{82.24}$$

2. Set the partial derivatives of S with respect to the n constants of integration equal to n new constants:

$$\frac{\partial S}{\partial a_i} = \beta_i. \tag{82.25}$$

3. Solve these equations for the n position coordinates q_i, obtaining them in the form

$$q_i = f_i(a_1, \ldots, a_n, \beta_1, \ldots, \beta_n, t). \tag{82.26}$$

This procedure amounts to a *complete solution of the integration problem*, since the mechanical variables are expressed as explicit functions of the time t and $2n$ constants of integration which can be adjusted to arbitrary initial conditions.

Problem. Assume H to be independent of t and put

$$S = -Et + S_1(q_1, \ldots, q_n; a_1, \ldots, a_{n-1}; E).$$

Show that the general procedure applied to this form for S leads back to the integration method discussed first, which was valid for conservative systems.

Summary. Instead of trying to integrate the canonical equations directly, we adopt a transformation procedure. If the system is conservative, we look for a canonical transformation which transforms the Hamiltonian function H into one of the new variables. If the system is rheonomic, we look for a time-dependent canonical transformation which transforms H into zero. In both cases the transformation solves the problem of motion since in the new system of coordinates the canonical equations can be directly integrated. The desired transformation is accomplished if we can find a complete solution of the partial differential equation of Hamilton-Jacobi.

---

3. Solution of the partial differential equation by separation. We do not possess any general method for the solution of a partial differential equation. Yet under certain special conditions a complete solution of the Hamilton-Jacobi equation is possible. This special class of problems turned out to be of great importance in the development of theoretical physics, because some of the fundamental problems of Bohr's atomic theory belonged to this class. In such problems the one partial differential equation in n variables can be replaced by n ordinary differential equations in a single variable. These equations are completely integrable. Such problems are called "separable problems."

The method of separation consists in the following procedure. We try to solve the given differential equation by setting S equal to a sum of functions, each of which depends on a single variable only:

$$S = S_1(q_1) + S_2(q_2) + \ldots + S_n(q_n). \tag{83.1}$$

Whether or not we shall succeed under such a special assumption cannot be seen in advance. The purely algebraic form of the Hamiltonian function does not reveal its separable or non-separable character. Levi-Civita[1] has shown how to test a given Hamiltonian function as to its separability, but in a given case it is simpler to assume S to be of the form (83.1) and test this assumption in the partial differential equation.

The characteristic feature of a solution of the form (83.1) is that now the momentum

$$p_k = \frac{\partial S}{\partial q_k} = \frac{\partial S_k(q_k)}{\partial q_k} \tag{83.2}$$

becomes a function of q_k alone. Let us write down the equation

$$H(q_1, \ldots, q_n, p_1, \ldots, p_n) - E = 0, \tag{83.3}$$

and solve it for p_k. In view of (83.2) this must yield p_k as a function of q_k alone, while the actual elimination will give it as a function of all the other q_i and p_i too. The contradiction can only be removed if *certain combinations of the other variables reduce to mere constants.* Hence we can see in advance that a separation is possible only if the equation (83.3) can be considered as a consequence of n relations of the form

$$p_k = f_k(q_k, a_1, \ldots, a_{n-1}, E), \tag{83.4}$$

where $a_1, \ldots, a_{n-1}$ are arbitrary constants, obtained by the process of separation. Usually not all the a_i will enter into each of

[1] T. Levi-Civita, *Mathematische Annalen* (1904), LIX, 383.

the equations (83.4). Frequently the separation in a properly chosen first variable brings in one constant, then the separation in the next variable one additional constant, and so on, until we have finally n constants of separation $a_1, \ldots, a_n$, and the energy constant E is some function of these constants:

$$E = \phi(a_1, \ldots, a_n). \tag{83.5}$$

Now we can eliminate a_n from this relation, expressing it as a function of $a_1, \ldots, a_{n-1}$ and E. The general equations (83.4) are thus wide enough to include all separable systems and, on the other hand, if our system is separable, n equations of the general form (83.4) must hold.

If we replace p_k by dS_k/dq_k, according to (83.2), we obtain, by a straightforward quadrature,

$$S_k = \int f_k(q_k, a_1, \ldots, a_{n-1}, E) dq_k + C_k. \tag{83.6}$$

The additive constants C_k add up to an additive constant of the resulting function S, which is of no interest. The "constants of integration" are obtained by the technique of separation; the actual integration does not add any new constants.

As a typical example, consider the Kepler problem of planetary motion. If we use polar coordinates r, θ, ϕ, the line-element becomes:

$$ds^2 = dr^2 + r^2 d\theta^2 + r^2 \sin^2 \theta d\phi^2, \tag{83.7}$$

and the Hamiltonian function becomes:

$$H = \frac{1}{2m}\left(p_r^2 + \frac{p_\theta^2}{r^2} + \frac{p_\phi^2}{r^2 \sin^2 \theta}\right) - \frac{k^2}{r}. \tag{83.8}$$

The equation $H = E$ can be separated as follows:

$$p_\phi = \text{const.} = a,$$
$$p_\theta^2 + \frac{a^2}{\sin^2\theta} = \text{const.} = \beta^2, \tag{83.9}$$
$$\frac{1}{2m}\left(p_r^2 + \frac{\beta^2}{r^2}\right) - \frac{k^2}{r} = E.$$

The technique of separation brings in automatically the right number of constants. The actual integration does not produce any new constants.

Problem 1. Separate the problem of motion in a homogeneous field of gravity, $V = mgz$, in rectangular coordinates.

Problem 2. Separate the problem of the "anisotropic oscillator"

$$V = \tfrac{1}{2} m (k_1^2 x^2 + k_2^2 y^2 + k_3^2 z^2),$$

in rectangular coordinates.

Problem 3. Separate the problem of the Stark-effect

$$V = - k^2/r + Ez$$

(E = electric field strength acting in the z-direction), in parabolic coordinates (Epstein, 1916):

$$x = \sqrt{uv} \cos \phi,$$
$$y = \sqrt{uv} \sin \phi,$$
$$z = \tfrac{1}{2} (u - v).$$

Sometimes a separation is possible in more than one kind of coordinates. The older quantum theory called such systems "degenerate systems." For example, the Kepler-problem is separable in polar coordinates, but it is also separable in parabolic coordinates.

The equation (83.6) shows that separable systems allow a complete solution of the Hamilton-Jacobi partial differential equation by quadratures. We have the unusual situation that the conjugate variables q_k, p_k of each pair interact strictly with one another, without any interference from the other variables. The mechanical system of n degrees of freedom can be considered as a superposition of n systems of one degree of freedom. However, the actual equations of motion are contained in the equations

$$\frac{\partial S}{\partial a_i} = \beta_i,$$

$$\frac{\partial S}{\partial E} = t - \tau, \tag{83.10}$$

and these equations are *not separated* since generally a certain a_i—and also E—will be present in *more than one of the S_i.*

Problem 4. Obtain the function S and the complete solution of the Kepler problem (cf. example above).

Problem 5. Do the same for the problem of the homogeneous field of gravity, considered in Problem 1.

The separable nature of a problem constitutes no inherent feature of the physical properties of a mechanical system, but is entirely a matter of the *right system of coordinates*. A problem which is not separable in a given system of coordinates might become separable after a proper point transformation. Unfortunately, the finding of the right system of coordinates is to some extent a matter of chance, since we do not possess any systematic method of procedure.

Burgers[1] has shown that the problem of combined Stark-effect and Zeeman-effect (electron revolving around the nucleus, disturbed by external electric *and* magnetic fields) is not separable in any coordinates, but becomes separable after a proper canonical transformation.

Summary. Under favourable circumstances the differential equation of Hamilton-Jacobi is completely integrable by straight quadratures. This is the case if the energy equation can be separated into n equations, each containing just *one* pair of conjugate variables q_k, p_k. In that case the function S can be expressed as a sum of n functions, each containing but a single variable q_k. The constants of integration appear during the process of separation.

4. Delaunay's treatment of separable periodic systems. The method of separation—if it is applicable—provides us with the complete solution of the Hamilton-Jacobi equation which is required in Jacobi's integration theory. Now a "complete solution" of a partial differential equation of the first order can appear in many different forms. Let us assume that we have a certain complete solution

$$S = S(q_1, \ldots, q_n; a_1, \ldots, a_{n-1}, E). \qquad (84.1)$$

Let us apply an arbitrary point transformation to the constants of integration:

[1] J. M. Burgers, *Het Atoommodel van Rutherford-Bohr* (Leyden, 1918).

$$a_1 = f_1(\gamma_1, \ldots, \gamma_n),$$
$$\cdot \cdot \cdot \cdot \cdot \cdot \cdot \cdot \cdot \cdot \cdot \cdot \cdot \cdot$$
$$\cdot \cdot \cdot \cdot \cdot \cdot \cdot \cdot \cdot \cdot \cdot \cdot \cdot \cdot \qquad (84.2)$$
$$\cdot \cdot \cdot \cdot \cdot \cdot \cdot \cdot \cdot \cdot \cdot \cdot \cdot \cdot$$
$$a_{n-1} = f_{n-1}(\gamma_1, \ldots, \gamma_n),$$
$$E = f_n(\gamma_1, \ldots, \gamma_n).$$

If we introduce these relations in the function (84.1), we obtain S in the form

$$S = S(q_1, \ldots, q_n; \gamma_1, \ldots, \gamma_n). \qquad (84.3)$$

This S is likewise a complete solution of the Hamilton-Jacobi equation, provided that we add to our solution the equation

$$E = f_n(\gamma_1, \ldots, \gamma_n) \qquad (84.4)$$

as an auxiliary condition.

Just as we considered the separation constants $a_1, \ldots, a_{n-1}, E$ as the transformed variables $Q_1, \ldots, Q_n$, so also can we consider the new constants $\gamma_1, \ldots, \gamma_n$. Then we obtain a canonical transformation from the q_i, p_i to the Q_i, P_i, which transforms the Hamiltonian function H into

$$H' = f_n(Q_1, \ldots, Q_n). \qquad (84.5)$$

Although this form of the Hamiltonian function is more complicated than the simpler form $H' = Q_n$, nevertheless the canonical equations are still explicitly integrable in the new coordinates. For, since H' depends on only *one* set of variables, without involving the P_i at all, the variables of our problem are still all "ignorable." We have

$$\dot{Q}_i = \frac{\partial H'}{\partial P_i} = 0 , \qquad (84.6)$$

which gives $\qquad\qquad Q_i = \text{const.} = \gamma_i; \qquad (84.7)$

while the second set of canonical equations gives

$$-\dot{P}_i = \frac{\partial H'}{\partial Q_i} = \text{const.} = \nu_i. \qquad (84.8)$$

The constancy of the right-hand side follows from the fact that all the Q_i are constants, so that we get

$$-P_i = \nu_i t + \delta_i, \qquad (84.9)$$

and the canonical equations are completely integrated.

The advantage of this more general form of the solution of a separable system is as follows. It is possible that the original constants $a_1, \ldots, a_{n-1}$, which appear as the result of the separation process, have no direct physical significance, while the new constants $\gamma_1, \ldots, \gamma_n$ may be adjusted to the nature of the physical problem and thus may be much better suited for theoretical conclusions than the original set of constants.

This is the leading idea of a remarkable method invented by the French astronomer Delaunay[1] (1816-1872) for the treatment of a certain class of separable problems. At first sight Delaunay's theory seems rather technical and involved. Yet it was this procedure, originally developed for purely astronomical purposes, which opened the eyes of physicists to the power of the Hamiltonian methods.

In view of the separation, we know that the conjugate variables in each pair interact with each other, without any interference on the part of the other variables. Hence we can draw the stream lines in the q_k, p_k plane, on the basis of the equation (83.4), considering the a_i as constants. In the classical problems of mechanics H is a *quadratic* function of the p_k, and thus the solution of the equation $H = E$ for a certain p_k must lead to the solution of a quadratic equation. This gives generally *two* solutions, so that to the same q_k two different p_k can be found. We assume that the discriminant of the solution is positive only for a definite finite range of q_k. In this case q_k must oscillate between two definite limits a_k and b_k, and the stream lines of the associated two-dimensional phase-fluid must be *closed*.

There is still another possibility for stream lines to be closed, without a double-valued relation between p_k and q_k. It is possible that a certain variable q_k has a definite range from a_k to b_k *on account of the geometrical connectivity of space.* For example, if the coordinate is an angle ϕ limited to the interval 0 to 2π, an apparently open stream line, as in Fig. 14, is actually closed because the two end points a_k and b_k coincide. We could illustrate this by bending the plane of the figure into a cylinder and

[1]Ch. E. Delaunay, *Sur une nouvelle théorie analytique du mouvement de la lune* (1846).

glueing the two end-ordinates together. The moving C-point jumps back from M'' to M' and continues its motion along the same stream line.

No matter whether for the first or for the second reason, we assume that the stream lines in *all* the q_k, p_k planes are *closed*.

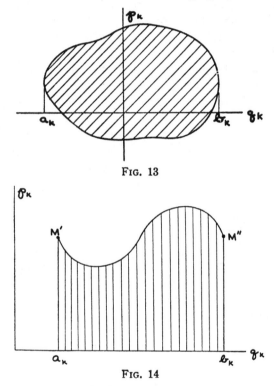

FIG. 13

FIG. 14

In that case the moving fluid particle returns to the same point and repeats its motion in identical fashion. Hence we have a periodic motion. This, however, holds only for the *path* of the moving point when projected on the q_k, p_k planes. It does not hold for the motion in *time*. The velocity with which the point starts its second revolution is not identical with the original velocity because $\dot{q}_k$, $\dot{p}_k$ will in general depend on *all* the q_i, p_i, and thus the return of *one* pair of variables q_k, p_k to their initial values is not a sufficient condition for the motion to be periodic. However, the motion does contain n independent periodicities, but these are spread over all the variables in a non-separable manner. It is the virtue of Delaunay's method that it shows how the frequencies of the motion can be obtained by studying the properties of the two basic functions: the Hamiltonian function H and the generating function S. A proper transformation exhibits the multiply-periodic nature of the given separable system and determines its frequencies explicitly. The process requires nothing but quadratures and eliminations.

We know that the constants of separation a_i and the energy constant E are the new position coordinates Q_i of a canonical transformation. We shall now change these variables into a new set of variables J_i, called "action variables," since they have the dimensions of "action"—i.e. of p times q.

We start with the equations (83.4) which, according to our assumption, give closed lines in the q_k, p_k planes. We evaluate the following definite integrals, extended over a complete stream line:

$$J_k = \oint p_k dq_k = \oint f_k(q_k, a_1, \ldots, E) dq_k. \qquad (84.10)$$

This line integral is equal to the area enclosed by the stream line. This area is a function of the a_i and E, which occur as parameters during the process of integration:

$$J_k = J_k(a_1, \ldots, a_{n-1}, E). \qquad (84.11)$$

From these equations we can eliminate the a_i and E, expressing them in terms of the J_i:

$$a_i = a_i(J_1, \ldots, J_n), \qquad (84.12)$$

$$E = E(J_1, \ldots, J_n). \qquad (84.13)$$

On account of these relations we can introduce the new "action" variables in the generating function S, and obtain S as a function of the q_i and the J_i:

$$S = S(q_1, \ldots, q_n, J_1, \ldots, J_n). \qquad (84.14)$$

Our transformation is now determined. The two basic functions of the new system of variables are the functions E and S. The function E is actually the Hamiltonian function H of the new system.

The action variables J_i are the position coordinates Q_i of the new reference system. The conjugate momenta P_i are called *angle variables*. They are pure numbers without physical dimensions. We prefer to use the negative P_i instead of the P_i themselves and denote them by $\omega_1, \ldots, \omega_n$. According to the general transformation scheme we have

$$- P_i = \omega_i = \frac{\partial S}{\partial J_i} \, . \tag{84.15}$$

We shall investigate these equations somewhat more closely. We are interested in the functional relation between the original q_i and the new ω_i, keeping the J_i as constants. We can then speak of the mapping of the n-dimensional q-space on the n-dimensional ω-space, and vice versa. The relations (84.15), which express the ω_i in terms of the J_i, are not single-valued. At first we notice that the configuration space of the q_i in which the motion takes place is a *limited* portion of the n-dimensional space. The coordinate q_k varies between a certain minimum and a certain maximum. Therefore, if the q_i are plotted as rectangular coordinates of an n-dimensional space, the entire motion must be restricted to a certain "rectangular box" of that space, on account of the separation. For example, in the Kepler problem the separation equations (83.9) show that ϕ varies between 0 and 2π, θ between a certain minimum value θ_0 and a maximum $\pi - \theta_0$, and r between a certain minimum r_1 (perihelion) and a certain maximum r_2 (aphelion). This means, for the physical space, that the motion is restricted to a region limited by a double cone and two concentric spheres.

We now pick out one stream line of the q_k, p_k plane and follow it for a complete revolution; q_k then changes from its minimum to its maximum and back to its minimum, while all the other q_i remain fixed. Let us see what happens to the associated ω_i during this process. The equations (84.15) give

$$\Delta\omega_i = \oint d\omega_i = \oint \frac{\partial^2 S}{\partial J_i \partial q_k} \, dq_k. \tag{84.16}$$

Now the J_i are mere parameters during the process of integration. It is hence permissible to bring the operation of differentiation with respect to J_i outside the integral sign, making use of the rule for differentiating with respect to a parameter. This gives

$$\Delta\omega_i = \frac{\partial}{\partial J_i} \oint \frac{\partial S}{\partial q_k} dq_k$$
$$= \frac{\partial}{\partial J_i} \oint p_k dq_k. \tag{84.17}$$

The last integral is by definition equal to J_k. Therefore,

$$\Delta\omega_i = \frac{\partial J_k}{\partial J_i} = \delta_{ik}, \tag{84.18}$$

where δ_{ik} is Kronecker's symbol.

We see that the passage of q_k through a complete cycle causes no change in the ω_i, *except for the variable ω_k which changes by one.*

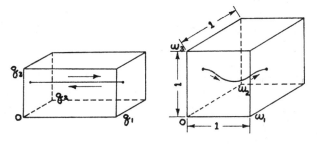

FIG. 15

This result gives a definite clue as to the relationship between the ω- and the q-variables. The ω-space has the character of a "mirror-cabinet." Let us imagine a cube of the n-dimensional space with an edge of length one in every direction. The walls of this cube are reflecting mirrors. We consider an arbitrary point inside the cube. An infinite number of images is formed in every direction perpendicular to the reflecting walls. The unit cube is repeated an infinite number of times. But all the images of a point Ω inside the unit cube belong to one and the same point Q of the q-space.

This geometrical picture has a corresponding analytical counterpart. Let us reverse the relationship between the ω- and the q-spaces by expressing the q_i as functions of the ω_i. The fact that the change of an arbitrary ω_i by one does not change the values of the q_i means that the q_i are *periodic functions of all the angle variables* $\omega_1, \ldots, \omega_n$ *with the period one.* This again means that the q_i can be expressed as multiple Fourier series, i.e. as infinite series of sines and cosines with the arguments

$$2\pi(k_1\omega_1 + k_2\omega_2 + \ldots + k_n\omega_n), \qquad (84.19)$$

where $k_1, k_2, \ldots, k_n$ are arbitrary integers. The amplitudes of these terms are constants.

Fig. 15 shows that the transformation from the q-space to the ω-space is double-valued (one q-point is transformed into two ω-points), in accordance with the back-and-forth oscillation of q_k between two limits, as represented by Fig. 13. If the conditions of Fig. 14 hold, then the transformation from the q-space to the ω-space is single-valued, because the original q_k is already an angle-variable. The transformation from the ω-space to the q-space is always single-valued (one ω-point is transformed into one q-point).

This definite information about the relation between the q- and ω-variables helps us to get a rather concrete picture concerning the complete integration of the problem of motion. So far we have not taken into account the canonical equations; we have merely investigated a certain canonical transformation. But now we solve the canonical equations in the new coordinate system of the J_i, ω_i variables. We know that the J_i correspond to the Q_i, the $-\omega_i$ to the P_i. The Hamiltonian function is $E(J_1, \ldots, J_n)$. The first set of canonical equations yields [cf. (84.6)]

$$\dot{J}_i = -\frac{\partial E}{\partial \omega_i} = 0, \qquad (84.20)$$

which gives
$$J_i = \text{const.} \qquad (84.21)$$

The constancy of J_i was already assumed in our previous considerations.

The second set of equations gives [cf. (84.7)]

$$\dot{\omega}_i = \frac{\partial E}{\partial J_i} = \text{const.} = \nu_i, \qquad (84.22)$$

and hence
$$\omega_i = \nu_i t + \delta_i. \qquad (84.23)$$

If these values of ω_i are introduced in the arguments (84.19) of the multiple Fourier series, we notice that the entire motion is analytically represented as a *superposition of simple harmonic motions, with the fundamental frequencies* ν_i. Thus the frequencies of the motion are obtained from the total energy, expressed in terms of the action variables. The partial derivatives of E with respect to the J_i give directly the frequencies of the system.

Problem. Consider the problem of the anharmonic oscillator, already treated in Problem 2 of section 3. Applying Delaunay's method, show that

$$E = \frac{1}{2\pi}(k_1 J_1 + k_2 J_2 + k_3 J_3),$$

which gives the three frequencies $\nu_i = k_i/2\pi$.

However, our analysis can be pursued still further. We can follow the path of the moving point in the ω-space. The equations (84.23) show that the path becomes a straight line, traversed with uniform velocity. The corresponding curve in the q-space may be a very complicated Lissajou-figure.

Let us make use of the fact that the ω-space is reduced to a cube whose edges are of unit length. If the straight line reaches the right-hand wall of the unit cube, it jumps back to the corresponding point of the left-hand wall and continues in the same direction. This jumping back and forth occurs again and again, as is illustrated (in two-dimensional projection) in Fig. 16.

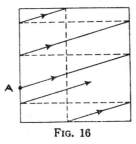

FIG. 16

Eventually the line may close, but it may equally well happen that it never returns to its starting point. The criterion is whether or not we can find some linear relations of the form

$$k_1\nu_1 + k_2\nu_2 + \ldots + k_n\nu_n = 0, \tag{84.24}$$

with *integral values* of the k_i. If $n - 1$ relations of this nature exist, then the line does return after a finite time to its starting point and so the motion is strictly periodic. If no relation of the form (84.24) exists, then the line *fills the entire unit-cube*, coming eventually arbitrarily near to every point of the cube. If m conditions of the form (84.24) exist, the moving point will stay on a definite $(n - m)$-dimensional plane of the cube and come arbitrarily near to any point of that subspace. These results can be directly transferred to the original configuration space of the variables q_i, although straight lines are changed into curves and a plane subspace into a curved subspace of the same number of dimensions.

Delaunay's method throws a new light on the "degenerate systems" of the older quantum theory. If the motion completely fills the permissible portion of the configuration space, the system is non-degenerate and the separation can occur in coordinates of only one kind. This corresponds to the case of *no* integral relations between the frequencies. If one or more conditions of the form (84.24) exist, the motion fills only parts of the configuration space. The system is then simply or multiply degenerate.

For example, in the case of the Kepler ellipses, E takes the form

$$E = - \frac{2\pi^2 m k^2}{(J_1 + J_2 + J_3)^2}. \tag{84.25}$$

Here
$$\nu_1 = \nu_2 = \nu_3, \tag{84.26}$$

and the motion continues along the same ellipse. The introduction of a weak magnetic field (Zeeman effect) removes one degeneracy by letting the plane of the ellipse precess slowly about the polar axis. The second degeneracy could be removed by assuming that the attracting force varies inversely as a power of the distance, which power differs slightly from 2. Actually this degeneracy is removed by the fact that the motion occurs according to relativistic instead of classical mechanics. This causes the perihelion of the ellipse to precess slowly in its own plane. These effects together give a motion which fills the permissible portion of the configuration space completely.

Delaunay's method originated from astronomical perturbation problems. Yet it was eminently adapted to the problems of the earlier quantum theory. The quantum theory of Bohr assumed that only certain orbits were allowed for the revolving electron. These orbits were entirely free from energy losses, so that the motion in them occurred according to the ordinary laws of mechanics. Hence the quantum theory accepted the principles of mechanics—and thus also the canonical equations—without modification. It merely added certain additional *restrictions* to the initial conditions. The $2n$ constants of integration were no longer arbitrary quantities, but were "quantized" according to certain rules. For these "quantum conditions" Delaunay's treatment of multiply-periodic systems was eminently suitable. The second set of constants, the "phase-angles" δ_i which enter when integrating the second set of canonical equations in Delaunay's transformation [cf. (84.23)] were left arbitrary. However, the first set of constants, the action variables, were quantized. The quantum conditions require that the action variables of Delaunay shall be equal to integral multiples of Planck's fundamental constant h:

$$J_k = n_k h. \qquad (84.27)$$

(The integers n_k are called "quantum numbers.") And thus Delaunay's method, originally conceived for planetary perturbation problems, found its most important applications in the realm of atomic physics.

The conditions (84.27) are called the "Sommerfeld-Wilson quantum conditions" (1915). They did not answer the question as to what would happen in the case of non-separable systems. Moreover, the quantization depended on the coordinates employed, and led to entirely different mechanical paths if the separation coordinates were changed. Einstein (in the year 1917) gave an astonishingly imaginative new interpretation of the Sommerfeld-Wilson quantum conditions by abandoning the stream lines of the q_k, p_k planes and operating with the S-function itself. Notice that the "phase-integrals"(84.10) can be replaced by ΔS_k—i.e. the change of S_k in a complete revolution—on account of the relation (83.2). Hence the quantum conditions enunciate something about the multiple-valued nature of S_k. Einstein now took the *sum* of all the quantum conditions:

$$\Sigma J_k = \Sigma \Delta S_k = \Delta S = nh, \qquad (84.28)$$

where $n = \Sigma n_k$ is again an integer. This one equation cannot replace, of course, the original set of equations. But Einstein handled this equation as a *principle* rather than an equation, by requiring that the multiple-valuedness of S shall be such that for *any* closed curve of the configuration space the change in S for a complete revolution shall be a multiple of h. Taking these curves to be $q_k =$ const. in the case of separable systems, the quantum conditions (84.27) are immediate consequences of Einstein's principle. Einstein's invariant formulation of the quantum conditions led de Broglie (in 1924) to his fundamental discovery of matter waves.

Summary. Delaunay invented a beautiful method for treating separable systems which satisfy the additional condition that the stream lines of the separated phase planes (q_k, p_k) are closed lines. He considers a canonical transformation whose position coordinates are the "action variables" J_k defined by the areas enclosed by the stream lines. The J_k are constants for the actual motion while the negatives of the conjugate momenta, the "angle variables" ω_k, change linearly with the time t. The partial derivatives of E with respect to the J_i give n new constants which are the frequencies ν_i of the motion. Each q_k can be written as a multiple Fourier series containing all the frequencies ν_i and their overtones. For this reason the system is called multiply-periodic.

5. The role of the partial differential equation in the theories of Hamilton and Jacobi. We have pointed out in the last chapter (chapter VII, section 9) that it was Hamilton who discovered the fundamental partial differential equation of analytical mechanics. It was likewise he who first conceived the idea of a fundamental function which can provide all the mechanical paths by mere differentiations and eliminations. Yet Hamilton's original scheme was practically unworkable. And, moreover, Hamilton's principal function satisfies two partial differential equations. This second differential equation is an unnecessary complication from the standpoint of integration theory. On the other hand, Jacobi's theory requires a complete solution of the basic differential equation and nothing more. In the case of separable systems such a solution can actually be obtained. Thus, superficially, we get the impression that Jacobi freed the Hamiltonian theory from unnecessary complication and converted it into a practicable scheme.

This is actually true if we presume that the principal goal of mechanics is nothing more than an integration of the equations of motion. But this very limited viewpoint would not do justice to the far-sighted investigations of Hamilton. Granted that we cannot obtain the principal function of Hamilton directly, but have to make use of Jacobi's approach, yet the principal function remains an important and interesting function which serves much more fundamental purposes than the mere integration of the canonical equations. Thus the problem of comparing Hamilton's W-function with Jacobi's S-function deserves more than passing attention. If we perceive all the implications of Hamilton's theory, we come to the conclusion that the *two* partial differential equations in Hamilton's theory are logically just as necessary and natural as the *one* partial differential equation in Jacobi's theory.

Let us first concentrate on the problem of a *conservative* mechanical system. This is actually the general case because, if we add the time t to the mechanical variables by defining the configuration space as the q-t space and change the phrase "independent of the time t" to "independent of the parameter τ," every mechanical system can be made conservative. Now we connect the point $\bar{q}_1, \ldots, \bar{q}_n$ with the point $q_1, \ldots, q_n$ by a path which makes the integral

$$A = \int_{\tau_1}^{\tau_2} \sum_{k=1}^{n} p_k q'_k d\tau = \int_{\tau_1}^{\tau_2} \sum_{k=1}^{n} p_k dq_k \qquad (85.1)$$

stationary, under the auxiliary condition that the path is restricted to lie on the energy surface

$$H(q_1, \ldots, q_n, p_1, \ldots, p_n) - E = 0. \qquad (85.2)$$

Then we express the value of the integral A, taken for the stationary path, as a function of the coordinates q_i and $\bar{q}_i$, obtaining Hamilton's principal function W:

$$W = W(q_1, \ldots, q_n, \bar{q}_1, \ldots, \bar{q}_n). \tag{85.3}$$

Hamilton shows that this function W satisfies the partial differential equation

$$H\left(q_1, \ldots, q_n, \frac{\partial W}{\partial q_1}, \ldots, \frac{\partial W}{\partial q_n}\right) - E = 0. \tag{85.4}$$

So far the analogy to Jacobi's S-function, which satisfies the same differential equation, seems to be perfect. Both functions depend on $2n$ variables, and each can be considered as the generating function of a canonical transformation. From the standpoint of integrating the partial differential (85.4), both functions contain n constants of integration.

And yet we need not bring in the second partial differential equation to discover a fundamental difference between the two functions. In Jacobi's theory the energy constant E was *one of the new variables Q_n*. Aside from the energy constant E, the solution contained but $n - 1$ *constants of integration*. In Hamilton's theory *all* variables are on an equal footing and the energy constant E plays the role of a given constant and *not* the role of a variable. Hamilton's solution of the partial differential equation is thus not a complete but an *over-complete* solution which contains *one more* constant of integration than a complete solution. *This homogeneity in all the n variables* is the characteristic property which differentiates Hamilton's W-function from Jacobi's S-function. It endows the W-function with transformation properties which are completely different from those of the S-function.

Jacobi's S-function accomplishes the transformation of the energy surfaces $H = E$ into the planes $Q_n = E$. The significance of the partial differential equation is here that the Hamiltonian function is transformed into one of the new variables Q_n. In the Hamiltonian case the situation is entirely different. Far from transforming the energy surface into a plane, the entire Hamiltonian construction is strictly limited to the energy surface $H = E$ and never leaves that surface. In the Jacobian case we have a *regular* transformation which is solvable for both the q_k, p_k and the Q_k, P_k. No identity exists here between the coordinates because the partial differential equation establishes a relation between the q_k, p_k *and* Q_n, and not one between the q_k, p_k alone. In the Hamiltonian case the transformation cannot be regular because the *space* q_k, p_k is mapped on to a *surface*. The equation

$$p_i = \frac{\partial W}{\partial q_i} \tag{85.5}$$

are not solvable for the $\bar{q}_i$, because the canonical transformation which transforms the q_i, p_i into the $\bar{q}_i$, $\bar{p}_i$ exists only if the point q_i, p_i of the phase space is chosen somewhere on the energy surface $H = E$. This shows that the functional determinant of the equations (85.5) with respect to the $\bar{q}_j$ must be zero, which gives the following determinantal condition for the function W:

$$\left\| \frac{\partial^2 W}{\partial q_i \partial \bar{q}_j} \right\| = 0. \tag{85.6}$$

This is a characteristic property of the W-function which has no analogue in Jacobi's theory. The partial differential equation of Hamilton brings about an identity between the q_k, p_k which forces them to remain on a definite surface of the phase space.

We illustrate the situation by the example of a linear transformation which maps the ordinary three-dimensional space on itself. The coordinates x, y, z belong, say, to Space I; the coordinates X, Y, Z, to Space II. If the determinant of the transformation is different from zero, we have an ordinary point-to-point transformation. But now let us assume that the determinant of the transformation vanishes. Then the coordinates x, y, z which are expressed as linear functions of the X, Y, Z, satisfy an identity of the form

$$Ax + By + Cz = 0, \tag{85.7}$$

which means that the point x, y, z is confined to a plane. We wish to rotate our axes and normalize this plane to $z = 0$. Hence we set up our singular transformation in the following form:

$$\begin{aligned} x &= X + aZ, \\ y &= Y + \beta Z, \\ z &= 0. \end{aligned} \tag{85.8}$$

The transformation (85.8) has the property that Space I is reduced to the plane $z = 0$, which is transformed into the entire Space II. Consequently, the transformation cannot be a point-to-point correspondence. Going from Space II to Space I, we have a point-to-point transformation. But going from Space I to Space II we have a point-to-line transformation. The point p at the foot of the straight line L is transformed into the entire line.

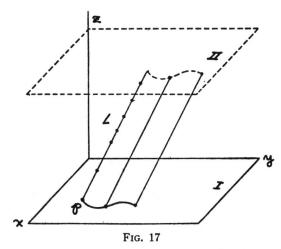

FIG. 17

This point-to-line transformation is chracteristic of a singular transformation, and explains why the transformation generated by the principal function must be singular. Somehow the *motion* has to come in, i.e. the transformation of the point $\bar{q}_k$, $\bar{p}_k$ into all its later positions q_k, p_k; but this means a point-to-line transformation.

In the Jacobian theory the point-to-line transformation arises in an entirely different manner. There the variable Q_n is distinguished from the other variables. The point Q_i, P_i is fixed, *except* that the coordinate P_n is *not* constant. Hence we need to solve the $n - 1$ equations:

$$P_k = -\frac{\partial S}{\partial Q_k} \ (k = 1, \ldots, n - 1) \tag{85.9}$$

for the q_i, obtaining $n - 1$ equations between the n variables q_i; this is a *line*. In the Hamiltonian case *all* the P_k, *including* P_n, are constants, and the point-to-line transformation is generated by the singularity of the transformation.

But now we can go one step further and see that the singularity of the transformation in *one* direction cannot suffice. Since the motion occurs on the surface $H = E$, it would not help to transform the points of that surface into the surrounding space. That transformation would have nothing to do with the problem of motion. Hence in our illustrative example we assume that the angle of inclination of the parallel straight lines L to the z-plane becomes smaller and smaller until eventually the lines L lie in the plane $z = 0$. The transformation is now singular in

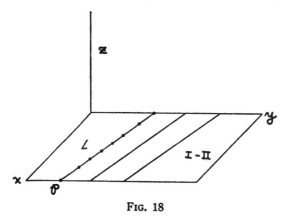

FIG. 18

both directions. Space I is reduced to the surface $z = 0$, but so also is Space II. The point-to-line transformation now holds quite symmetrically, going from Space I to Space II *and also* from Space II to Space I.

This is the picture of the transformation generated by Hamilton's principal function. The surface $H = E$ is transformed into itself because the points q_k, p_k are transformed into lines which lie on that surface. This means that not only the point q_k, p_k, but also the point $\bar{q}_k$, $\bar{p}_k$, lies on the surface $H = E$—which is exactly the statement of the second partial differential equation of Hamilton.

Just as the equations (85.5) cannot be solved for the $\bar{q}_i$ because the functional determinant vanishes, so also the equations

$$\bar{p}_i = -\frac{\partial W}{\partial \bar{q}_i} \tag{85.10}$$

cannot be solved for the q_i because here too the functional determinant vanishes. Indeed, this functional determinant is identical with the previous functional determinant (85.6) in view of the symmetry of the expression in q_i and $\bar{q}_i$. Thus we see that the singularity of the transformation in one direction *involves* already the singularity of the transformation in the other direction. The one partial differential equation cannot exist without the other, the common tie between the two being provided by the determinantal condition (85.6).

We now proceed to the general rheonomic case. In view of the similarity of the situation we can restrict ourselves to a brief exposition. We add the time t to the mechanical variables and require that the "extended action"

$$A = \int_{\tau_1}^{\tau_2} \sum_{i=1}^{n+1} p_i q_i' \, d\tau = \int_{\tau_1}^{\tau_2} \left(\sum_{i=1}^{n} p_i \, dq_i + p_t dt \right) \tag{85.11}$$

be stationary, under the auxiliary condition that the C-point of the extended phase space remains on the "extended energy surface"

$$p_t + H(q_1, \ldots, q_n, t, p_1, \ldots, p_n) = 0. \tag{85.12}$$

Again we evaluate the integral (85.11) taken between the two points $\bar{q}_1, \ldots, \bar{q}_n, \bar{t}$ and $q_1, \ldots, q_n, t$ of the extended configuration space, expressing it as a function of these two sets of coordinates:

$$W = W(q_1, \ldots, q_n, t; \bar{q}_1, \ldots, \bar{q}_n, \bar{t}). \tag{85.13}$$

Problem 1. Show that this definition of Hamilton's principal function is equivalent to the following procedure:

i. Obtain a complete solution of the Lagrangian equations in the form

$$q_i = f_i(a_1, \ldots, a_{2n}, t).$$

ii. Evaluate the definite integral

$$A = \int_{\bar{t}}^{t} L dt$$

for this solution, obtaining A as a function of t, $\bar{t}$ and the $2n$ constants of integration a_i.

iii. Eliminate the constants of integration by solving the $2n$ equations:

$$q_i = f_i(a_1, \ldots, a_{2n}, t),$$
$$\bar{q}_i = f_i(a_1, \ldots, a_{2n}, \bar{t}),$$

for the a_i.

iv. Substitute these solutions for the a_i in A. The function so constructed gives Hamilton's principal function W.

Problem 2. Consider the dynamical problem of a uniform gravitational field, $V = mgz$, using rectangular coordinates. Here the Lagrangian equations are integrable by elementary means. Construct Hamilton's W-function on the basis of the previous procedure, obtaining

$$W = \frac{m}{2}\left[\frac{(x - \bar{x})^2 + (y - \bar{y})^2 + (z - \bar{z})^2}{t - \bar{t}} - g(z + \bar{z})(t - \bar{t}) - \frac{1}{12}g^2(t - \bar{t})^3\right].$$

Once more the transformation equations hold;

$$p_i = \frac{\partial W}{\partial q_i}, \qquad \bar{p}_i = -\frac{\partial W}{\partial \bar{q}_i}$$

$$p_t = \frac{\partial W}{\partial t}, \qquad p_{\bar{t}} = -\frac{\partial W}{\partial \bar{t}}. \tag{85.14}$$

The auxiliary condition (85.12), written down for both points (q_i, t, p_i, p_t) and $(\bar{q}_i, \bar{t}, \bar{p}_i, \bar{p}_{\bar{t}})$, yields the two partial differential equations that the W-function satisfies [cf. (79.20)].

Here again the functional determinant of the transformation vanishes, which makes the transformation singular. We cannot express the q_i, t, p_i, p_t in terms of the $\bar{q}_i, \bar{t}, \bar{p}_i, \bar{p}_{\bar{t}}$, or vice versa. However, if we agree to choose the first and second sets of co-ordinates in harmony with the condition (85.12), we can omit the last members of the two sets of equations (85.14), because they are consequences of the previous equations. The remaining equations

$$p_i = \frac{\partial W}{\partial q_i}, \qquad \bar{p}_i = -\frac{\partial W}{\partial \bar{q}_i}, \tag{85.15}$$

are solvable for the q_i and p_i and we obtain a transformation of the following form:

$$q_i = f_i(t, \bar{q}_1, \ldots, \bar{q}_n, \bar{p}_1, \ldots, \bar{p}_n, \bar{t}), \tag{85.16}$$
$$p_i = g_i(t, \bar{q}_1, \ldots, \bar{q}_n, \bar{p}_1, \ldots, \bar{p}_n, \bar{t}).$$

This gives the complete solution of the problem of motion, adjustable to arbitrary initial conditions. Compared with Jacobi's solution, the number of integration constants is not $2n$ but $2n + 1$, in agreement with the over-completeness of the W-function.

Summary. While Jacobi's transformation theory flattens out the energy surfaces to planes and the curves of motion to straight lines, the transformation established by Hamilton's principal function is of an entirely different nature. The transformation is limited to the energy surface $H = E$ and has thus a singular character. The motion is here a consequence of the point-to-line nature of the transformation, caused by the vanishing of the functional determinant.

6. Construction of Hamilton's principal function with the help of Jacobi's complete solution. In spite of the different viewpoints which characterize Hamilton's and Jacobi's integration theories, there is a definite relationship between the W-function and the S-function. Jacobi's integration theory includes the over-determined W-function in an indirect way. If

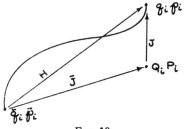

FIG. 19

we know Jacobi's S-function, we can construct Hamilton's principal function by differentiation and elimination.

In order to show this we go back to the theory of infinitesimal canonical transformations, developed before (cf. chapter VII, section 7). We consider Jacobi's S-function as the generating function of an infinite succession of continuously changing canonical transformations, physically realized by the motion of the phase fluid. An arbitrary fixed point Q_i, P_i of the phase space is transformed into the point q_i, p_i which moves with the time t. We find the transformed point in the position $\bar{q}_i$, $\bar{p}_i$ at the time $\bar{t}$, and in the position q_i, p_i at the time t. Hamilton's W-function on the other hand can be conceived as the generating function of a canonical transformation which eliminates the point Q_i, P_i and transforms the point $\bar{q}_i$, $\bar{p}_i$ *directly* into q_i, p_i.

We now make use of the *group property* of canonical transformations (cf. chapter VII, section 7). The transformation from $\bar{q}_i$, $\bar{p}_i$ to Q_i, P_i, and then from Q_i, P_i to q_i, p_i are both canonical, and thus the resulting transformation from $\bar{q}_i$, $\bar{p}_i$ to q_i, p_i is likewise canonical. The generating function of the resulting transformation is equal to the *difference* between the two generating functions which are employed in the two separate transformations [cf. (77.3)]. Thus the following remarkable relation holds:

$$W = \Delta S = S(q_1, \ldots, q_n, Q_1, \ldots, Q_n, t)$$
$$- S(\bar{q}_1, \ldots, \bar{q}_n, Q_1, \ldots, Q_n, \bar{t}) \qquad (86.1)$$

However, our construction is not finished, since we have to consider the equations

$$P_i = \frac{\partial S(q, Q, t)}{\partial Q_i}, \qquad P_i = \frac{\partial S(\bar{q}, Q, \bar{t})}{\partial Q_i}, \qquad (86.2)$$

as auxiliary conditions. If we subtract these two equations, we obtain the n conditions

$$\frac{\partial \Delta S}{\partial Q_i} = 0. \qquad (86.3)$$

These n conditions can be used to eliminate the Q_i, so that finally W appears as a function of the q_i, t and $\bar{q}_i$, $\bar{t}$ alone.

Hence we arrive at the following procedure which allows us to obtain Hamilton's principal function from a complete solution $S(q_1, \ldots, q_n, t, a_1, \ldots, a_n)$ of the Hamilton-Jacobi partial differential equation:

1. Take the difference
$$\Delta S = S(q_1, \ldots q_n, t, a_1, \ldots, a_n) - S(\bar{q}_1, \ldots, \bar{q}_n, \bar{t}, a_1, \ldots, a_n).$$

2. Solve the equations

$$\frac{\partial \Delta S}{\partial a_i} = 0$$

for the a_i.

3. Substitute these a_i in ΔS, thus obtaining $W = \Delta S$ as a function of $q_1, \ldots, q_n, t$ and $\bar{q}_1, \ldots, \bar{q}_n, \bar{t}$. *This gives Hamilton's principal function W.*

Problem 1. Solve the partial differential equation of Hamilton-Jacobi by separation for the case of a uniform gravitational field (cf. Problem 2, section 5). Obtain from this solution Hamilton's W-function, and show that the result agrees with that earlier result where the W-function was constructed on the basis of a preliminary complete solution of the equations of motion.

Problem 2. Assume a conservative system and suppose that the time independent W-function between two points of the configuration space:

$$W = W(q_1, \ldots, q_n, \bar{q}_1, \ldots, \bar{q}_n, E)$$

is known. Show that the time-dependent W-function can now be obtained as follows:

$$W_1 = -E(t - \bar{t}) + W,$$

provided that we eliminate the energy constant E with the help of the equation

$$\frac{\partial W}{\partial E} = t - \bar{t}.$$

Summary. Although Hamilton's principal function is over-determined, complete solution of the equation of Jacobi's integration theory provides Hamilton's principal function by mere differentiation and elimination.

---

7. Geometrical solution of the partial differential equation. Hamilton's optico-mechanical analogy. In our previous considerations we have assumed that we possess a *complete* solution of the partial differential equation of Hamilton-Jacobi. We now assume much less. We are satisfied if we have a *particular* solution of the given partial differential equation

$$H\left(q_1, \ldots, q_n, \frac{\partial S}{\partial q_1}, \ldots, \frac{\partial S}{\partial q_n}\right) = E, \qquad (87.1)$$

without any constants of integration.

Analytically, such a particular solution can be exploited for integrating *one half* of the complete set of canonical equations. We can express the p_i in terms of the q_i by means of the equations

$$p_i = \frac{\partial S}{\partial q_i}. \tag{87.2}$$

If we introduce these p_i in the first set of canonical equations

$$\dot{q}_i = \frac{\partial H}{\partial p_i}, \tag{87.3}$$

our task is reduced to the solution of these n differential equations of the first order, instead of the original set of $2n$ equations. In the following treatment we pursue a geometrical, rather than an analytical, method of solution.

We can construct a particular solution of the partial differential equation (87.1) by geometrical means. The physical interpretation of this construction leads to a new insight into the nature of mechanical problems and elucidates one of Hamilton's most startling discoveries, the analogy between optical rays and mechanical paths. Through present theories of wave-mechanics this analogy has become one of the leading ideas of modern atomic research.

In the following discussion we assume that we are dealing with the motion of a particle in a field of potential energy V. As coordinates we employ the ordinary rectangular coordinates x, y, z. However, the next section will show that all our conclusions are applicable to the motion of arbitrary mechanical systems.

The energy-equation of our present problem takes the form

$$\frac{1}{2m} (p_1{}^2 + p_2{}^2 + p_3{}^2) + V(x, y, z) = E. \tag{87.4}$$

This leads to the following partial differential equation for the function S:

$$\left(\frac{\partial S}{\partial x}\right)^2 + \left(\frac{\partial S}{\partial y}\right)^2 + \left(\frac{\partial S}{\partial z}\right)^2 = 2m(E - V). \tag{87.5}$$

We consider S as a function of x, y, z *without* any constants of integration:

$$S = S(x, y, z). \tag{87.6}$$

This means that we assume a *particular* solution of the differential equation (87.5) instead of the complete solution which would contain two arbitrary constants of integration a_1 and a_2. Such a particular solution of the equation (87.5) permits a simple geometrical interpretation.

We know that the gradient of a function S has the direction of the normal to the surface

$$S(x, y, z) = \text{const.} \tag{87.7}$$

We now consider two neighbouring surfaces $S = C$ and $S = C + \epsilon$.

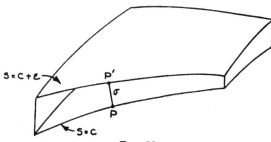

FIG. 20

From an arbitrary point P of the first surface we arrive at the point P' of the second surface by proceeding along the normal, i.e. along the gradient of S. Now the absolute value of grad S can be defined as the directional derivative of S:

$$|\text{grad } S| = \frac{\partial S}{\partial \nu}, \tag{87.8}$$

taken in the direction of the normal. Relative to Fig. 20 we can write

$$|\text{grad } S| = \frac{\epsilon}{\sigma}. \tag{87.9}$$

But then equation (87.5) can be rewritten as follows:

$$\sigma = \frac{\epsilon}{\sqrt{2m(E - V)}}. \tag{87.10}$$

This equation gives a method of constructing the surface $S = C + \epsilon$ if the surface $S = C$ is given. For this purpose we lay off the infinitesimal distance σ, evaluated from (87.10), perpendicular to the given surface, at every point of the surface.

This construction provides a successive infinitesimal procedure by which a particular solution of the Hamilton-Jacobi partial differential equation can be obtained. We start out with any given basic surface of the three-dimensional space. We attach to this surface the value $S = 0$. We now construct the neighbouring surface $S = \epsilon$, from this the surface $S = 2\epsilon$, then the surface $S = 3\epsilon$, and so on. Eventually a certain finite portion of the three-dimensional space will be filled by surfaces $S = $ const., one passing through each point of space. This means that we have a function S in a certain region which satisfies the partial differential equation (87.5).

The surfaces $S = $ const. stand in a remarkable relation to the problem of motion. To solve the canonical equations of motion with the help of the generating function S is not possible in the manner of Jacobi's integration theory, because we do not know how the function S depends on the variables Q_i. However, instead of making use of the *second* set of transformation equations, we can now utilize the *first* set

$$p_i = \frac{\partial S}{\partial q_i}. \tag{87.11}$$

This relation can be written in the form of the vector equation

$$\mathbf{P} = \text{grad } S, \tag{87.12}$$

and since the momentum $\mathbf{P} = m\mathbf{v}$ has the direction of the tangent to the path, we obtain the theorem that *the mechanical path of the moving point is perpendicular to the surfaces $S = $ const.* We obtain a family of possible mechanical paths by *constructing the orthogonal trajectories of the surfaces $S = $ const.*

This theorem has the following significance. Let us consider a family of mechanical paths which all have the same total energy E and which all start perpendicular to a given surface $S = 0$. Then these paths have the property that *one can find an infinite family of surfaces $S = $ const. to which they remain perpendicular.* We say that these mechanical paths "have the ray property," because they behave exactly like the light rays of optics. These

light rays are characterized by the fact that they are the orthogonal trajectories of the wave surfaces. The same holds now for
the mechanical paths of a conservative system. They can be
considered as the orthogonal trajectories of the surfaces $S =$
const.

The orthogonality of the light rays to the wave surfaces does not hold in
crystal optics. Nor is a mechanical path always perpendicular to the surfaces
$S =$ const. An electron moving in a magnetic field does not cross the surfaces
$S =$ const. perpendicularly. In all these cases, however, the orthogonality
is restored if we think not in Euclidean terms but in terms of a geometry which
is intrinsically connected with the dynamical structure of the configuration
space, as Section 9 of the present chapter will show. We shall see that in the
case of a line-element of the Jacobian form $\overline{d\sigma} = \sqrt{E - V}\, \overline{ds}$ the difference
remains latent, because here the intrinsic orthogonality leads to orthogonality
in the ordinary sense. In other cases, however, the orthogonality of rays
and wave surfaces, taken in the intrinsic sense, does not involve orthogonality
in the ordinary Euclidean sense.

The "ray property" of a selected family of mechanical paths
is by no means trivial. An arbitrary continuous family of curves
in a space of more than two dimensions cannot be considered as
orthogonal trajectories of a family of surfaces. The ray property
of mechanical paths holds only because they are derivable from
a variational principle. Without the principle of least action,
the ray property of mechanical paths could not be established.

It is important to emphasize the condition under which the ray property
of mechanical paths holds, since we made a definite *selection* amongst all
possible mechanical paths. The light rays of a given optical field form a two-
dimensional manifold of curves, while the totality of mechanical paths in a
given mechanical field form a five-dimensional manifold. The paths could
start from a given basic surface with arbitrary initial velocities. *We select
those paths which start with the same total energy and are perpendicular to the
given surface.* The ray property is established for these paths. Mechanical
paths are isolated lines which do not intersect with one another. If we single
out artificially a family of mechanical paths through a certain condition, there
is no reason why we could not add a "stray" path which does not belong to that
family. An optical ray, however, cannot exist as an isolated unit but is always
part of a *field*.

In the "electron microscope," moving electrons produce the optical image.
All the electrons leave the heated cathode with practically the same total
energy because the temperature of the cathode is everywhere the same.

Moreover, on account of the strong electric forces which act on the cathode, the electrons leave the cathode practically perpendicularly. To the extent to which these two conditions are violated, the optical image becomes blurred. The violation of the energy condition has an optical analogy in "chromatic aberration," as we shall see below.

The ray property of mechanical paths is only part of the picture in the close analogy between optics and mechanics. The construction of the wave-surfaces on the basis of Huygens' principle has likewise a mechanical counterpart. In fact, the infinitesimal formulation of Huygens' principle coincides with Hamilton's partial differential equation for the realm of optics.

Huygens' principle, if restricted to infinitesimal distances, can be formulated as follows   We start with an arbitrary surface on which there is a light disturbance at a certain time $t = 0$. We then construct the neighbouring surface at a time $t = \epsilon$ by using the value of the velocity of light at every point of the basic surface $t = 0$. The construction is entirely similar to that of Fig. 20.   If v is the velocity of light at the point P, we have

$$\sigma = v\epsilon. \tag{87.13}$$

We assume that our medium is optically heterogeneous so that the velocity of light changes from point to point.   We then have

$$v = \frac{c}{n(x, y, z)}, \tag{87.14}$$

where n is the "refractive index" of the medium and c the velocity of light *in vacuo*. If n is given everywhere, we can construct the infinitesimal distance

$$\sigma = \frac{c}{n} \epsilon \tag{87.15}$$

at every point of the basic surface, and then continue this infinitesimal construction further and further.   Eventually a certain portion of space will be filled by surfaces which may be denoted by

$$\phi(x, y, z) = C. \tag{87.16}$$

This ϕ-function corresponds to our previous S-function.   The significance of ϕ is that it is equal to the time that light requires to travel from the basic surface $\phi = 0$ to the given point x, y, z.

In view of the relation which exists between the infinitesimal distance σ separating two neighbouring surfaces $\phi = $ const. and the gradient of ϕ [cf. equ. (87.9)], we obtain for ϕ the partial differential equation

$$\left(\frac{\partial \phi}{\partial x}\right)^2 + \left(\frac{\partial \phi}{\partial y}\right)^2 + \left(\frac{\partial \phi}{\partial z}\right)^2 = \frac{n^2}{c^2}. \qquad (87.17)$$

This basic differential equation of geometrical optics, which expresses Huygens' principle in infinitesimal form, was discovered by Hamilton in his fundamental researches in the realm of geometrical optics.

In mechanical problems we have encountered the partial differential equation (87.17) in the form (87.5). This differential equation can now be considered as an analogue of Huygens' principle. Comparing the right-hand sides of the equations (87.5) and (87.17), we can correlate a definite optical problem to a given mechanical problem by defining the refractive index of a hypothetical optical medium according to the law

$$\frac{n}{c} = \alpha \sqrt{2m(E - V)}, \qquad (87.18)$$

where α is an arbitrary constant.

It is historically a startling fact that, at the early dawn of the physical sciences, when the methods of higher mathematics were barely established, John Bernoulli treated the motion of a particle in the field of gravity as an optical problem, assuming a fictitious refractive index proportional to $\sqrt{E - V}$.

In this discussion we have emphasized the wave-surfaces, in both optics and mechanics, and considered both optical and mechanical paths as the orthogonal trajectories of these surfaces. But now we wish to show how the minimum principle, by which the paths can be defined, is obtainable from the partial differential equation which determines the wave surfaces.

In Fig. 21 we consider light travelling along the orthogonal trajectory T which originates in the point M of the wave surface $\phi = 0$ and terminates in the point N of the wave surface $\phi = n\epsilon$. Together with this path we consider another path, C, terminating in the same end points M and N, which is not an orthogonal

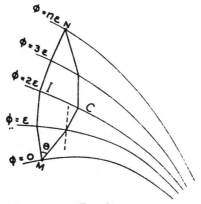

FIG. 21

trajectory. Then it follows from the geometrical construction of the surfaces ϕ = const. that the time required to travel from M to N along the orthogonal trajectory T is

$$t = \lim_{\epsilon \to 0} (\epsilon + \epsilon + \epsilon + \ldots), \qquad (87.19)$$

while the time required to travel along C is

$$t' = \lim_{\epsilon \to 0} \left(\frac{\epsilon}{\cos \theta_1} + \frac{\epsilon}{\cos \theta_2} + \ldots \right), \qquad (87.20)$$

where θ is the angle between the tangent to the path and the normal. Since the sum (87.19) is formed of purely positive quantities and the terms of (87.20) are all either equal to or larger than the corresponding terms of (87.19), we get

$$t' > t. \qquad (87.21)$$

This expresses Fermat's "principle of quickest arrival": *The path of a light ray is distinguished by the property that if light travels from one given point M to another given point N, it does so in the smallest possible time.*

FIG. 22

Our conclusion seems to indicate that Fermat's principle is a true minimum principle (and not merely a stationary value principle) if we make comparisons in the *local* sense, i.e. between the actual path and other paths arbitrarily near to it. However, our deduction requires that, all along the trajectory T, the wave surfaces shall be well defined, single-valued surfaces with definite normals. Now the following situation might arise (see Fig. 22). We consider the bundle of rays starting from the point M. These rays diverge in the beginning, but later on they may converge, and it is possible that the two neighbouring trajectories T and T' actually *intersect* at a certain point M'. In that case, the wave surface to which the point M' belongs degenerates into a line or a point. (In optical instruments to *every* point source M of optical waves there should correspond an "image" M' at which the wave surface degenerates into a point.) Our conclusion of a true local minimum holds only up to the point M' but cannot be extended *beyond*, because then our neighbouring path goes through a region in which it does not intersect any wave surfaces. Consequently the angle θ ceases to be real, and the conclusion $t' > t$ becomes illusory. In the corresponding mechanical situation M' is called the "kinetic focus" associated with M along the trajectory T. The principle of least action loses its "least" feature if we pass the kinetic focus.

Problem 1. Discuss the motion of a particle on a sphere, in the absence of external forces. Show that the action ceases to be a minimum if the moving point passes the "kinetic focus," which is here the antipole of M.

Problem 2. Find the kinetic focus in the field of gravity of the earth for particles which leave a certain point M in various directions with the same total energy E. Investigate the "least" feature of the principle of least action, in its relation to the kinetic focus.

Minimizing the time t means minimizing the integral

$$I = \int_{\tau_1}^{\tau_2} \frac{\overline{ds}}{v} = \int_{\tau_1}^{\tau_2} \frac{n}{c}\, \overline{ds}. \tag{87.22}$$

Hence we have *two* principles by which optical phenomena can be described. We can use the principle of the wave-surfaces, determining the function ϕ from the partial differential equation (87.17) and then constructing the orthogonal trajectories of these surfaces. Or we can use Fermat's principle, minimizing the integral (87.22). These two principles are entirely equivalent.

The same dualism appears in the phenomena of mechanics. We can define the mechanical paths as the orthogonal trajectories of the wave-surfaces $S = $ const., determining S by solving the partial differential equation (87.5). Or we can use a minimum principle which corresponds to Fermat's principle (87.22). In view of the correlation (87.18), this principle must take the form of minimizing the definite integral

$$A = \int_{r_1}^{r_2} \sqrt{2m(E - V)} \, ds. \tag{87.23}$$

This is nothing but Jacobi's principle (cf. chapter V, section 6), which again is equivalent to the principle of least action. If one assumes Jacobi's principle in mechanics and Fermat's principle in optics, the parallelism of mechanical and optical phenomena can be predicted. Jacobi's principle invites an optical interpretation through associating with a conservative mechanical system an optical medium whose refractive index changes proportionally to $\sqrt{E - V}$. The analogy can be exploited in both ways. On the one hand the canonical equations of Hamilton become available for optical problems. On the other hand, Huygens' construction of the wave surfaces can be transferred from optics to the realm of mechanics.

This gives a new significance to the function S, defined as an arbitrary particular solution of the partial differential equation of Hamilton-Jacobi. A complete solution of that differential equation was interpreted in abstract mathematical terms. It represented the generating function of a certain canonical transformation. But the S-function in the sense of a particular solution has a much more direct physical significance. The function S is optically paralleled by the function ϕ, which represents the time taken by light in passing from one wave surface to another. It can be defined by the integral (87.22), taken along the orthogonal trajectories of the wave surfaces. The corresponding integral in mechanics is the action (87.23), taken from a given basic surface along the actual mechanical paths which start perpendicularly from that surface with the same total energy E. If this action is expressed as a function of the end-point x, y, z of the path, the function S is obtained. The surfaces S = const. are thus *surfaces of equal action*, the action being measured from an arbitrary basic surface which is designated as the surface $S = 0$.

This interpretation of the S-function is strongly reminiscent of Hamilton's principal function W, the only difference being that in the case of the W-function we start from *a point* and not from *a surface*. Proceeding along an arbitrary mechanical path which starts from the point $\bar{x}, \bar{y}, \bar{z}$, the action is taken up to the point x, y, z. In the case of the previously considered S-function we can start from an arbitrary basic *surface* and then proceed along an arbitrary orthogonal trajectory to that surface.

One more feature of the optico-mechanical analogy deserves attention. An optical field can include at the same point (x, y, z) of space light vibrations of various frequencies. It can happen that the refractive index n is a function of the frequency. This is the phenomenon of "dispersion." The original wave-surface $\phi = 0$ is propagated in various ways for the different frequencies, and in the design of optical instruments we speak of a "chromatic aberration." Here too we can establish a corresponding mechanical analogy. The mechanical paths starting perpendicularly from a basic surface $S=0$ might be slightly inhomogeneous with respect to the total energy E. Such is the case, e.g., in the electron microscope where the thermal motion of the electrons causes slight variations in their total initial energy E. Accordingly we get dispersion, and a slight chromatic aberration in the picture produced by the electron microscope.

In the optico-mechanical analogy, the time of light propagation and the action are parallel quantities. Starting from a given initial surface, we can determine the infinite family of surfaces to which light pulses are propagated during successive small intervals of time. The corresponding surfaces in mechanics are the *surfaces of equal action*. It is of great importance that the optico-mechanical analogy holds only between the mechanical *paths* and light rays. The motion *in time* occurs according to entirely different laws. Particles starting in directions normal to a certain basic surface with the same total energy do *not* remain on the surfaces of equal action. This would be the case if the relation (87.13) between neighbouring wave surfaces were to hold. However, the quantity $\sqrt{2m(E - V)}$ is equal to the momentum mv. In view of the partial differential equation of Hamilton-Jacobi, the relation (87.13) has, therefore, to be written in mechanics as follows:

$$\sigma = \frac{\epsilon}{mv} . \tag{87.24}$$

The infinitesimal distance between two neighbouring wave surfaces is thus not directly, but *inversely*, proportional to the velocity. Hence the surfaces of simultaneous position of the moving particles are entirely different from the surfaces of equal action.

The decision as between the emission theory and the undulatory theory of light can not be made simply by observing the paths of optical rays. The laws of reflection and refraction can be explained on purely mechanical grounds. However, while the emission theory gives the law of refraction in the form

$$\frac{\sin i}{\sin r} = \text{const.} = \frac{v_2}{v_1} , \tag{87.25}$$

the corresponding law for the wave theory is

$$\frac{\sin i}{\sin r} = \text{const.} = \frac{v_1}{v_2} . \tag{87.26}$$

The decision between the two theories came through Foucault's famous experiment (in 1850) which showed that light travels in water more slowly than in air, while the emission theory demanded the opposite.

Problem 3. In the optico-mechanical analogy, normalize the mass of the moving particle to 1, and the velocity of light likewise to 1. Equation (87.18) then gives $n = v$. A sudden change in the refractive index between two optical media corresponds to a sudden change of the associated potential energy V. Replacing this discontinuity by a steep but continuous change of V in the boundary layer, explain the mechanical significance of the laws of reflection and refraction.

Problem 4. Assume a continuous change in the refractive index n from point to point, corresponding to a continuous field of force, of potential energy $V(x, y, z)$. Interpret the infinitesimal formulation of the law of refraction

$$n_1 \sin i = n_2 \sin r,$$

and show that it determines the curvature of the optical ray according to the mechanical law

$$mv^2/\rho = F_n,$$

where F_n is the component of the force along the principal normal and ρ is the radius of curvature of the path.

Summary. The mechanical paths connected with a conservative system may be obtained by finding a particular solution of the partial differential equation of Hamilton-Jacobi and then constructing the orthogonal trajectories of the surfaces S = const. This construction is parallel to that of the wave surfaces in geometrical optics. The surfaces of equal time in optics are paralleled by the surfaces of equal action in mechanics. Fermat's principle of least time is paralleled by the principle of least action, or Jacobi's principle. Both optical and mechanical phenomena can be described in wave terms as well as in particle terms. In wave terms we have an infinite family of surfaces, controlled by Hamilton's partial differential equation. In particle terms we have the orthogonal trajectories of these surfaces, controlled by Fermat's and Jacobi's principles. The parallelism extends to the paths of mechanical particles only and does not include the manner in which the motion occurs in time. Moreover, this analogy selects amongst all possible mechanical paths those which start the motion normal to a given surface.

8. The significance of Hamilton's partial differential equation in the theory of wave motion. We have pointed out that Hamilton's partial differential equation (87.17) in optics expresses Huygens' principle in infinitesimal form. Although Huygens' principle is based on undulatory assumptions, nevertheless the construction of the successive wave-fronts with the help of this principle is a method of geometrical, and not of physical, optics. In order to examine more deeply the relation of Hamilton's partial differential equation to the principles of physical optics, we modify somewhat the definition of wave surfaces. We considered these surfaces in connection with the propagation of elementary light waves in geometrical optics.

However, these surfaces have no less significance in connection with the propagation of a light wave of a definite frequency in physical optics. The wave surfaces can then be defined as *surfaces of equal phase*, and the velocity with which light is propagated is also the velocity with which the phase angle, $\overline{\phi}$ say, is propagated, normal to the wave surfaces.

The propagation of a wave of definite frequency may be described by means of the equation

$$u = A e^{i\, 2\pi\, [\nu t - \overline{\phi}\, (x,\, y,\, z)]}\, ,\qquad (88.1)$$

where u represents the disturbance, A is the amplitude, ν the frequency and $\overline{\phi}$ the phase angle. In this picture the wave surfaces are characterized by the equation

$$\overline{\phi}(x,\, y,\, z) = \text{const.}\qquad (88.2)$$

The propagation of a definite phase requires the relation

$$\nu dt - d\overline{\phi} = 0,\qquad (88.3)$$

or

$$d\overline{\phi} = \nu dt.\qquad (88.4)$$

This relation shows that the time taken for light to pass between neighbouring surfaces $\overline{\phi}$ = const. is $d\overline{\phi}/\nu$, whereas the corresponding time for the surfaces ϕ = const. of the previous section [cf. (87.16)] is $d\phi$. Hence the differential equation satisfied by $\overline{\phi}$ is derived from (87.17) by putting

$$\overline{\phi} = \nu\phi,\qquad (88.5)$$

and the partial differential equation of Hamilton becomes

$$\mid \text{grad}\ \overline{\phi}\ \mid^2 = \nu^2 n^2/c^2 = 1/\lambda^2\, ,\qquad (88.6)$$

if λ denotes the wave-length.

The interpretation of $\overline{\phi}$ as a phase function gave rise to de Broglie's fundamental discovery of matter waves. Let us assume that there is physical truth in the optico-mechanical analogy, notwithstanding the obvious discrepancy which exists between the propagation of matter and the propagation of light waves. Let us forget about the electron as a particle and consider only the *path* of the moving electron as a kind of light ray. This path is a closed orbit around the nucleus. Now if there exists a vibration along a closed path, then this vibration, coming back again and again to the same point, will add up to a resultant amplitude. This amplitude is zero if the phase angle of the returning vibration is not in resonance with the original phase angle. On the other hand, the amplitude can become arbitrarily large if the new phase angle differs from the original one by 2π or a multiple of 2π. In view of the fact that the phase angle is $2\pi\bar{\phi}$, we obtain the following selection principle for those paths which are not blotted out by interference:

$$\Delta\bar{\phi} = n, \tag{88.7}$$

where n is an integer. On the other hand, Einstein's invariant formulation of the quantum conditions [cf. (84.28)] gives

$$\Delta S = nh. \tag{88.8}$$

In the optico-mechanical analogy the phase angle $\bar{\phi}$ and the action S are corresponding quantities. The resonance condition (88.7) shows that we can obtain a natural and adequate interpretation of Einstein's quantum condition if we conceive the action function S as a phase function $\bar{\phi}$, in accordance with the correlation:

$$\bar{\phi} = S/h. \tag{88.9}$$

Now, if we compare the partial differential equations (87.5) and (88.6) which determine the functions S and $\bar{\phi}$, respectively, we obtain the result that the moving electron is associated with a definite wave-length according to the law

$$\frac{1}{\lambda^2} = \frac{2m(E - V)}{h^2} = \frac{m^2v^2}{h^2} . \tag{88.10}$$

This gives the celebrated "de Broglie wave-length,"

$$\lambda = h/mv, \tag{88.11}$$

which has influenced so decisively the course of contemporary atomic physics.

Hamilton's partial differential equation in optics is equivalent to an infinitesimal formulation of Huygens' principle. But Huygens' principle is only an *approximate* consequence of the true principles of physical optics. Light is a vectorial phenomenon and the adequate description of optical phenomena proceeds on the basis of Maxwell's electromagnetic field equations. Yet many optical phenomena are explainable on the basis of the simpler scalar theory of Fresnel. Here we consider a field function $\phi(x, y, z, t)$ which satisfies Fresnel's wave equation:

$$\frac{\partial^2\phi}{\partial x^2} + \frac{\partial^2\phi}{\partial y^2} + \frac{\partial^2\phi}{\partial z^2} - \frac{n^2}{c^2}\frac{\partial^2\phi}{\partial t^2} = 0. \tag{88.12}$$

This wave equation is a linear differential equation of the second order in contrast to Hamilton's partial differential equation (88.6) which is of the first order but of the second degree. The connection between these two basic equations can be found as follows.

Let us assume that the optical vibration occurs with a definite frequency ν. Then a separation with respect to the time becomes possible by writing ϕ in the form

$$\phi = e^{2\pi i \nu t} \psi (x, y, z). \tag{88.13}$$

ψ is called the "amplitude-function," and the differential equation which determines it is the "amplitude equation"

$$\frac{\partial^2 \psi}{\partial x^2} + \frac{\partial^2 \psi}{\partial y^2} + \frac{\partial^2 \psi}{\partial z^2} + \frac{4\pi^2}{\lambda^2} \psi = 0. \tag{88.14}$$

The comparison of (88.13) with (88.1) shows that we have the following connection between the wave function ψ and the phase function $\bar\phi$:

$$\psi = e^{-2\pi i \bar\phi} \tag{88.15}$$

If we substitute the expression (88.15) in the amplitude equation (88.14), we obtain the following differential equation for $\bar\phi$:

$$\left| \text{grad } \bar\phi \right|^2 = \left(\frac{1}{\lambda} \right)^2 - \frac{i}{2\pi} \left(\frac{\partial^2 \bar\phi}{\partial x^2} + \frac{\partial^2 \bar\phi}{\partial y^2} + \frac{\partial^2 \bar\phi}{\partial z^2} \right). \tag{88.16}$$

The last term of this differential equation becomes negligible as λ tends to zero. Hence the interesting conclusion can be drawn that Hamilton's partial differential equation in optics is equivalent to Fresnel's wave equation for the case of infinitely small wave lengths or infinitely large frequencies. For small but finite wave lengths, Hamilton's partial differential equation is only an approximation and should be replaced by the correct equation (88.16), or better still by the amplitude equation for the function ψ.

Schroedinger had in 1927 the original idea of going beyond the analogy between geometrical optics and mechanics, established by Hamilton's partial differential equation, and changing over from the phase function $\bar\phi$ to the wave function ψ. If we introduce de Broglie's wave-length (88.10) into the amplitude equation (88.14), we obtain Schroedinger's famous differential equation

$$\frac{\partial^2 \psi}{\partial x^2} + \frac{\partial^2 \psi}{\partial y^2} + \frac{\partial^2 \psi}{\partial z^2} + \frac{8\pi^2 m}{h^2} \left(E - V \right) \psi = 0, \tag{88.17}$$

which is the basis of modern wave mechanics. The great development from classical mechanics to wave mechanics is thus characterized by the following landmarks: Delaunay's treatment of separable multiply-periodic mechanical systems; the Sommerfeld-Wilson quantum conditions; Einstein's invariant formulation of the quantum conditions; de Broglie's resonance interpretation of Einstein's quantum condition; Schroedinger's logarithmic transformation from the phase function S to the wave function ψ.

Summary. The wave surfaces of light propagation can be defined as surfaces of equal phase. Hamilton's partial differential equation in optics then determines the space distribution of the phase angle in a stationary optical field. This differential equation is closely related to Fresnel's wave equation. It is an approximate consequence of that equation which becomes exact for the case of infinitely small wavelengths or infinitely large frequencies.

9. The geometrization of dynamics. Non-Riemannian geometries. The metrical significance of Hamilton's partial differential equation. Again and again we have found how much the pictorial language of geometry helps in the deeper understanding of the problems of mechanics. The configuration space with its Riemannian metric made it possible for us to picture a mechanical system, however complicated, as a single point of a properly defined many-dimensional space. The principles which govern the motion of a single particle could be extended to the motion of arbitrary mechanical systems.

The advantage of a geometrical language is particularly conspicuous if the mechanical system is not subject to external forces. In that case the mechanical path of the system can be conceived as a geodesic of the configuration space (Hertz's principle of the straightest path). Furthermore, if the potential energy does not depend on the time t, we can introduce an auxiliary line element

$$\overline{d\sigma} = \sqrt{E - V}\ \overline{ds}, \tag{89.1}$$

where $\overline{ds}$ is the line element of the original configuration space, and once more obtain all the mechanical paths which belong to the same total energy E as the geodesics of this manifold.

However, not all mechanical systems are conservative. Nor is it always true that the work function of the external forces is a function of the coordinates alone. The work function of the forces which act on an electron in the presence of an external electromagnetic field depends on the velocities $\dot{q}_i$ and may also depend on the time t. Hence the ordinary conditions of a work function independent of time and velocity do not hold here. Moreover, if we change from classical to relativistic mechanics, the ordinary form of the kinetic energy responsible for the Riemannian structure of the line element can no longer be retained.

The close relation of dynamics to geometry continues to hold even under these more general conditions. The Riemannian geometry is not the only possible form of metrical geometry. Riemann's geometry is distinguished by the unique feature that space flattens out in the neighbourhood of an arbitrary point of the manifold, and thus the customary Euclidean geometry holds at least in infinitesimal regions. But the constructions of geometry, based on straight lines and angles, do not require this restriction. In view of the general problems of dynamics a more general form of geometry deserves attention, in which the line element $\overline{ds}$ is defined in a more general manner than the Riemannian line element.

For the following discussion the time t must abandon the preferred position which it occupies in the customary problems of mechanics. We add the time t to the position coordinates q_i by putting

$$t = q_{n+1}, \tag{89.2}$$

and we consider all the variables q_i on an equal footing. The subscript i, if not stated otherwise, will always run from 1 to $n + 1$. The configuration space has thus $n + 1$, instead of n, dimensions. We introduce a line element $\overline{ds}$ in this space by the definition

$$\overline{ds} = F(q_1, \ldots, q_{n+1}, dq_1, \ldots, dq_{n+1}). \tag{89.3}$$

The function F is here an arbitrary function of the $2n + 2$ variables q_i and dq_i, except for the natural restriction that it is supposed to be a *homogeneous differential form of the first order* in the variables dq_i. This means that

$$F(q_1, \ldots, q_{n+1}, a\,dq_1, \ldots, a\,dq_{n+1}) =$$
$$a F(q_1, \ldots, q_{n+1}, dq_1, \ldots, dq_{n+1}). \qquad (89.4)$$

Let us consider an arbitrary curve of this manifold, given in parametric form

$$q_i = f_i(\tau). \qquad (89.5)$$

Due to the condition (89.4) the line element of this curve can be written

$$\overline{ds} = F(q_1, \ldots, q_{n+1}, q'_1, \ldots, q'_{n+1})d\tau. \qquad (89.6)$$

Hence the problem of the geodesic lines, i.e. the problem of minimizing the length of a curve between two points τ_1 and τ_2, leads to the variational problem of minimizing the definite integral

$$A = \int_{\tau_1}^{\tau_2} F(q_1, \ldots, q_{n+1}, q'_1, \ldots, q'_{n+1})d\tau. \qquad (89.7)$$

The same problem, however, can be stated in a slightly different form by choosing the last variable $q_{n+1} = t$ as the parameter τ. Then the equations of the geodesic appear in the form

$$q_i = f_i(t), \qquad (89.8)$$

and the functions $f_i(t)$ are found by minimizing the integral

$$A = \int_{t_1}^{t_2} F(q_1, \ldots, q_n, t, \dot{q}_1, \ldots, \dot{q}_n, 1)dt. \qquad (89.9)$$

Here we recognize the standard problem of Lagrangian mechanics, namely that of finding the motion of a mechanical system by minimizing the time-integral of the Lagrangian function L. *The function F can thus be interpreted mechanically as the Lagrangian function L of analytical mechanics.*

Conversely, if we start with a mechanical problem, characterized by a given Lagrangian function L,

$$L = L(q_1, \ldots, q_n, t, \dot{q}_1, \ldots, \dot{q}_n), \qquad (89.10)$$

we can define the line element $\overline{ds}$ of an $(n + 1)$-dimensional space by writing

$$\overline{ds} = L\left(q_1, \ldots, q_{n+1}, \frac{dq_1}{dq_{n+1}}, \ldots, \frac{dq_n}{dq_{n+1}}\right) dq_{n+1}$$

$$= F(q_1, \ldots, q_{n+1}, dq_1, \ldots, dq_{n+1}), \tag{89.11}$$

and transform the problem of motion to the problem of finding the geodesics of this $(n + 1)$-dimensional manifold. Hence we see that *the problem of solving the equations of dynamics and the problem of finding the geodesics of a certain—in general non-Riemannian—manifold are equivalent.*

In this geometrical interpretation of dynamics, the "principal function" of Hamilton is particularly significant. Since the "action" is now geometrically interpreted as "arc length," and "least action" means "least length," Hamilton's W-function simply means the distance between the two points $\overline{q}_1, \ldots, \overline{q}_{n+1}$ and $q_1, \ldots, q_{n+1}$ of the manifold.[1]

In the light of this interpretation we can see directly why Hamilton's principal function must be conditioned by some kind of differential equation. We cannot expect that an arbitrary function of the coordinates of two points of a manifold could be prescribed as the "distance" between these two points. "Distance" actually means "least distance," and the word "least" would not be applicable to an arbitrary definition of distance.

Obviously we can expect some fundamental geometrical significance to be attached to Hamilton's partial differential equation if we attack the problem from the geometrical angle. In order to derive this new interpretation, we start out by transforming the variational problem (89.7) into the canonical form. We introduce the momenta

$$p_i = \frac{\partial F}{\partial q'_i}, \tag{89.12}$$

[1] In view of this interpretation, Hamilton's notation S for his principal function assumes a prophetic significance. The reason why the present treatise did not adhere to Hamilton's notation is that the contemporary literature uses the S-function in the sense of Jacobi's integration theory, i.e. as a generating function of a general canonical transformation. Considering the deep-seated differences which exist between the theories of Hamilton and Jacobi (cf. section 5), a distinctive notation for Hamilton's principal function seemed justified.

and proceed to the construction of the Hamiltonian function H. However, since F is a homogeneous function of the first order in the variables q'_i, we have

$$H = \sum_{i=1}^{n+1} \frac{\partial F}{\partial q'_i} q'_i - F = 0. \tag{89.13}$$

We know that the vanishing of the Hamiltonian function is always compensated for by the existence of an identity between the q_i, p_i:

$$K(q_1, \ldots, q_{n+1}, p_1, \ldots, p_{n+1}) = 0. \tag{89.14}$$

This identity has to be considered as an auxiliary condition of the variational problem which requires the stationary value of

$$A = \int_{\tau_1}^{\tau_2} \sum_{i=1}^{n+1} p_i dq_i. \tag{89.15}$$

Problem 1. Show that in the case of a Euclidean line-element

$$\overline{ds} = \sqrt{dx^2 + dy^2 + dz^2},$$

the p_i are the "direction cosines" a_i of an arbitrary direction. The identity (89.14) now takes the form

$$\Sigma a_i^2 = 1.$$

Hence we can conceive of the "momenta" p_i, associated with an arbitrary metrical manifold, as the "generalized direction cosines" of that manifold. These direction cosines are not independent of one another, but are connected by the identity (89.14).

We now consider an arbitrary surface

$$f(q_1, \ldots, q_{n+1}) = 0 \tag{89.16}$$

of our manifold. Our aim is to find the normal to this surface. For this purpose, we drop a perpendicular on the surface from a point $\bar{q}_1, \ldots, \bar{q}_{n+1}$ outside. We then have the problem of finding the shortest distance of the point $\bar{q}_i$ from an arbitrary point of the surface. Hence we have to minimize the function $W(q_i, \bar{q}_i)$ under the auxiliary condition (89.16). The solution of this minimum problem is given by the equations

$$\frac{\partial W}{\partial q_i} - \lambda \frac{\partial f}{\partial q_i} = 0. \tag{89.17}$$

Making use of the fundamental properties of Hamilton's principal function we get

$$p_i = \lambda \, \frac{\partial f}{\partial q_i} \, . \qquad (89.18)$$

This equation expresses the direction cosines of the normal in terms of the gradient of f. The factor λ has the following significance. We can write equation (89.13) in the form

$$\Sigma p_i dq_i = \overline{ds}, \qquad (89.19)$$

and thus obtain

$$\overline{ds} = \lambda \, \Sigma \frac{\partial f}{\partial q_i} \, dq_i = \lambda df, \qquad (89.20)$$

from which

$$\frac{1}{\lambda} = \frac{df}{d\overline{s}} = | \, \text{grad} f \, | \, . \qquad (89.21)$$

Hence $1/\lambda$ is equal to the *directional derivative* of f, taken along the normal, or is *the steepest rate of change of the function f* at the given point. The equation (89.18) of the normal now becomes

$$p_i = \frac{1}{| \, \text{grad} f \, |} \, \frac{\partial f}{\partial q_i} \, . \qquad (89.22)$$

In order to evaluate λ, we substitute the expression (89.18) for p_i in our identity (89.14), obtaining

$$K \left(q_1, \ldots, q_{n+1}, \, \lambda \, \frac{\partial f}{\partial q_1}, \ldots, \lambda \, \frac{\partial f}{\partial q_{n+1}} \right) = 0. \qquad (89.23)$$

From this condition the undetermined factor λ can be found.

Problem 2. Assume a Riemannian line element

$$\overline{ds} = \sqrt{\Sigma a_{ik} dx_i dx_k},$$

and obtain for the steepest rate of change of a function $f(x_i)$ the law

$$| \, \text{grad} f \, | = \sqrt{\Sigma b_{ik} \frac{\partial f}{\partial x_i} \frac{\partial f}{\partial x_k}},$$

where the b_{ik} are the elements of a matrix which is the reciprocal of the matrix a_{ik}.

We notice that this procedure for determining the factor λ, and with it the steepest rate of change of f, reminds us of the partial differential equation of Hamilton. There we replaced p_i by $\partial W/\partial q_i$ and obtained a condition on the function W. In the present problem the function f can be an *arbitrary* function, and the condition (89.23) serves to obtain the steepest rate of change of f. Hamilton's partial differential equation is thus equivalent to the condition

$$\lambda = 1, \tag{89.24}$$

which means

$$| \operatorname{grad} W | = \left(\frac{dW}{ds}\right)_{max} = 1. \tag{89.25}$$

This property of the principal function has a very plausible geometrical significance. The surfaces $W = $ const. are by definition concentric spheres with the common centre $\bar{q}_i$. The radius vector $\mathbf{R}$ is everywhere perpendicular to the surface of the sphere. Hence, to proceed in the direction of the radius vector means to proceed in the direction of the normal—and, by the definition of W as the length of the radius vector $\mathbf{R}$:

$$| \operatorname{grad} W | = \left(\frac{dW}{ds}\right)_{max} = \frac{dr}{dr} = 1. \tag{89.26}$$

We notice that the perpendicularity of the radius vector to the surface of a sphere is not a characteristic property of Euclidean geometry alone. *It is an invariant property of any kind of metrical geometry*.

The condition (89.26) is equivalent to Hamilton's partial differential equation

$$K \left(q_1, \ldots, q_{n+1}, \frac{\partial W}{\partial q_1}, \ldots, \frac{\partial W}{\partial q_{n+1}} \right) = 0. \tag{89.27}$$

This differential equation is sufficient to determine Hamilton's principal function W if we add the proper boundary conditions. These boundary conditions follow from the definition of W as the distance between the two points $\bar{q}_i$ and q_i of our manifold. Let us assume that the two points come *arbitrarily near* to one another, i.e.

$$q_i = \bar{q}_i + dq_i. \tag{89.28}$$

The distance between two neighbouring points is simply the line element $\overline{ds}$ which was given in the form (89.3). Hence we obtain the condition

$$W(q_1, \ldots, q_{n+1}, q_1 - dq_1, \ldots, q_{n+1} - dq_{n+1})$$
$$= F(q_1, \ldots, q_{n+1}, dq_1, \ldots, dq_{n+1}). \tag{89.29}$$

The differential equation (89.27), together with the boundary condition (89.29), determines Hamilton's principal function uniquely.

The boundary condition (89.29) permits us to obtain an explicit expression for the distance-function W in the form:

$$W = F(q_1, \ldots, q_{n+1}; q_1 - \bar{q}_1, \ldots, q_{n+1} - \bar{q}_{n+1}), \qquad (89.30)$$

provided that the point $\bar{q}_i$ is near to the point q_i. A *complete integration* becomes possible in the special case where F depends on the dq_i alone, without involving the q_i:

$$\overline{ds} = F(dq_1, \ldots, dq_{n+1}). \qquad (89.31)$$

The problem of the geodesics is then completely integrable because all the q_i are now ignorable variables. The result is

$$W = F(q_1 - \bar{q}_1, \ldots, q_{n+1} - \bar{q}_{n+1}). \qquad (89.32)$$

which gives the distance between *any* two points of the manifold by mere substitution, without any integration. In this geometry, space is homogeneous, i.e. the properties of space around every point are the same. At the same time, space is *not* isotropic, i.e. the properties of space depend on the direction. Figures can be freely translated, but generally not rotated, in this geometry.

Problem 3. Obtain the distance between two points in Euclidean geometry from the line element $\overline{ds} = \sqrt{\Sigma dx_i^2}$, on the basis of the principle (89.32).

Problem 4. Obtain by the same method the principal function of a free particle of mass m which moves in the absence of external forces. Show that the result agrees with the expression obtained in Problem 2, section 5, if we put $g = 0$.

Problem 5. Show that the determinantal relation (85.6) is satisfied for this geometry.

In Section 7 we dealt with the remarkable analogy between optical and mechanical phenomena. We constructed the wave surfaces S = const., ϕ = const., and we saw that the orthogonal trajectories of these wave surfaces are, respectively, the mechanical paths and the optical rays. Our considerations were restricted to the case of a single particle acted upon by a conservative external force which did not depend on the velocity of the particle. We now abandon this restriction and show that our results remain valid for arbitrary mechanical systems *pro-*

vided that we associate each system with its own natural geometry.

The fundamental differential equation which the function S has to satisfy can be formulated geometrically, in the light of (89.24) and (89.25), as follows:

$$| \text{ grad } S | = 1. \tag{89.33}$$

This equation has a striking geometrical significance. *The surface $S = $ const. has everywhere the same constant distance from a basic surface $S = 0$.* Hence the surfaces $S = $ const. represent a *family of parallel surfaces.* The wave surfaces, which before were "surfaces of equal action," now become "surfaces of equal distance," i.e. parallel surfaces.

The orthogonality of the wave surfaces $S = $ const. to the mechanical (or optical) paths can once more be established. In Section 7 of this chapter we obtained the mechanical paths from the wave surfaces by solving the first set of transformation equations

$$p_i = \frac{\partial S}{\partial q_i} . \tag{89.34}$$

If we write these equations in the form

$$p_i = \frac{1}{| \text{ grad } S |} \frac{\partial S}{\partial q_i} , \tag{89.35}$$

which is permissible in view of (89.33), we see that we are concerned here with the equations of the normals to the surfaces $S = $ const. Thus we can establish once more the fact that *the mechanical paths are orthogonal trajectories of the wave surfaces.*

This theorem applies now in a much more general sense than before. In our earlier treatment the orthogonality of path and wave surface was understood in the sense of ordinary Euclidean geometry. Actually the theorem belongs to the non-Euclidean— and in general even non-Riemannian—geometry which is intrinsically associated with the given mechanical problem. That our earlier treatment sufficed is due to the somewhat accidental fact that the line element of the intrinsic geometry was proportional to the Euclidean line element [cf. (89.1)].

Problem 6. Show that the equation of the normal remains invariant with respect to a multiplication of $\overline{ds}$ by some factor $a(q_1, \ldots, q_{n+1})$.

In more general cases—such as the motion of an electron in a magnetic field, non-conservative systems, relativistic mechanics, optical rays within a crystal—the proportionality of the intrinsic $\overline{ds}$ and the ordinary ds is no longer true. The orthogonality of paths and wave surfaces then still holds in the intrinsic sense, although it ceases to hold in the ordinary sense.

The relation of the wave surfaces to the mechanical paths is more significant if we regard the problem from the point of view of its own intrinsic geometry. We then have, quite generally, the same situation that we encounter in optics when light is propagated in an optically *homogeneous* medium. The light rays are then *straight lines*, i.e. shortest lines or geodesics. The elementary waves in Huygens' construction are *spheres*, not only in infinitesimal, but even in finite regions. The envelopes of these spheres, i.e. the wave surfaces, are *parallel surfaces*, and the optical rays—or mechanical paths—are *orthogonal trajectories* of this family of parallel surfaces. All this remains true for arbitrary optical and mechanical systems, provided that we operate in a properly defined metrical space.

It was Gauss who, in his immortal "general investigations of curved surfaces" (*Disquisitiones generales circa superficies curvas*, 1827), first discovered that the *orthogonal trajectories of an arbitrary family of parallel surfaces are always geodesics*. His investigations were naturally restricted to the Riemannian form of metric. But actually we have here a theorem which holds equally in *any* kind of metrical geometry.

An amazing form of non-Riemannian geometry is realized in nature in the optical phenomena associated with crystals. In this geometry the line element is defined as follows:

$$\overline{ds} = \frac{\sqrt{dx^2 + dy^2 + dz^2}}{v} \qquad (89.36)$$

where v—the velocity of light—is obtained by solving "Fresnel's equation"

$$\frac{dx^2}{v^2 - v_1^2} + \frac{dy^2}{v^2 - v_2^2} + \frac{dz^2}{v^2 - v_3^2} = 0. \qquad (89.37)$$

The constants v_1, v_2, v_3 are called the "principal velocities." Since this equation leads to a quadratic equation for v, we have here a *double-valued* metric, i.e. the coexistence of two different kinds of geometries in the same problem. As a consequence, a light wave striking the crystal separates into two (differently polarized) waves and we obtain the phenomenon of "double refraction." The elementary wave surfaces of Huygens' theory form a double family of complicated surfaces of the fourth order, although in the proper geometrical setting they remain spheres.

Problem 7. In Fresnel's equation, put $v_1 = v_2$. We then have rotational symmetry around the z-axis and the crystal becomes "uniaxial" instead of "biaxial." Show that now one form of metric degenerates into the Euclidean type, so that wave surfaces of one kind become spheres in the Euclidean sense ("ordinary ray"). The distance-expression for the other kind of geometry (which belongs to the "extraordinary ray") becomes

$$W = \frac{(x - \bar{x})^2 + (y - \bar{y})^2 + (z - \bar{z})^2}{\sqrt{v_3^2[(x - \bar{x})^2 + (y - \bar{y})^2] + v_1^2(z - \bar{z})^2}} .$$

Summary. The problems of dynamics can be formulated in completely geometrical language if we associate the given dynamical problem with the proper form of metrical geometry. This geometry is generally not of the Riemannian type. The configuration space includes now the time with the other variables on an equal footing. The mechanical paths are shortest lines or geodesics of this manifold while the wave surfaces become parallel surfaces. The geodesics can be constructed as the orthogonal trajectories of the wave surfaces. The dynamical phenomena correspond to the propagation of light in an optically homogeneous medium.

CHAPTER IX

RELATIVISTIC MECHANICS

To some-one who could view the universe from a unified standpoint, the entire creation would appear as a unique truth and necessity. D'ALEMBERT, L'Encyclopédie (1751)

1. Historical introduction. Ever since the time of Faraday's "lines of force" concept the question has been asked: What happens to these lines if the bodies are set in motion? Is an electric field generated by material bodies carried along rigidly, if the bodies move in some rigid fashion? H. Hertz, the first demonstrator of electromagnetic waves, answered the question in the affirmative. But Fizeau's experiment with running water demonstrated that the velocity of propagation of light in the water is not $c + v$, but only $c + (1 - 1/n^2)v$, where n is the index of refraction of water. Lorentz explained the "convection coefficient" $1 - 1/n^2$ on the basis of a "stationary ether," which is not affected by the electric charges moving relative to it. On the other side, the assumption of a stationary ether led to the conclusion that on the earth (moving relative to the ether on account of the yearly revolution around the sun) certain optical effects should be expected which are proportional to v^2/c^2, where v is the velocity of the earth around the sun, and c is the velocity of light. The experimental demonstration of the non-existence of these phenomena brought theoretical physics to an impasse from which it was rescued by Einstein's paper of 1905 "On the Electro-dynamics of Moving Bodies." Both H. A. Lorentz and H. Poincaré realized, before Einstein, that a satisfactory theory should be in harmony with the observed non-measurability of an absolute velocity, but did not find a satisfactory answer to the problem. It was Einstein who realized that we should put the equivalence of all reference systems in uniform motion relative to each other at the head of our speculations, in the form of a universal *postulate* which we accept without trying to prove it. To this postulate he added a second postulate, demonstrated by the negative result of the Michelson-Morley and many other experiments, which demands that in all legitimate reference systems the velocity of light must have the same constant value c in every direction. On the basis of our ordinary kinematic concepts these two postulates can not hold simultaneously since they contradict each other. But Einstein showed that they can be reconciled if we abandon our common-sense notion of the existence of an "absolute time." He found the relation that must hold between the space and time measurements of two observers who are in uniform motion relative to each other. These equations show that the time t ceases to be an absolute quantity but has to be added to the three space coordinates. The time t has changed from an *in*variant to a *co*variant quantity, whereas the light velocity c has changed from a *co*variant to an *in*variant quantity.

Einstein could demonstrate that these transformation equations were in full harmony with all the known first-order and second-order effects, giving a perfect explanation of all the phenomena that come about by the motion of the light source relative to the observer, or vice versa. Moreover, the two basic postulates demanded a modification of the Newtonian equations of motion, and a new dynamical law came into being. The most decisive result of the new theory was, however, that the two previously independent concepts of mass and energy became united in the form of the celebrated equation $E = mc^2$, which Einstein discovered in incomplete form in 1905, and in complete form in 1907.

The objection that Einstein did not give a truly physical explanation of the relativistic effects, but transformed away the problem by a maze of physically unmotivated mathematical formulae, was removed a few years later, when H. Minkowski (in 1908–09) discovered the deeper significance of the two postulates of Einstein. He found that these postulates express a new *geometrical* structure realized in the universe, which adds time to space, thus forming an extended manifold of four dimensions: the space-time world. The relativistic phenomena are thus direct consequences of the geometry realized in nature, independently of any physical assumptions.

The overwhelming evidence provided by the demonstration of all the conclusions of the relativistic postulates by means of the most painstaking experiments led within a few years to their universal acceptance and made the theory of relativity one of the best-documented chapters of mathematical physics. The only dissenting voice was from Einstein himself, who felt that the first postulate of relativity was not sufficiently general in restricting the class of legitimate reference systems to *uniformly* moving systems, instead of including *all* possible reference systems. Reference systems are by their very nature auxiliary constructions which should have no absolute significance, and the notion of the legitimacy of reference systems should disappear altogether from the realm of mathematical physics. The postulate of the equivalence of *all* reference systems is called the "principle of general relativity," in contradistinction to "special relativity," which restricts the equivalence of reference systems to uniformly moving systems.

Even more decisive was Einstein's observation that in the presence of gravity the principle of the constancy of the velocity of light cannot be maintained. As a consequence, the geometrical structure of the world could not be of the Minkowskian type but had to be generalized to a structure which was Minkowskian only in the small, but had a "curvature" in finite dimensions. This meant that the geometry of the world remained metrical but changed from a four-dimensional Euclidean to a four-dimensional Riemannian structure. The new theory gave a perfect account of all the gravitational phenomena, together with a new interpretation of physical "matter" in purely geometrical terms. While special relativity united space and time in one entity, general relativity united *space*, *time*, and *matter* in one geometrical entity: a metrical geometry of the Riemannian type, established in a world of four dimensions.

General relativity cannot be properly treated without the field concept, which belongs to the realm of the mechanics of continua. Our following exposition is thus restricted to the kinematics and dynamics of Special Relativity. However, the dynamical law of General Relativity falls within the scope of Lagrangian and Hamiltonian dynamics and will be included in our discussions.

2. Relativistic kinematics. In his fundamental paper Einstein[1] introduced the following two postulates:

(*a*) All reference systems that move relative to each other with constant velocity are equally legitimate for the description of nature, without any preference for any particular one of them.

(*b*) The velocity of light *c* is the same universal constant, measured in any of these systems.

The second postulate leads to the following apparent absurdity. Consider an observer *B* moving relative to *A* with the velocity of light *c*. He emits a light signal in the direction of his own motion which moves relative to him with the velocity *c*. According to our common notions the velocity of this signal relative to *A* should be 2*c*, while the second postulate of relativity demands that it is *c*. Hence *c* = 2*c*, which is self-contradictory. This is no longer so, however, if we give up the intuitive notion of a time *t* which is the same for both observers *A* and *B*.

According to Einstein we should not prejudice in advance the transformation equations which exist between the coordinates of the system *A* and those of the system *B*. Assuming that the motion of *B* relative to *A* occurs in the *X*-direction, we put

$$x = \alpha x' + \beta t',$$
$$t = \gamma x' + \delta t',$$
(92.1)

where α, β, γ, δ are constants, since we do not want to lose the *linearity* of the transformation. But by abandoning the equation $t = t'$ (which corresponds to the choice $\gamma = 0$, $\delta = 1$) we have gained the freedom of two additional constants.

All formulae of special relativity gain greatly in simplicity, if we agree to measure the time in such units that the velocity

[1]Ann. der Physik *17*, 891 (1905).

of light becomes unity (this means that the usual time t is replaced by t/c). If we want to return to the formulae involving our ordinary time t, we have merely to change all t to ct and all v to v/c. We shall frequently make use of this simplification and thus agree that the invariant velocity of light has the numerical value $c = 1$.

We consider (x, t) as the coordinates of the system associated with the observer A, and (x', t') as the coordinates associated with the observer B. The points of the system B, if measured by A, move with the velocity v (<1). This means, considering x' as fixed, that

$$dx = \beta dt',$$
$$dt = \delta dt',$$

and thus

$$\frac{dx}{dt} = v = \frac{\beta}{\delta}. \tag{92.2}$$

Also, from the second postulate, if $x = t$, then $x' = t'$, because if A finds that a signal is propagated relative to him with the velocity of light (unit velocity), B must find the same relative to his own system. This yields the relation

$$\alpha + \beta = \gamma + \delta. \tag{92.3}$$

But light can also be sent in the opposite direction: $x = -t$ which must correspond to $x' = -t'$. This yields the new relation

$$\alpha - \beta = \delta - \gamma. \tag{92.4}$$

As a result of these relations we can now put

$$x = k(x' + vt'),$$
$$t = k(vx' + t'), \tag{92.5}$$

with the constant k still undetermined.

We may, however, invert this transformation, expressing the coordinates (x', t') in terms of (x, t):

$$x' = \frac{1}{k} \frac{x - vt}{1 - v^2},$$
$$t' = \frac{1}{k} \frac{-vx + t}{1 - v^2}. \tag{92.6}$$

Now we make use of the fact that the systems A and B must show no preference with respect to each other. The change of v to $-v$ is natural because, if the system B moves relative to A with the velocity v, system A moves relative to B with the velocity $-v$. But no other change is permitted if the two systems are to be entirely equivalent. This gives

$$\frac{1}{k(1-v^2)} = k, \tag{92.7}$$

which determines the constant k as

$$k = \frac{1}{\sqrt{1-v^2}}. \tag{92.8}$$

We have thus obtained the relation between the coordinates of the two systems:

$$x = \frac{x' + vt'}{\sqrt{1-v^2}}, \qquad x' = \frac{x - vt}{\sqrt{1-v^2}},$$
$$t = \frac{vx' + t'}{\sqrt{1-v^2}}, \qquad t' = \frac{-vx + t}{\sqrt{1-v^2}}. \tag{92.9}$$

To this we shall add

$$y = y', \qquad z = z', \tag{92.10}$$

since we have no reason to assume that the motion in the X-direction has any influence on the y and z coordinates.

The addition of velocities. Let us consider a signal travelling with the velocity u in the system B, that is

$$x' = ut'. \tag{92.11}$$

Then

$$x = \frac{v + u}{1 + vu} t, \tag{92.12}$$

and thus the velocity of the signal, measured by the observer A, becomes

$$w = \frac{v + u}{1 + vu}, \tag{92.13}$$

at variance with the Newtonian equation $w = v + u$. This is Einstein's celebrated *addition theorem of velocities.*

If the signal in B moves perpendicularly to the X'-axis, we have

$$x' = 0,$$
$$y' = ut', \tag{92.14}$$

and thus

$$x = vt,$$
$$y = \sqrt{1 - v^2}\, ut. \tag{92.15}$$

This does not mean that the addition of vectors according to the usual parallelogram construction is no longer valid. This construction is put out of action only if we are interested in the resultant of two vectors which have been *measured in two different reference systems*.

In Fizeau's experiment light travels in water with the velocity

$$u = \frac{1}{n} \tag{92.16}$$

(n = refractive index), while the water has the velocity v. The resultant velocity of light, measured by A, becomes

$$w = \frac{v + 1/n}{1 + v/n}. \tag{92.17}$$

In view of the smallness of v we can put $(1 + v/n)^{-1} = 1 - v/n$, and if quantities that are higher than first order in v are neglected, we obtain

$$w = \left(1 - \frac{v}{n}\right)\left(v + \frac{1}{n}\right) = \frac{1}{n} + v\left(1 - \frac{1}{n^2}\right). \tag{92.18}$$

This result is in full harmony with the experimental facts and agrees with Fresnel's "convection coefficient."

The aberration of light. Let us consider light that is emitted in system B in a direction which forms the angle θ' with the X-axis. Hence

$$x' = t' \cos \theta',$$
$$y' = t' \sin \theta'. \tag{92.19}$$

Substituting in (92.9), we obtain

$$x = t' \frac{v + \cos \theta'}{\sqrt{1 - v^2}},$$

$$y = t' \sin \theta',$$

(92.20)

and thus

$$\frac{y}{x} = \tan \theta = \frac{\sin \theta'}{v + \cos \theta'} \sqrt{1 - v^2}.$$

(92.21)

The last factor is very near to 1 under astronomical conditions; hence we can then put

$$\tan \theta = \frac{\sin \theta'}{v + \cos \theta'}.$$

(92.22)

Therefore the direction in which a star appears in consequence of the motion of the earth is practically the same as that derived from the elementary vector addition of $\mathbf{v}$ and $\mathbf{c}$ (although this is not true of the length of the resulting vector, which remains c instead of $|\mathbf{v} + \mathbf{c}|$).

The Doppler Effect. We consider a plane wave, once more propagating in system B at the angle θ' to the X-axis. We represent it by the wave function

$$C \, e^{2\pi i \nu' (t - x' \cos \theta' - y' \sin \theta')}.$$

(92.23)

Making use of the transformation formulae (92.9), we express the same wave, now measured in system A, obtaining:

$$C \, e^{2\pi i \nu (t - x \cos \theta - y \sin \theta)},$$

(92.24)

where

$$\nu = \frac{\nu'}{\sqrt{1 - v^2}} (1 + v \cos \theta'),$$

(92.25)

$$\sin \theta = \frac{\sin \theta'}{1 + v \cos \theta'} \sqrt{1 - v^2},$$

$$\cos \theta = \frac{v + \cos \theta'}{1 + v \cos \theta'}.$$

(92.26)

Hence

$$\cos \theta' = \frac{-v + \cos \theta}{1 - v \cos \theta},$$

(92.27)

and substituting in (92.25), we obtain

$$\nu = \nu' \frac{\sqrt{1 - v^2}}{1 - v \cos \theta},\tag{92.28}$$

which up to quantities of first order agrees with the usual expression

$$\nu = \nu'(1 + v \cos \theta).\tag{92.29}$$

But the relativistic formula (92.28) shows that even a light ray coming from a light source passing perpendicularly ($\cos \theta = 0$) suffers a slight red-shift. This so-called "transversal Doppler effect" has been experimentally investigated and its agreement with the theoretically predicted value demonstrated.[1]

The clock paradox. One of the "paradoxes" of relativity, which released a flood of literature in half-learned magazines (from the pens of less than half-learned authors), is the so-called "clock paradox," which was first posed around 1918 and completely analysed and elucidated by Einstein.[2] In view of the transformation equations (92.9), if an observer in system B compares the readings of his clock with those of system A, he finds that the clocks of system A move faster (that this is not caused by any real change in the rates of the clocks is demonstrated by the fact that an observer in A would find exactly the same, if he were to compare the readings of his clock with those of system B). If the relative velocity v is almost the velocity of light, his clock may register a time interval of let us say 1 second, while the clocks he encounters of system A may register 1 year. This may also be expressed in another form. Suppose that a man is being shot out of a cannon with almost the velocity of light and travels to the star Sirius, and is then shot back to earth again with almost the velocity of light. He will arrive back at his starting point after, let us say, 16 seconds according to his own time—that is without aging at all—whereas the inhabitants of the earth will have aged by 16 years. Although this result appears highly paradoxical according to our "common-sense" notions—which are based on the wrong assumption of

[1]H. E. Ives and G. R. Stilwell, J. Opt. Soc. Amer. *28*, 215 (1938) and *31*, 369 (1941).
[2]Cf. Die Naturwissenschaften *6*, 697 (1918).

an absolute time—there is in fact nothing self-contradictory in it. The traveller to Sirius and back passed through entirely different parts of the space-time world than the inhabitants of the earth and thus there is no reason why they should age by the same amount. The alleged "paradox" is evident in the following kinematic formulation of the experiment. A says: "I see B moving to the right with the velocity v and returning with the same velocity to the starting point." The observation of B relative to A is exactly the same, except that "right" is changed to "left." Why then should there be a dissymmetry in the aging of A and B? This purely kinematic description of the events, in fact, omits essential elements of the situation and is thus physically incomplete. If the two participants A and B each carry an accelerometer, the accelerometer of A shows zero permanently, while the accelerometer of B gives a very strong reading at the moment of time when B arrives on Sirius and is kicked back again. The principle of symmetry is not violated because the physical states of A and B are completely different.

An experimental verification of the clock paradox became possible in experiments with very short-lived particles, the so-called mu-mesons, which are created by cosmic-ray showers at very high altitudes. The lifetime of these particles is known from laboratory measurements, and is such that these particles, which move with more than 99.5 per cent of the velocity of light, would not be able to penetrate more than 600 metres before they disintegrated, if it were not for the relativistic time dilatation. This dilatation makes it possible for them to penetrate 6000 metres and more, thus arriving at sea level. This fact could never have been explained without the transformation formulae of relativity. The effect on which the so-called "clock paradox" is based has thus been fully verified by experiments.[1]

3. Minkowski's four-dimensional world. The transformation formulae (92.9–10) have the property that they leave the expression

$$C(x^2 + y^2 + z^2 - t^2) = C(x'^2 + y'^2 + z'^2 - t'^2) \quad (93.1)$$

[1]Cf. W. F. G. Swann, Report on Progress in Physics *10*, 1 (1946), particularly p. 16.

invariant. The two reference systems to which they pertain are special cases of a much more general class of reference systems, which have equal validity for the description of natural events. Such reference systems allow a free translation of the origin (3 parameters), a free translation of the origin of time (1 parameter), a free rotation of the space axes (3 parameters), and a translatory motion with an arbitrary velocity v (<1) (3 parameters)—all in all a class of reference systems which allow the freedom of 10 parameters. This entire family of legitimate reference systems can be characterized by one single statement, namely that their coordinates are such that they leave the following algebraic quantity invariant:

$$C[(x - \bar{x})^2 + (y - \bar{y})^2 + (z - \bar{z})^2 - (t - \bar{t})^2] \quad (93.2)$$

(the last term should be multiplied by c if we are dealing with time units that have not been normalized). We have to add the condition that the relation between the two coordinate systems be *linear*. In Newtonian physics the equivalence of reference systems was restricted to the invariance of

$$s^2 = (x - \bar{x})^2 + (y - \bar{y})^2 + (z - \bar{z})^2, \quad (93.3)$$

which is the square of the distance between the two points (x, y, z) and $(\bar{x}, \bar{y}, \bar{z})$. The equivalence of reference systems came about from the fact that the space of physics was considered to be a Euclidean space of three dimensions, in which all points are equivalent (freedom of translation), and all directions are equivalent (freedom of rotation).

The new realization that the equivalence of the reference systems demands the invariance of the quadratic form

$$s^2 = C[(x - \bar{x})^2 + (y - \bar{y})^2 + (z - \bar{z})^2 - c^2(t - \bar{t})^2] \quad (93.4)$$

demonstrates that the world of physics demands a metrical geometry in which the distance between two points is expressed by (93.4), instead of (93.3). This means that the time t joins the space coordinates x, y, z, to form a Euclidean space of *four dimensions*. This space is not strictly Euclidean, on account of the minus sign of the last term. We can make it *formally* Euclidean by introducing a set of four mathematical coordinates, defined as follows:

$$x_1 = ix, \quad x_2 = iy, \quad x_3 = iz, \quad x_4 = ct, \tag{93.5}$$

in which case the expression for the distance between the two points x_1, x_2, x_3, x_4 and $\bar{x}_1$, $\bar{x}_2$, $\bar{x}_3$, $\bar{x}_4$ becomes

$$s^2 = -C[(x_1 - \bar{x}_1)^2 + (x_2 - \bar{x}_2)^2 + (x_3 - \bar{x}_3)^2 + (x_4 - \bar{x}_4)^2]. \tag{93.6}$$

Newton's "absolute space" and "absolute time" cease to exist as independent categories but become amalgamated into a higher unity: the "absolute space-time world" of relativity.[1] In this interpretation all the relativistic effects of the previous section have to be considered as *geometrical* effects, caused by the fact that the phenomena of nature are put into a metrical space of four dimensions of a quasi-Euclidean structure.

The constant C can be dropped since we take it for granted that the distance between two points will depend on the units in which distances are measured. But it is not immaterial whether C is normalized to $+1$ or -1. The choice $+1$ is dictated if we want the distance between two physically connectable points to come out as a *real* number. Then we put

$$s^2 = c^2(t - \bar{t})^2 - (x - \bar{x})^2 - (y - \bar{y})^2 - (z - \bar{z})^2, \tag{93.7}$$

or in differential form:

$$ds^2 = c^2dt^2 - dx^2 - dy^2 - dz^2. \tag{93.8}$$

The indefinite signature $(+, -, -, -)$ of this quadratic form leads to fundamental deviations from the properties of a truly Euclidean space of four dimensions, whose signature is $(+, +, +, +)$. "Zero distance" between two points, in Euclidean terms, means that the two points collapse into one. "Zero distance" in the Minkowskian world means that

$$c^2(t - \bar{t})^2 - (x - \bar{x})^2 - (y - \bar{y})^2 - (z - \bar{z})^2 = 0, \tag{93.9}$$

which represents the geometrical locus of all the points of a sphere which, starting at a certain time moment $\bar{t}$, expands with the velocity of light from the centre $(\bar{x}, \bar{y}, \bar{z})$. To this have to be

[1]H. Minkowski, Phys. Zeitschr. *10*, 104 (1909).

added the points of a sphere which in the past *contracted* towards the centre $(\bar{x}, \bar{y}, \bar{z})$, reaching the radius zero at the time moment $\bar{t}$. The invariance of the distance with respect to rectangular coordinate transformations has the immediate consequence that if $s = 0$ in *one* legitimate coordinate system, it remains zero in *all* legitimate coordinate systems. This means that the propagation of a light wave is an *absolute phenomenon* which remains the same in all legitimate reference systems. Einstein's postulate of the invariance of the velocity of light is thus absorbed in the more comprehensive principle of the invariance of the line-element (93.8).

The impossibility of any material velocity of more than the velocity of light is now a natural geometrical consequence of the fact that such a velocity belongs to a distance which becomes *imaginary*, because its square becomes negative. An arbitrary point $(\bar{x}, \bar{y}, \bar{z}, \bar{t})$ of the physical world can only be in physical communication with such points (x, y, z, t) for which

$$c^2(t - \bar{t})^2 - (x - \bar{x})^2 - (y - \bar{y})^2 - (z - \bar{z})^2 \geqslant 0. \quad (93.10)$$

The equation

$$c^2(t - \bar{t})^2 - (x - \bar{x})^2 - (y - \bar{y})^2 - (z - \bar{z})^2 = 0 \quad (93.11)$$

is called the "light cone" associated with the point $(\bar{x}, \bar{y}, \bar{z}, \bar{t})$. The points for which the "greater than zero" sign of the inequality (93.10) holds are referred·to as "lying inside the light cone" with the centre $(\bar{x}, \bar{y}, \bar{z}, \bar{t})$.

4. The Lorentz transformations. The transformation (92.9) is a special case of a wider group of transformations which has received (with little historical justification) the name "Lorentz transformations." They are characterized by the general four-dimensional rotations of a Euclidean space of four dimensions, whose coordinates are $x_1 = ix$, $x_2 = iy$, $x_3 = iz$, and $x_4 = ct$. But we can avoid the use of imaginary quantities if we define the Lorentz transformations by the group of linear transformations of the four real variables (x, y, z, t) which leave the quadratic form

$$c^2t^2 - x^2 - y^2 - z^2 \quad (94.1)$$

invariant.

Although at the time that Hamilton invented his Quaternion Calculus, the idea of a four-dimensional universe was not yet known, it so happens that his quaternions are exceptionally well adapted to the study of the general nature of an arbitrary Lorentz transformation. Hamilton's quaternion is defined as a hypercomplex number of the form

$$Q = q_1\mathbf{i} + q_2\mathbf{j} + q_3\mathbf{k} + q_4 \qquad (94.2)$$

in which the fourth component (the "time-part" of the quaternion) behaves algebraically like a pure number (scalar), and the hypercomplex units $\mathbf{i}$, $\mathbf{j}$, $\mathbf{k}$ satisfy the following operational rules:

$$\begin{aligned}
\mathbf{ij} &= -\mathbf{ji} = \mathbf{k}, \\
\mathbf{jk} &= -\mathbf{kj} = \mathbf{i}, \\
\mathbf{ki} &= -\mathbf{ik} = \mathbf{j}, \\
\mathbf{i}^2 &= \mathbf{j}^2 = \mathbf{k}^2 = -1.
\end{aligned} \qquad (94.3)$$

Multiplication is generally not commutative ($AB \neq BA$), but it is associative. With the exception of the commutative rule of multiplication, all the other rules of ordinary algebra are preserved, including the non-factorability of zero. We shall define the "conjugate" of a quaternion by putting

$$\tilde{Q} = -q_1\mathbf{i} - q_2\mathbf{j} - q_3\mathbf{k} + q_4. \qquad (94.4)$$

Then $Q + \tilde{Q}$ is a mere scalar and so is

$$|Q|^2 = Q\tilde{Q} = \tilde{Q}Q = q_1{}^2 + q_2{}^2 + q_3{}^2 + q_4{}^2, \qquad (94.5)$$

which commutes with any quaternion. The following operational rule holds:

$$(AB)^{\tilde{}} = \tilde{B}\tilde{A}. \qquad (94.6)$$

Hence

$$|AB|^2 = (AB)(AB)^{\tilde{}} = AB\tilde{B}\tilde{A} = (A\tilde{A})(B\tilde{B}) = |A|^2|B|^2. \qquad (94.7)$$

A quaternion can be interpreted as a vector (4-vector) of a space of four dimensions. Certain properties of the electromagnetic field (for example, the Maxwell equations) can be adequately expressed in the language of quaternions.[1] But the study of the Lorentz transformations becomes particularly facilitated by the utilization of quaternion calculus.

[1]Cf. L. Silberstein, *Theory of Relativity* (Macmillan, 1924), p. 46.

Let us put

$$R = x_1\mathbf{i} + x_2\mathbf{j} + x_3\mathbf{k} + x_4 \qquad (94.8)$$

and choose

$$A = a_1\mathbf{i} + a_2\mathbf{j} + a_3\mathbf{k} + a_4 \qquad (94.9)$$

subject to the condition

$$A\tilde{A} = a_1{}^2 + a_2{}^2 + a_3{}^2 + a_4{}^2 = 1. \qquad (94.10)$$

Then the equation

$$R' = AR \qquad (94.11)$$

represents a linear transformation of the coordinates which, according to (94.7), satisfies the condition

$$R'\tilde{R}' = R\tilde{R}. \qquad (94.12)$$

This means that *the length of the vector R has not changed.* Hence the transformation (94.11) represents a *rotation* of a four-dimensional Euclidean space. The same is true of the transformation

$$R' = RB \qquad (94.13)$$

if the quaternion B satisfies the condition

$$B\tilde{B} = 1. \qquad (94.14)$$

In view of the group property of rotations, the quaternion product

$$R' = ARB \qquad (94.15)$$

is once more a rotation in four dimensions. The two quaternions of unit length A and B allow 6 degrees of freedom, in harmony with the 6 degrees of freedom of an arbitrary rotation (without reflection) in four dimensions. Hence (94.15) can be considered as the most general proper rotation of a space in four dimensions.

We shall now assume that the quaternions A and B have generally *complex* coefficients. We shall denote by an asterisk the "complex conjugate" of a complex number. The vector R has a real time part and an imaginary space part in a Minkowskian space. This is uniquely characterized by the property

$$\tilde{R} = R^*. \qquad (94.16)$$

In order to obtain a rotation of four-space which has physical significance (that is which transforms the real form (94.1) into itself), it is necessary and sufficient that the condition (94.16) shall remain preserved in the new reference system:

$$\tilde{R}' = R'^*. \tag{94.17}$$

This means

$$\tilde{B}\tilde{R}\tilde{A} = A^*R^*B^*$$

or

$$\tilde{B}R^*\tilde{A} = A^*R^*B^*. \tag{94.18}$$

This condition is satisfied if we choose

$$\tilde{B} = A^*. \tag{94.19}$$

Then also

$$\tilde{A} = B^*. \tag{94.20}$$

Hence a general (proper) Lorentz transformation can be characterized by the quaternion product

$$R' = AR\tilde{A}^*, \tag{94.21}$$

where the eight components of the complex quaternion

$$A = A' + iA''$$

satisfy the two scalar conditions

$$A'\tilde{A}' - A''\tilde{A}'' = 1,$$
$$A'\tilde{A}'' + A''\tilde{A}' = 0.$$

If A is real, then $A^* = A$ and

$$A\tilde{A}^* = A\tilde{A} = 1.$$

Hence $R = x_4$ is transformed into

$$R' = x_4,$$

which means that the *time axis remains preserved*. The Lorentz transformation is then reduced to a mere *rotation* in ordinary space. This rational representation of an arbitrary rotation of three-space with the help of a real quaternion of length 1 was known to and utilized by Euler.

In an ordinary three-dimensional rotation we are particularly interested in the *axis* of rotation, i.e. a straight line which is transformed into itself. More generally we have to ask for the "principal axes" of our transformation, characterized by the condition

$$R' = \lambda R, \qquad (94.22)$$

where λ is a mere (not necessarily real) number, called the "principal value" of the matrix associated with the linear transformation (94.21). Since our transformation has the property that for *all* positions of the radius vector

$$|R'|^2 = |R|^2, \qquad (94.23)$$

the condition (94.22) for the principal axes yields

$$(\lambda^2 - 1)|R|^2 = 0, \qquad (94.24)$$

which shows that either

$$\lambda = \pm 1 \qquad (94.25)$$

or

$$|R|^2 = 0. \qquad (94.26)$$

The condition (94.26) means that the corresponding principal axis lies on the null cone

$$x^2 + y^2 + z^2 - c^2t^2 = 0 \qquad (94.27)$$

and thus the principal axes of a Lorentz transformation can belong to real space in spite of the fact that λ is different from ± 1, which could not happen in a Euclidean space. Indeed, an even-dimensional Euclidean space might not contain *any* straight line which has been transformed into itself (at variance with an odd-dimensional space, where at least *one* real axis of rotation must exist). But a Lorentz transformation—although the λ-values are usually all different from 1—has always *two* principal axes which belong to the real null cone (94.27)[1] (in the limit these two axes may collapse into one), while the other two principal axes are likewise on the null cone (94.27) but belonging to complex values of the coordinates.

[1] For a detailed study of the null cone, cf. J. L. Synge, *Relativity, the Special Theory* (North-Holland, Amsterdam, 1956), pp. 89ff.

We shall generate an arbitrary Lorentz transformation in terms of the two real principal axes, which we shall briefly call the "null axes," since they lie on the real null cone. We shall denote them, considered as quaternions, by P and Q. Since they lie on the null cone, their length is zero:

$$P\tilde{P} = Q\tilde{Q} = 0, \tag{94.28}$$

while the reality of these axes demands the fulfilment of the condition (94.16). We shall put

$$\begin{aligned} P &= p_4[1 + i(p_1\mathbf{i} + p_2\mathbf{j} + p_3\mathbf{k})] = p_4(1 + i\mathbf{P}), \\ Q &= q_4[1 + i(q_1\mathbf{i} + q_2\mathbf{j} + q_3\mathbf{k})] = q_4(1 + i\mathbf{Q}), \end{aligned} \tag{94.29}$$

where $\mathbf{P}$ and $\mathbf{Q}$ can be conceived as ordinary 3-vectors, but also as quaternions whose scalar part is zero.

The condition (94.28) now yields

$$\mathbf{P}\tilde{\mathbf{P}} = p_1^2 + p_2^2 + p_3^2 = 1 \tag{94.30}$$

and likewise

$$\mathbf{Q}\tilde{\mathbf{Q}} = q_1^2 + q_2^2 + q_3^2 = 1.$$

We shall normalize the lengths of the two vectors $\mathbf{P}$ and $\mathbf{Q}$ by the conditions

$$p_4 = q_4 = p \tag{94.31}$$

and

$$\begin{aligned} P\tilde{Q} + Q\tilde{P} &= p_4(1 + i\mathbf{P})q_4(1 - i\mathbf{Q}) \\ &\quad + q_4(1 + i\mathbf{Q})p_4(1 - i\mathbf{P}) \\ &= 2p^2(1 - \cos\theta) = 1, \end{aligned} \tag{94.32}$$

where θ denotes the angle between $\mathbf{P}$ and $\mathbf{Q}$ (we exclude for the time being the possibility that the two vectors collapse into one).

We shall now generate the quaternion A of the transformation (94.21) as follows:

$$\begin{aligned} A &= \alpha P\tilde{Q} + \frac{1}{\alpha}Q\tilde{P}, \\ \tilde{A}^* &= \alpha^*\tilde{Q}P + \frac{1}{\alpha^*}\tilde{P}Q. \end{aligned} \tag{94.33}$$

(In the second equation we made use of the condition (94.16).)

First of all we shall show that the condition (94.10) is satisfied, no matter what (generally complex) values we may assign to α:

$$A\tilde{A} = \left(\alpha P\tilde{Q} + \frac{1}{\alpha}Q\tilde{P}\right)\left(\alpha Q\tilde{P} + \frac{1}{\alpha}P\tilde{Q}\right)$$

$$= (P\tilde{Q})^2 + (Q\tilde{P})^2$$

$$= (P\tilde{Q} + Q\tilde{P})P\tilde{Q} + (Q\tilde{P} + P\tilde{Q})Q\tilde{P}$$

$$= (P\tilde{Q} + Q\tilde{P})^2 = 1. \tag{94.34}$$

The vectors **P** and **Q** (in view of the normalization of their lengths according to (94.30–32)) represent four degrees of freedom, added to which are the two degrees of freedom of the complex constant α.

We shall demonstrate (by a method of complementation similar to that used in the proof of (94.34)) that P and Q are indeed the principal axes of our transformation:

$$P' = \left(\alpha P\tilde{Q} + \frac{1}{\alpha}Q\tilde{P}\right)P\left(\alpha^*\tilde{Q}P + \frac{1}{\alpha^*}\tilde{P}Q\right)$$

$$= \alpha\alpha^* P\tilde{Q}P\tilde{Q}P = \alpha\alpha^*(P\tilde{Q} + Q\tilde{P})P(\tilde{Q}P + \tilde{P}Q)$$

$$= \alpha\alpha^* P. \tag{94.35}$$

Hence

$$P' = \alpha\alpha^* P, \qquad \lambda_1 = \alpha\alpha^*, \tag{94.36}$$

and by exactly the same method

$$Q' = \frac{1}{\alpha\alpha^*}Q, \qquad \lambda_2 = \frac{1}{\alpha\alpha^*}. \tag{94.37}$$

We can equally show the principal axis character of the two quaternions PQ and QP:

$$\left(\alpha P\tilde{Q} + \frac{1}{\alpha}Q\tilde{P}\right)PQ\left(\alpha^*\tilde{Q}P + \frac{1}{\alpha^*}\tilde{P}Q\right)$$

$$= \frac{\alpha}{\alpha^*}P\tilde{Q}PQ\tilde{P}Q$$

$$= \frac{\alpha}{\alpha^*}(P\tilde{Q} + Q\tilde{P})PQ(\tilde{P}Q + \tilde{Q}P) = \frac{\alpha}{\alpha^*}PQ. \tag{94.38}$$

Hence

$$(PQ)' = \frac{\alpha}{\alpha^*} (PQ), \qquad \lambda_3 = \frac{\alpha}{\alpha^*}, \qquad (94.39)$$

$$(QP)' = \frac{\alpha^*}{\alpha} (QP), \qquad \lambda_4 = \frac{\alpha^*}{\alpha}. \qquad (94.40)$$

We have thus obtained the four principal axes of an arbitrary Lorentz transformation. The first pair (96.36–37) belong to real eigenvalues (which are reciprocal to each other); they lie on the real null cone. The second pair (94.39–40) lie on the complex null cone. Their eigenvalues are conjugate complex, and once more they are reciprocal to each other.

Limiting cases. A general Lorentz transformation has four distinct principal values. Hence it cannot happen that the transformation leaves more than four straight lines unchanged. Yet we have seen in dealing with the special Lorentz transformation (92.9–10) that an *entire plane* may remain untouched by the transformation. This can only happen if two eigenvalues collapse into one. Our above classification shows that this may happen in one of two ways:

(a) If α is chosen to be *real*. Then

$$\lambda_3 = \lambda_4 = 1. \qquad (94.41)$$

(b) If α is chosen to be complex but of *absolute value one*. Then

$$\lambda_1 = \lambda_2 = 1. \qquad (94.42)$$

(Both conditions together can only hold for $\alpha = 1$, but then we get the identity transformation $R' = R$ and every point remains unaltered. Apart from this extreme case it can never happen that more than a plane remains untouched by the transformation.)

1. Let us first deal with case (a), which represents a 5-parameter family of transformations since α was restricted by one condition only. This class of transformations is distinguished by the property that *all four eigenvalues are real* and thus all the principal axes must likewise be real. This is not revealed by the two expressions (94.39) and (94.40) since both PQ and QP are

complex quantities. But in view of the collapse of the two eigen-values λ_3 and λ_4 we can now take any linear combination of these two axes and we see from the property (94.16) that the following two 4-vectors must be real:

$$C_1 = PQ + QP, \qquad C_2 = i(PQ - QP). \qquad (94.43)$$

Let us obtain the scalar (94.5) for these two vectors:

$$(PQ \pm QP)(\tilde{Q}\tilde{P} \pm \tilde{P}\tilde{Q}) = \pm(PQ\tilde{P}\tilde{Q} + QP\tilde{Q}\tilde{P}).$$

Considering the factor i in C_2 we see that $|C_1|^2 = |C_2|^2$ and if we express P and Q in the form (94.29), we obtain

$$|C_1|^2 = |C_2|^2 = -2p^2 \sin^2 \theta = -(1 + \cos \theta). \qquad (94.44)$$

Hence both principal axes lie *outside* the null cone and thus outside the physical space which is transformed by the Lorentz transformation. The same is true of the invariant plane which passes through these two 4-vectors. The Y–Z plane of the special transformation (92.9–10) lies outside the null cone and thus belongs to this class of transformations. In particular we find in this case:

$$\cos \theta = -1, \qquad p = \tfrac{1}{2}, \qquad \alpha = \sqrt[4]{\frac{1-v}{1+v}},$$
$$P = \tfrac{1}{2}(1 + i\mathbf{i}), \qquad Q = \tfrac{1}{2}(1 - i\mathbf{i}),$$
$$A = \frac{1}{2}\frac{\sqrt{1-v} + \sqrt{1+v}}{\sqrt[4]{1-v^2}} + \frac{1}{2}\frac{\sqrt{1-v} - \sqrt{1+v}}{\sqrt[4]{1-v^2}}i\mathbf{i}.$$

2. We now come to case (*b*), which is once more a 5-parameter class of Lorentz transformations, since α is now restricted by the condition that it must be of the form $e^{i\gamma}$. The invariant plane is now characterized by a linear combination of the two principal axes P and Q:

$$C = aP + bQ, \qquad (94.45)$$

and we obtain

$$|C|^2 = (aP + bQ)(a\tilde{P} + b\tilde{Q}) = ab(P\tilde{Q} + Q\tilde{P}) = ab. \qquad (94.46)$$

Those points of the invariant plane for which the factors a and b have the same sign lie inside the null cone (real space) and those for which their sign is opposite, outside the null cone (imaginary space). The two cases (a) and (b) totally exclude each other.

3. There is a point of tangency between these two classes of transformations, which yields a 4-parameter class of Lorentz transformations,[1] obtainable by letting the two vectors P and Q infinitely approach each other and at the same time go with α to the limit 1. Let us put

$$\alpha = 1 + \epsilon, \tag{94.47}$$

considering ϵ as an infinitesimal parameter. Then the quaternion (94.33) becomes

$$A = 1 + \epsilon(P\tilde{Q} - Q\tilde{P}). \tag{94.48}$$

If we put

$$\begin{aligned} P &= p(1 + i\mathbf{P}_0 + i\mu\mathbf{P}_1), \\ Q &= p(1 + i\mathbf{P}_0 - i\mu\mathbf{P}_1), \end{aligned} \quad |\mathbf{P}_0| = |\mathbf{P}_1| = 1, \quad \mathbf{P}_1 \perp \mathbf{P}_0 \tag{94.49}$$

we obtain (neglecting quantities of second order in μ):

$$P\tilde{Q} - Q\tilde{P} = 4p^2 i\mu(\mathbf{P}_1 + i\mathbf{P}_0\mathbf{P}_1), \tag{94.50}$$

$$\cos\theta = 1 - \mu^2. \tag{94.51}$$

Considering the condition (94.32), we now have

$$2p^2 = \frac{1}{\mu^2} \tag{94.52}$$

and thus

$$\epsilon(P\tilde{Q} - Q\tilde{P}) = \frac{2\epsilon}{\mu}i(\mathbf{P}_1 + i\mathbf{P}_0\mathbf{P}_1). \tag{94.53}$$

The ratio $2\epsilon/\mu$ of the two infinitesimal parameters may approach any finite value β. Hence we can express the result of our calculations in the following form. The complex quaternion A, which according to (94.21) generates a Lorentz transformation, becomes:

$$A = 1 + i\beta(\mathbf{P}_1 + i\mathbf{P}_2), \tag{94.54}$$

[1]The author is indebted to Professor J. L. Synge for introducing him to this class of transformations.

where
$$|\mathbf{P}_1| = |\mathbf{P}_2| = 1, \qquad \mathbf{P}_1 \perp \mathbf{P}_2.$$
The principal axes of this transformation become $\mathbf{P}_2$ and $1 + i\mathbf{P}_0$ where
$$\mathbf{P}_0 = \mathbf{P}_1 \times \mathbf{P}_2. \qquad (94.55)$$
Although the real eigenvalue $\lambda = 1$ is four-fold, we obtain only *two* principal axes because the two null axes P and Q collapse into one; moreover, the first axis C_1 of the pair (94.43) likewise collapses into P (since $P^2 = 2pP$). Hence the null axis becomes a *three-fold axis*, whereas C_2 approaches in the limit the 3-vector $\mathbf{P}_2$.

The invariant plane, determined by the two vectors $\mathbf{P}_2$ and $1 + i\mathbf{P}_0$, lies completely outside the null cone but touches the null cone in the line $a(1 + i\mathbf{P}_0)$.

As an example of such a transformation let us choose
$$\mathbf{P}_0 = \mathbf{i}, \qquad \mathbf{P}_1 = \mathbf{j}, \qquad \mathbf{P}_2 = \mathbf{k},$$
$$A = 1 + \beta i(\mathbf{j} + i\mathbf{k}).$$
Then
$$R' = (1 + \beta i\mathbf{j} - \beta\mathbf{k})R(1 + \beta i\mathbf{j} + \beta\mathbf{k}),$$
or written out in coordinates
$$
\begin{aligned}
x_1' &= (1 - 2\beta^2)x_1 + 2\beta x_2 && + 2i\beta^2 x_4, \\
x_2' &= - 2\beta x_1 + x_2 && + 2i\beta x_4, \\
x_3' &= x_3, \\
x_4' &= 2\beta^2 i x_1 - 2\beta i x_2 && + (1 + 2\beta^2)x_4.
\end{aligned}
$$
That $\mathbf{P}_2 = \mathbf{k}$ is a principal axis is seen at once. We can likewise demonstrate that $x_1 = i$, $x_2 = x_3 = 0$, $x_4 = i$ yields the second principal axis.

Infinitesimal Lorentz transformations. The choice $\alpha = 1$ yields the identity transformation. If we choose α very near to 1, we obtain an "infinitesimal transformation," corresponding to an infinitesimal rotation of the axes. An arbitrary finite rotation can be generated as a succession of infinitesimal rotations.

We have now once more the relation (94.48), and since $P\tilde{Q} - Q\tilde{P}$ is a mere space quaternion (whose scalar part is zero), we can put

$$A = 1 + \tfrac{1}{2}\epsilon i(\mathbf{b} + i\mathbf{a}), \tag{94.56}$$

where the two 3-vectors $\mathbf{a}$ and $\mathbf{b}$ are arbitrary and ϵ is a (real) infinitesimal parameter. Then

$$\begin{aligned} R' = AR\tilde{A}^* &= [1 + \tfrac{1}{2}\epsilon i(\mathbf{b} + i\mathbf{a})]R[1 + \tfrac{1}{2}\epsilon i(\mathbf{b} - i\mathbf{a})] \\ &= R - \tfrac{1}{2}\epsilon(\mathbf{a}R - R\mathbf{a}) + \tfrac{1}{2}\epsilon i(\mathbf{b}R + R\mathbf{b})]. \end{aligned} \tag{94.57}$$

This means, if written out in components:

$$\begin{aligned}
x_1' &= x_1 & &+ \epsilon a_3 x_2 &- \epsilon a_2 x_3 &+ i\epsilon b_1 x_4, \\
x_2' &= - \epsilon a_3 x_1 &+ x_2 & &+ \epsilon a_1 x_3 &+ i\epsilon b_2 x_4, \\
x_3' &= \epsilon a_2 x_1 & &- \epsilon a_2 x_2 &+ x_3 &+ i\epsilon b_3 x_4, \\
x_4' &= - i\epsilon b_1 x_1 &- i\epsilon b_2 x_2 &- i\epsilon b_3 x_3 &+ x_4.
\end{aligned} \tag{94.58}$$

5. Mechanics of a particle. The variational principles of mechanics permit us to derive the equations of motion of an arbitrary mechanical system, if one fundamental quantity, the "Lagrangian function" L, is given. In Newtonian mechanics space and time played separate roles and the time t served as the independent variable. In relativity this is no longer possible. Time is not more than one of the coordinates, on equal footing with the three space coordinates. The physical events take place in a four-dimensional world, which has a very definite metrical structure. This structure demands that we should have no privileged orientation in the space-time world. Equations which demand such a privileged orientation contradict the principle of relativity and have to be discarded, or corrected in such a way that the final system will reflect the proper metrical structure of the physical world.

The existence of a Lagrangian greatly facilitates this task. If the Lagrangian function L is a true scalar of the four-dimensional world (that is, a quantity which is invariant with respect to an arbitrary Lorentz transformation or, as we say briefly, "Lorentz invariant"), then also the equations derived from this Lagrangian are relativistically correct, by being likewise Lorentz invariant.

We have seen how the dynamical problems of Newtonian mechanics were solvable with the help of a Lagrangian function

$L = T - V$ (cf. 51.7) where T is the kinetic and V the potential energy of the system. Let us restrict ourselves to the consideration of a single particle, characterized by mass and position. The kinetic energy

$$T = \tfrac{1}{2}mv^2, \tag{95.1}$$

which was responsible for the inertia of the particle, could be admitted as a true scalar of the Newtonian world. It is no longer a scalar, however, in the united world of space-time. In the unified world of space and time the motion of a particle can be pictured as a line—called "world-line"—of the four-dimensional space, whose line-element is given by the expression

$$\overline{ds}^2 = -dx^2 - dy^2 - dz^2 + c^2 dt^2$$
$$= dx_1^2 + dx_2^2 + dx_3^2 + dx_4^2. \tag{95.2}$$

In this picture we see the *tangent* of the world-line as representing the velocity of the particle. This tangent is a vector of four components:

$$\frac{dx_1}{ds}, \quad \frac{dx_2}{ds}, \quad \frac{dx_3}{ds}, \quad \frac{dx_4}{ds}. \tag{95.3}$$

Its relation to our traditional concepts becomes evident if we realize the extremely high value of the velocity of light compared with the usual velocities of Newtonian mechanics. The line-element (95.2) may be written in the form

$$ds = \sqrt{c^2 dt^2 - dx^2 - dy^2 - dz^2}$$
$$= c\sqrt{1 - \left(\frac{v}{c}\right)^2}\, dt, \tag{95.4}$$

and we see that the replacement of ds by cdt involves a very small error only. In particular we see that to a very high degree of accuracy we can put

$$ds = c\left(1 - \frac{1}{2}\frac{v^2}{c^2}\right)dt \tag{95.5}$$

and thus

$$\int_{t_1}^{t_2} mc\, ds = \int_{t_1}^{t_2} mc^2\, dt - \int_{t_1}^{t_2} T\, dt. \tag{95.6}$$

The fact that the first term on the right side of (95.6) does not appear under ordinary circumstances is explainable (in spite of the tremendous constant c^2) by the extraordinarily high stability of the mass m. If m is a constant, the variation of the first term gives zero and it can be omitted.

We thus come to the conclusion that the kinetic part of the action integral:

$$\int_{t_1}^{t_2} T dt \tag{95.7}$$

must be replaced by

$$- c \int_{t_1}^{t_2} m ds \tag{95.8}$$

in order to make it relativistically invariant.

From this form of the inertial term in the action integral a number of important consequences follow. Let us assume that we designate the time t as the independent variable. Then the Lagrangian L of the action integral (95.8) becomes

$$L = - c^2 m \sqrt{1 - \frac{1}{c^2}(\dot{q}_1^2 + \dot{q}_2^2 + \dot{q}_3^2)} \tag{95.9}$$

if the coordinates x, y, z are considered as the dynamical variables q_i. Now we know from the general principles of the variational treatment of mechanics (cf. 53.4) that the "momenta" of a mechanical system can be defined as the partial derivatives of L with respect to the $\dot{q}_i$. This now yields:

$$p_i = \frac{m\dot{q}_i}{\sqrt{1 - \frac{v^2}{c^2}}} \tag{95.10}$$

or, in vector form,

$$\mathbf{p} = \frac{m\mathbf{v}}{\sqrt{1 - \frac{v^2}{c^2}}}. \tag{95.11}$$

Hence the correction compared with the Newtonian definition of the momentum is very slight for small velocities, but it becomes of decisive importance for velocities which approach the velocity of light.

Another fundamental quantity is the *energy* of the system. This can also be defined in terms of the Lagrangian function L, according to the definition (53.12):

$$E = \sum_{i=1}^{3} p_i \dot{q}_i - L. \tag{95.12}$$

In our case we obtain

$$E = \frac{mv^2}{\sqrt{1 - v^2/c^2}} + mc^2 \sqrt{1 - v^2/c^2}$$

$$= \frac{mc^2}{\sqrt{1 - v^2/c^2}}. \tag{95.13}$$

This celebrated result of Einstein represents one of the most important discoveries of theoretical physics. Whereas in Newtonian physics the kinetic energy of a particle appeared in the form $\frac{1}{2}mv^2$, which indicates that a mass achieves energy only by its motion, the new equation (95.13) puts the enormous quantity mc^2 in front of the Newtonian term, thus demonstrating that a mass, by its very existence, is the seat of a tremendous amount of energy, compared with which the ordinary kinetic energy is usually negligibly small. Considering the ease with which the various forms of energy may be transformed into each other, the possibility of converting the latent energy, associated with a given mass, into other energy forms appeared on the horizon. The successful construction of the atom bomb tragically confirmed this conclusion of relativity.

Since $c = 3 \times 10^{10}$ cm./sec., the conversion of one gram of matter into energy amounts to 9×10^{20} ergs = 9×10^{13} joules = 25 million kilowatt-hours = 21,500 million kilogram-calories.

A further fundamental conclusion can be drawn from the canonical integrand (cf. 64.3), which can be rewritten in mathematical coordinates:

$$p_1 dx + p_2 dy + p_3 dz - H dt$$

$$= \frac{p_1}{i} dx_1 + \frac{p_2}{i} dx_2 + \frac{p_3}{i} dx_3 - \frac{H}{c} dx_4$$

$$= \sum_{i=1}^{4} \bar{p}_i dx_i. \tag{95.14}$$

The four components $\bar{p}_i$ of the 4-momentum must have invariant significance by being the four components of a 4-vector. Indeed, we obtain from (95.11) and (95.13):

$$\bar{p}_1 = -mc\frac{dx_1}{ds}, \qquad \bar{p}_2 = -mc\frac{dx_2}{ds},$$

$$\bar{p}_3 = -mc\frac{dx_3}{ds}, \qquad \bar{p}_4 = -mc\frac{dx_4}{ds}, \qquad (95.15)$$

which establishes the 4-vector $\bar{p}_i$ as the four-dimensional *tangent vector* (95.3) to the world-line of the particle, multiplied by the constant $-mc$.

We see that the two Newtonian concepts *momentum* and *energy*, which in the Newtonian form of physics represent two separate entities, form relativistically an inseparable union; *the components of the momentum* (divided by i) *plus the negative energy* (divided by c) *form together the components of a 4-vector of the Minkowskian space-time world.*

In the early relativistic literature it was customary to write the equation (95.11) in the form

$$\mathbf{p} = \frac{m_0}{\sqrt{1 - v^2/c^2}}\,\mathbf{v} \qquad (95.16)$$

and to interpret

$$m = \frac{m_0}{\sqrt{1 - v^2/c^2}} \qquad (95.17)$$

as the "relativistic mass" of the particle, which changes with the velocity and increases to infinity as v approaches the velocity of light. This is useful for the demonstration that no finite force can ever accelerate a material body up to the velocity of light since the inertia of the body grows to infinity if v becomes c. But in actual fact the correction factor of the denominator belongs to dt, because we have to differentiate with respect to the arc-length s, which has an invariant significance, and not with respect to t, which is not more than one of the coordinates. Instead of the arc-length s we can introduce a more physical quantity by defining the "proper time" of a particle. This is the time measured by a clock which is carried along by the particle and which is thus attached to the world-line of the particle. If this proper time is denoted by σ, we have

$$ds = c d\sigma \tag{95.18}$$

since the clock is permanently in a frame of reference in which the particle is at rest. In terms of this time we have

$$\frac{c dx}{ds} = \frac{dx}{d\sigma} = \frac{dx}{\sqrt{1 - v^2/c^2}\, dt} \tag{95.19}$$

and if we agree that we define the velocity of a particle by the time rate of change of position, *the time being measured in proper time*, the definition (95.11) of the momentum becomes

$$\mathbf{p} = m\mathbf{v} \tag{95.20}$$

and Newton's definition of the momentum remains unchanged.

6. The Hamiltonian formulation of particle dynamics. In the previous section we have made use of the time t as the independent variable, which is not in the spirit of relativity since the time should be handled with the other coordinates on equal footing. This can be done by introducing a parameter τ into our discussion, considering the four mathematical coordinates $q_i = x_i$ ($i = 1, 2, 3, 4$) as functions of τ. We have followed this method in the Hamiltonian treatment of the equations of motion (cf. chap. VI, section 10).

In the Hamiltonian form of dynamics the fundamental function is no longer the Lagrangian L but the Hamiltonian function

$$H = \sum_{i=1}^{n} p_i \dot{q}_i - L \tag{96.1}$$

(cf. 62.3), expressed in terms of the q_i and p_i. Now in our present problem

$$L = -mc \frac{ds}{d\tau} = -mc \sqrt{\sum_{i=1}^{4} \dot{q}_i^{\,2}}. \tag{96.2}$$

Hence

$$p_i = \frac{\partial L}{\partial \dot{q}_i} = -mc \frac{\dot{q}_i}{\sqrt{\sum \dot{q}_i^{\,2}}} = -mc \frac{dq_i}{ds} \tag{96.3}$$

and thus

$$H = \sum_{i=1}^{4} p_i \dot{q}_i - L = -mc \sum_{i=1}^{4} \frac{dq_i^2}{ds \, d\tau} + mc \frac{ds}{d\tau} = 0. \quad (96.4)$$

Hence we are faced with the same situation that we have encountered before in (610.5). Although the canonical Lagrangian is

$$L = \sum_{i=1}^{4} p_i \dot{q}_i - H \quad (96.5)$$

and H vanishes in our case, we have no difficulty with the Hamiltonian treatment because the role of the Hamiltonian function is taken over by the *identity* which must exist between the p_i. Indeed, the expression (96.3) shows that we cannot express the $\dot{q}_i$ in terms of the p_i, on account of the identity

$$\sum_{i=1}^{4} p_i^2 - m^2 c^2 = 0, \quad (96.6)$$

which has to be considered as an *auxiliary condition* of our variational problem. We thus have the principle of minimizing the canonical integral

$$A = \int_{\tau_1}^{\tau_2} \sum_{i=1}^{4} p_i \dot{q}_i d\tau \quad (96.7)$$

under the auxiliary condition (96.6). This means that the resulting L becomes

$$L' = \sum_{i=1}^{4} p_i \dot{q}_i - \lambda \left(\sum_{i=1}^{4} p_i^2 - m^2 c^2 \right). \quad (96.8)$$

The appearance of the undetermined factor λ is in harmony with the freedom of the parameter τ. We can normalize λ to 1, which merely normalizes the parameter τ to a definite quantity. The essential difference compared with the customary form of dynamics is the fact that the summation extends over *four*, instead of three, coordinates.

7. The potential energy V. We return to the Lagrangian of Newtonian physics: $L = T - V$. We have seen before how to modify the kinetic energy T in order to harmonize it with the principle of relativity. Now we consider the variational integral associated with the potential energy V:

$$A = -\int_{t_1}^{t_2} V dt. \tag{97.1}$$

If we consider V as a given scalar of the four-dimensional world:

$$V = V(x_1, \ldots, x_4), \tag{97.2}$$

we have to modify (67.1) in order to obtain an integral which has four-dimensional significance. We can do this by replacing dt by the differential of the "proper time" $d\sigma$. We then obtain the corrected action integral in the form

$$-\int_{t_1}^{t_2} V \frac{ds}{c} \tag{97.3}$$

and the entire action integral, which combines kinetic and potential energy, becomes

$$A = -\int_{t_1}^{t_2} \left(mc + \frac{V}{c} \right) ds$$
$$= -c \int_{t_1}^{t_2} \left(m + \frac{V}{c^2} \right) ds. \tag{97.4}$$

We see that the presence of a scalar potential energy is equivalent to an *increase of the mass of the particle* by the amount of V/c^2. If we wind up a watch, thus increasing its latent potential energy by a certain amount of elastic energy, we change the mass of the watch. If a certain chemical compound changes its configuration to a more stable one of diminished potential energy —thus giving off the corresponding amount of chemical energy in the form of heat—the mass of the new compound has diminished. These differences are so small, however, that they remain unobservable, on account of the huge factor c^2 by which we have to divide.

The situation is different in nuclear disintegration or recombination processes where the change of mass can lead to spectacular events, on account of the huge number of atoms involved. The mass of the hydrogen nucleus (the proton) is 1.008146 (in the scale in which oxygen is 16), the mass of the neutron 1.00897, while the mass of the helium nucleus is 4.003879. Hydrogen can be transformed into helium (at extreme temperatures) in the presence of neutrons (and carbon), since the helium nucleus is formed of a combination of two protons and two neutrons. But the mass of the resulting helium particle is smaller by 0.03136, that is by about 1 per cent of the atomic mass. The corresponding energy is released in the form of heat and gives rise to the destructive effects of the hydrogen bomb.

8. Relativistic formulation of Newton's scalar theory of gravitation. Let us write down the Lagrangian equations which are associated with the action principle (67.4). We put all the four variables $q_i = x_i$ on equal footing by introducing a parameter τ as independent variable and considering the q_i as functions of τ. Then

$$-L = \left(m + \frac{V}{c^2}\right)\sqrt{\sum_{i=1}^{4} \dot{q}_i^2} \tag{98.1}$$

and the Lagrangian equations (52.1) give

$$\frac{d}{d\tau}\left[\left(m + \frac{V}{c^2}\right)\frac{dq_i}{ds}\right] = \frac{1}{c^2}\frac{\partial V}{\partial q_i}\sqrt{\sum_{i=1}^{4} \dot{q}_i^2}, \tag{98.2}$$

which may be put in the form

$$c\frac{d}{ds}\left[\left(m + \frac{V}{c^2}\right)c\frac{dq_i}{ds}\right] = \frac{\partial V}{\partial q_i}, \tag{98.3}$$

that is

$$\frac{d}{d\sigma}\left[\left(m + \frac{V}{c^2}\right)\frac{dq_i}{d\sigma}\right] = \frac{\partial V}{\partial q_i}. \tag{98.4}$$

Apart from the fact that the operation d/dt of Newtonian physics is replaced by the invariant operation $d/d\sigma$, we observe only the slight correction of m to $m + V/c^2$. It is true that in Newtonian physics the right side has a negative sign (since $L = T - V$). But this is explained by the fact that our coordinates q_1, q_2, q_3 are not the Newtonian x, y, z, but ix, iy, iz. If we introduce our customary coordinates, the right side becomes multiplied by -1.

In addition, we get a surplus equation if we identify the subscript i with 4. This equation expresses the *conservation law of energy*. Let us assume that V is independent of t (and thus also of x_4, i.e. the external field is static). Then the fourth equation gives

$$\frac{mc^2 + V}{\sqrt{1 - v^2/c^2}} = \text{const.} \tag{98.5}$$

which agrees—up to quantities which are of second order in v/c—with the law of conservation of energy in Newtonian mechanics:

$$\tfrac{1}{2}mv^2 + V = \text{const.} \tag{98.6}$$

The same problem may be treated by the Hamiltonian method. For this purpose we can utilize our earlier result, obtained in connection with Jacobi's principle (cf. pp. 188–191), which is formally equivalent with the action principle (67.4) if $\sqrt{[2(E - V)]}$ is replaced by $m + V/c^2$. Hence we obtain the canonical Lagrangian

$$L = \sum_{i=1}^{4} p_i \dot{q}_i - H, \tag{98.7}$$

where the differentiation is taken with respect to a parameter τ, while the Hamiltonian function H is given by

$$H = \frac{\sum\limits_{i=1}^{4} p_i^2}{(m + V/c^2)^2} \tag{98.8}$$

with the added condition that for the actual motion the energy constant in $H = \text{const.}$ is to be chosen as 1.

Newton explained the planetary orbits on the basis of a scalar field function, the "gravitational potential." In the early years of relativity Poincaré (1905) and later Minkowski (1908) attempted to modify the Newtonian theory, in order to harmonize it with the four-dimensional structure of the world. Hence they changed the Newtonian equations of motion to the system (68.4). Such attempts have been discarded with the advent of Einstein's Theory of General Relativity (1916), which demonstrates with overwhelming conclusiveness that the problem of gravity demands a much more radical revision of our traditional concepts (cf. section 11 of this chapter).

9. Motion of a charged particle. Let us assume that the scalar function V is replaced by a vectorial function A_i, called the "vector potential." In harmony with the demands of relativity we have to assume that this vector A_i has *four* components (the space part being purely imaginary, if mathematical coordinates are used), thus forming a 4-vector of the space-time world. We consider this vector as a field quantity, given as a function of the four coordinates $q_i = x_i$. If we form the scalar product of this vector with the 4-velocity, we obtain a true scalar of the Minkowskian space. Accordingly we replace the potential energy by the invariant

$$\frac{e}{c}\sum_{i=1}^{4} A_i \frac{dx_i}{ds}, \tag{99.1}$$

where the constant e is the electric charge of the particle. We now form the invariant action integral (omitting the minus sign, which is purely a matter of convention):

$$A = \int_{\tau_1}^{\tau_2}\left(mc + \frac{e}{c}\sum_{i=1}^{4} A_i \frac{dq_i}{ds}\right)ds. \tag{99.2}$$

Here again we want to keep the four coordinates q_i on equal footing and operate with a parameter τ as independent variable. Hence the Lagrangian function L becomes

$$L = mc\sqrt{\sum_{i=1}^{4} \dot{q}_i^{\,2}} + \frac{e}{c}\sum_{i=1}^{4} A_i\,\dot{q}_i. \tag{99.3}$$

We form the four components of the momentum:

$$p_i = \frac{\partial L}{\partial \dot{q}_i} = mc\frac{dq_i}{ds} + \frac{e}{c}A_i \tag{99.4}$$

and obtain the equations of motion in the following form:

$$\frac{d}{d\tau}\left[mc\frac{dq_i}{ds} + \frac{e}{c}A_i\right] = \frac{e}{c}\sum_{\alpha=1}^{4}\frac{\partial A_\alpha}{\partial q_i}\frac{dq_\alpha}{d\tau} \tag{99.5}$$

with the two constants e and m representing the charge and mass of the particle respectively. Multiplying on both sides of the equation by $d\tau/d\sigma$, we can equally write

$$m\frac{d^2 q_i}{d\sigma^2} + \frac{e}{c}\frac{dA_i}{d\sigma} = \frac{e}{c}\sum_{i=1}^{4}\frac{\partial A_\alpha}{\partial q_i}\frac{dq_\alpha}{d\sigma}. \tag{99.6}$$

The second term on the left can be transferred to the right side and combined with the term there, on the basis of the relation

$$\frac{dA_i}{d\sigma} = \sum_{\alpha=1}^{4}\frac{\partial A_i}{\partial q_\alpha}\frac{dq_\alpha}{d\sigma}, \tag{99.7}$$

to yield the expression

$$\frac{e}{c}\sum_{\alpha=1}^{4} F_{i\alpha}\frac{dq_\alpha}{d\sigma}, \tag{99.8}$$

where we have put

$$\frac{\partial A_k}{\partial x_i} - \frac{\partial A_i}{\partial x_k} = F_{ik}. \tag{99.9}$$

The six quantities $F_{ik} = -F_{ki}$ form the components of an antisymmetric "tensor."

We have to remember that the three components A_1, A_2, A_3 are purely imaginary, but so are the coordinates $q_i = x_i$ ($i = 1, 2, 3$). Accordingly all the F_{ik} ($i, k = 1, 2, 3$) are *real*, while the three components F_{i4} ($i = 1, 2, 3$) are purely *imaginary*. The correlation of F_{ik} with the traditional quantities of an electromagnetic field occurs as follows:

$$\begin{aligned} F_{14} = iE_1, \qquad F_{24} = iE_2, \qquad F_{34} = iE_3, \\ F_{23} = H_1, \qquad F_{31} = H_2, \qquad F_{12} = H_3, \end{aligned} \tag{99.10}$$

where the vectors $\mathbf{E}$ and $\mathbf{H}$ are the traditional electric and magnetic field strengths.

In order to interpret the equation (99.6) in traditional form, we shall return to the time t as independent variable. We can do this by multiplying both sides of the equation by $d\sigma/dt$, obtaining

$$m\frac{d}{dt}\left(\frac{1}{\sqrt{1 - v^2/c^2}}\frac{dx_i}{dt}\right) = \frac{e}{c}\sum_{\alpha=1}^{4} F_{i\alpha}\frac{dx_\alpha}{dt}. \tag{99.11}$$

The left side can be expressed in Newtonian terms. If we restrict ourselves to the first three equations ($i = 1, 2, 3$) and divide the entire equation by i, the left side becomes the "time rate of change of the momentum" of the particle, the "momentum" being defined according to (95.11). The right side becomes the "moving force," if we make use of Newton's equation of motion, which says "time rate of change of momentum equals moving force":

$$\frac{d\mathbf{p}}{dt} = \mathbf{F},\tag{99.12}$$

where we get for the 3-vector $\mathbf{F}$, written in the traditional symbols of ordinary vector analysis:

$$\mathbf{F} = e\left(\mathbf{E} + \frac{1}{c}\mathbf{v} \times \mathbf{H}\right).\tag{99.13}$$

This force is called the "Lorentz force," acting on a charged particle. The last equation ($i = 4$) expresses once more the law of conservation of energy in the form: "the time rate of change of the kinetic energy equals the work of the moving force" (the kinetic energy being defined in the relativistic form (95.13)).

In consistent relativistic language we should only operate with quantities which have four-dimensional significance. It is preferable then to multiply the equation (99.11) by dt/ds and write

$$mc\frac{d^2x_i}{ds^2} = \frac{e}{c}\sum_{\alpha=1}^{4} F_{i\alpha}\frac{dx_\alpha}{ds},\tag{99.14}$$

where the four components of the 4-velocity are related to each other by the condition

$$\sum_{i=1}^{4}\left(\frac{dx_i}{ds}\right)^2 = 1.\tag{99.15}$$

That this condition is consistent with the equations of motion (99.14) can be seen by multiplying these equations by dx_i/ds and forming the sum. The right side gives zero, because of the anti-symmetry of $F_{ik} = -F_{ki}$. On the left side we get

$$\frac{1}{2}mc\frac{d}{ds}\sum_{i=1}^{4}\left(\frac{dx_i}{ds}\right)^2 = 0,$$

which means that the quantity (99.15) is a constant of the motion.

The equation (99.14) allows an interesting geometrical interpretation. We have seen in chapter VII, section 8, that the motion of the phase fluid could be considered as a continual succession of *infinitesimal canonical transformations*. Let us now focus our attention on the velocity vector

$$\frac{dx_i}{ds} = u_i \qquad (99.16)$$

and write the equation (99.14) in the form

$$\lim_{\epsilon \to 0} \frac{u_i(s + \epsilon) - u_i(s)}{\epsilon} = \frac{e}{mc^2} \sum_{\alpha=1}^{4} F_{i\alpha} u_\alpha, \qquad (99.17)$$

$$u_i(s + \epsilon) = u_i(s) + \frac{e}{mc^2} \epsilon \sum_{\alpha=1}^{4} F_{i\alpha}(s) u_\alpha(s). \qquad (99.18)$$

If we write out these four equations in detail, we find that they are in complete agreement with the equations (94.58) which gave the equations of an *infinitesimal Lorentz transformation*, the electric vector **E** taking the place of **a** and the magnetic vector **H** the place of **b**. Hence the motion of the velocity vector of an electron in an external electromagnetic field can be conceived as a *continual sequence of infinitesimal Lorentz transformations*, the electromagnetic tensor F_{ik} providing the components of this transformation.

An interesting limiting case arises if the electron moves in the field of a *plane wave*. Here $E = H$ and $\mathbf{E} \perp \mathbf{H}$. We then have a physical realization of that particular four-parameter class of Lorentz transformations studied earlier (cf. 94.47–55), in which all four eigenvalues collapse into one and three of the principal axes collapse into an axis of the null cone. This three-fold axis is given by

$$1 + i \frac{\mathbf{E} \times \mathbf{H}}{|E|^2},$$

whereas the other axis (which lies outside the null cone) has the direction of the magnetic field strength **H**.

We shall now treat the same problem of motion in Hamiltonian terms. Since the Lagrangian (99.3) is once more a homogeneous form of the first order in the $\dot{q}_i$, Euler's theorem (610.4) on homogeneous forms of the first order will again give $H = 0$. But once more the p_i will satisfy an identity which, treated as an auxiliary condition, will take the place of the Hamiltonian function.

Indeed, the four equations (99.4) give at once

$$\sum_{i=1}^{4} \left(p_i - \frac{e}{c} A_i \right)^2 = m^2 c^2 \tag{99.19}$$

and we obtain, in complete analogy to (96.8),

$$L' = \sum_{i=1}^{4} p_i \dot{q}_i - \sum_{i=1}^{4} \left(p_i - \frac{e}{c} A_i \right)^2. \tag{99.20}$$

We have normalized λ to 1, which merely normalizes the variable τ, with respect to which the derivatives are taken. The last term is equivalent to a Hamiltonian function

$$H = \sum_{i=1}^{4} \left(p_i - \frac{e}{c} A_i \right)^2. \tag{99.21}$$

It seems surprising that in this treatment the mass m of the particle does not appear at all. Hamilton's canonical equations become

$$\dot{q}_i = 2 \left(p_i - \frac{e}{c} A_i \right),$$

$$\dot{p}_i = 2 \frac{e}{c} \sum_{\alpha=1}^{4} \left(p_\alpha - \frac{e}{c} A_\alpha \right) \frac{\partial A_\alpha}{\partial q_i}, \tag{99.22}$$

and we can verify that these equations are equivalent to (99.14) with $\frac{1}{2}$ taking the place of the factor mc on the left side, and $d\tau$ taking the place of ds.

Now we multiply on both sides of the equation by $\dot{q}_i$ and form the sum. The right side gives zero, on account of the anti-symmetry of F_{ik}. Hence

$$\sum_{i=1}^{4} \dot{q}_i \ddot{q}_i = 0, \tag{99.23}$$

which means that

$$\sum_{i=1}^{4} \dot{q}_i{}^2 = \text{const.} \tag{99.24}$$

But now the auxiliary condition (99.19) demands that

$$\frac{1}{4} \sum_{i=1}^{4} \dot{q}_i{}^2 = m^2 c^2, \tag{99.25}$$

and since

$$\sum_{i=1}^{4} \left(\frac{dq_i}{ds} \right)^2 = 1 \tag{99.26}$$

we see that our—originally undetermined—parameter τ becomes *a posteriori* identified with

$$\tau = \frac{s}{2mc}, \tag{99.27}$$

and the previous equation (99.14) is fully restored.

10. Geodesics of a four-dimensional world. The motion of a particle that is free of external forces could be expressed by the action principle

$$\delta \int_{\tau_1}^{\tau_2} \overline{ds} = 0. \tag{910.1}$$

This principle resembles Jacobi's principle (56.12) if we omit the potential energy V. Moreover, the line-element $\overline{ds}$ is no longer the line-element of three-space but the line-element (93.8) of the Minkowskian four-space. Let us assume that we change from rectangular coordinates to arbitrary *curvilinear* coordinates by a point transformation. Then the same line-element $\overline{ds}$ will now appear in the form (15.16), but with the summation extending from 1 to 4:

$$\overline{ds} = \sqrt{ \sum_{i,k=1}^{4} g_{ik} dq_i dq_k }. \tag{910.2}$$

If we want to know how a force-free particle will move in this curvilinear reference system, we have to solve a variational problem with the Lagrangian

$$L = \sqrt{ \sum_{i,k=1}^{4} g_{ik} \dot{q}_i \dot{q}_k }, \tag{910.3}$$

where the differentiation is with respect to a certain parameter τ.

Once more we wish to change to a Hamiltonian problem, as we have done earlier in chapter VI, section 10. The Hamiltonian function vanishes, because L is once more a homogeneous form of the first order in the $\dot{q}_i$. But the p_i satisfy the identity (cf. 610.27)

$$\sum_{i,k=1}^{4} g^{ik} p_i p_k = 1, \qquad (910.4)$$

where the notation g^{ik} refers to the elements of a matrix which is the *reciprocal* of the original matrix of the g_{ik}. Hence the effective Hamiltonian function H of our problem becomes

$$H = \sum_{i,k=1}^{4} g^{ik} p_i p_k, \qquad (910.5)$$

where the g^{ik} are generally some given functions of the variables $q_1, \ldots, q_4$, and we have to restrict the energy constant in $H = \text{const.}$ to 1.

The equations of motion—which have now the significance of the straight lines of a Minkowskian manifold—become

$$\dot{q}_i = 2 \sum_{\alpha=1}^{4} g^{i\alpha} p_\alpha,$$
$$p_i = -\sum_{\alpha,\beta=1}^{4} \frac{\partial g^{\alpha\beta}}{\partial q_i} p_\alpha p_\beta. \qquad (910.6)$$

11. The planetary orbits in Einstein's gravitational theory. By pondering on the puzzling and yet obviously fundamental fact that *every inertial mass is also a heavy mass* and thus the source of a gravitational force, Einstein came to the conclusion that the constancy of the velocity of light cannot hold in the presence of a gravitational field. Since the Minkowskian line-element leads directly to the invariance of the light cone (93.11), Einstein's result demands that we should abandon the Minskowskian line-element in favour of a more general line-element which takes into account the existence of gravity. By an unprecedented boldness of mathematical imagination, which was nevertheless firmly rooted in the soil of observed physical phenomena, Einstein introduced the geometry of Riemann into the physical world and showed that on the basis of a curved geometry not only do space and time become united into one entity (as was done by Minkowski on the basis of a quasi-Euclidean four-dimensional line-element), but *space*, *time*, and *matter* become one geometrical entity, "matter" being interpretable as a certain curved portion of the four-dimensional world.

The exposition of the concepts of general relativity is beyond the scope of particle dynamics but we shall discuss the fundamental result of Einstein, which discarded the "force of gravity" as a separate force and explained planetary motion as a purely *geodesic* phenomenon, i.e. the forceless motion of a particle in a certain four-space of Riemannian structure.

The theory of Einstein generalized Newton's gravitational potential to a system of ten field quantities, which are the ten components $g_{ik} = g_{ki}$ of a four-dimensional Riemannian line-element. The scalar potential equation of Newton was generalized to the "Einsteinian field equations," which made it possible to obtain the gravitational field generated by the sun, under the assumption that this field is *spherically symmetric*. The result of the calculations is given by the "line-element of Schwarzschild," which in polar coordinates becomes

$$\overline{ds}^2 = \left(1 - \frac{\alpha}{r}\right)dx_4{}^2 - \frac{dr^2}{1 - \alpha/r} - r^2(d\theta^2 + \sin^2\theta\, d\phi^2). \quad (911.1)$$

The case $\alpha = 0$ leads to Minkowski's "flat" line-element (95.2), written in polar coordinates.

Hence the problem of obtaining the motion of the planets under the action of a central body becomes equivalent to the evaluation of the geodesics of a Riemannian space with the line-element (911.2). This again means the solution of a dynamical problem with the Hamiltonian function (910.5), which in our case becomes

$$H = \frac{p_4{}^2}{1 - \alpha/r} - \left(1 - \frac{\alpha}{r}\right)p_1{}^2 - \frac{1}{r^2}\left(p_2{}^2 + \frac{p_3{}^2}{\sin^2\theta}\right). \quad (911.2)$$

Actually we know in advance that, in consequence of spherical

symmetry, we can reduce our problem to a motion in the plane
$\phi = \frac{1}{2}\pi$. Hence it suffices to treat the Hamiltonian problem of
only *three* pairs of canonical variables, with the Hamiltonian
function

$$H = \frac{p_4^2}{1 - \alpha/r} - \left(1 - \frac{\alpha}{r}\right)p_1^2 - \frac{p_2^2}{r^2}. \tag{911.3}$$

We have the further advantage that H does not depend explicitly
on θ and x_4. Hence we have two ignorable coordinates which can
be eliminated in advance:

$$\begin{aligned} p_2 &= \text{const.} = -A, \\ p_4 &= \text{const.} = BA, \end{aligned} \tag{911.4}$$

thus reducing H to

$$H = \frac{B^2 A^2}{1 - \alpha/r} - \left(1 - \frac{\alpha}{r}\right)p_1^2 - \frac{A^2}{r^2}. \tag{911.5}$$

Of greatest interest to us is not how the motion occurs in
time, but what the *geometrical orbit* is in which the planets
revolve around the sun; this means *the relation between r and θ.*
Now we have

$$\dot{r} = \frac{\partial H}{\partial p_1} = -2\left(1 - \frac{\alpha}{r}\right)p_1, \tag{911.6}$$

$$\dot{\theta} = \frac{\partial H}{\partial p_2} = \frac{2A}{r^2},$$

and thus

$$\frac{dr}{d\theta} = -\frac{r^2}{A}\left(1 - \frac{\alpha}{r}\right)p_1. \tag{911.7}$$

To this can be added the energy theorem

$$H = 1, \tag{911.8}$$

which yields immediately p_1 as a function of r. The determination
of the planetary orbits is thus reducible to a *single quadrature:*

$$d\theta = \frac{dr}{r^2\sqrt{1 - \alpha/r}} \frac{1}{\sqrt{-1/A^2 - 1/r^2 + B^2/(1 - \alpha/r)}} \tag{911.9}$$

(we choose the sign of the square root in getting p_1 by the agreement that increasing values of r shall lead to increasing values of θ); or, introducing the reciprocal radius $r^{-1} = \rho$ as a new variable:

$$\theta = -\int \frac{d\rho}{\sqrt{1 - \alpha\rho}\sqrt{-1/A^2 - \rho^2 + B^2/(1 - \alpha\rho)}}. \qquad (911.10)$$

Since the denominator becomes the square root of a cubic function of ρ, the integration leads to an elliptic integral. The actual value of α in the solar system is so small, however, that we can expand $(1 - \alpha\rho)^{-1}$ in powers of $\alpha\rho$, neglecting third-order terms, which become immeasurably small. Hence we can replace the last term of the denominator by

$$B^2(1 + \alpha\rho + \alpha^2\rho^2) \qquad (911.11)$$

and we can similarly replace $(1 - \alpha\rho)^{-\frac{1}{2}}$ by $1 + \frac{1}{2}\alpha\rho$, since the next term gives a contribution which is entirely below observability. Then we obtain

$$\theta = -\int \frac{(1 + \frac{1}{2}\alpha\rho)d\rho}{\sqrt{-(A^{-2} - B^2) + \alpha B^2\rho - (1 - B^2\alpha^2)\rho^2}}, \qquad (911.12)$$

which is now integrable in terms of elementary functions. The quadratic function in the denominator may be resolved into its root factors:

$$-(1 - B^2\alpha^2)(\rho - \rho_1)(\rho - \rho_2), \qquad (911.13)$$

where ρ_1 and ρ_2 are determined by the conditions

$$\frac{\alpha B^2}{1 - \alpha^2 B^2} = \rho_1 + \rho_2, \qquad (911.14)$$

$$\frac{1}{A^2} - B^2 = \rho_1\rho_2. \qquad (911.15)$$

If we use the notations

$$\tfrac{1}{2}(\rho_1 + \rho_2) = \rho_0, \qquad \tfrac{1}{2}(\rho_1 - \rho_2) = b \qquad (911.16)$$

and put

$$\rho = \rho_0 + t \qquad (911.17)$$

the denominator becomes

$$\sqrt{1 - B^2\alpha^2}\sqrt{b^2 - t^2} \qquad (911.18)$$

while the numerator becomes

$$1 + \tfrac{1}{2}\alpha\rho_0 + \tfrac{1}{2}\alpha t. \qquad (911.19)$$

We thus have to integrate the function

$$\frac{1 + \tfrac{1}{2}\alpha\rho_0}{\sqrt{1 - B^2\alpha^2}} \frac{1}{\sqrt{b^2 - t^2}} + \frac{\alpha}{2} \frac{t}{\sqrt{b^2 - t^2}}, \qquad (911.20)$$

which can be accomplished by elementary functions. The quantity αB^2 becomes, according to (911.14), practically $\rho_1 + \rho_2$ and thus

$$\alpha^2 B^2 = 2\alpha\rho_0, \qquad (911.21)$$

which shows that the constant factor in front of the first term of (911.20) is replaceable by $1 + \tfrac{3}{2}\rho_0\alpha$. The integration of this term yields

$$\theta = (1 + \tfrac{3}{2}\rho_0\alpha) \arccos \frac{t}{b}, \qquad (911.22)$$

and going back to the original variable r we obtain

$$\frac{1}{r} = \frac{1}{r_0} + b \cos \frac{\theta}{1 + \tfrac{3}{2}\rho_0\alpha}, \qquad (911.23)$$

which is the *focal equation of an ellipse*, except for a small correction which amounts to a slow *precession* of the ellipse in its own plane. The change of angle from one minimum of r (perihelion) to the next one is not 2π but

$$2\pi + 3\pi\rho_0\alpha. \qquad (911.24)$$

The integration of the second term of (911.20) causes only a small *periodic* perturbation of the orbit (too small to be observed), which has no cumulative significance, while the advance of the perihelion occurs at *every* revolution and thus accumulates to a measurable quantity after hundreds of revolutions.

In order to obtain the physical value of α, we shall make use of the phenomenological theory of Poincaré–Minkowski, dealt with earlier in section 8. This theory was based on special relativity and adapted Newton's equations of planetary motion to the demands of special relativity, but agreeing with the old theory for small velocities. We established the Hamiltonian function of this theory in (98.8). In polar coordinates we obtain (for $\phi = $ const. $= \frac{1}{2}\pi$)

$$H = \frac{p_4^2}{(m + V/c)^2} - \frac{1}{(m + V/c)^2}\left(p_1^2 + \frac{p_2^2}{r^2}\right), \quad (911.25)$$

where V is the potential energy of the solar field (cf. p. 118, problem 2). If we compare this Hamiltonian with that of the Einstein theory (911.3), we see that we obtain agreement for small velocities by choosing $m = 1$. In that case the denominator of the first term becomes (neglecting quantities of second order):

$$1 - \frac{2fM}{c^2 r}. \quad (911.26)$$

Hence the integration constant of the Schwarzschild line-element (911.1) must be identified with

$$\alpha = \frac{2fM}{c^2} \quad (911.27)$$

where f is the gravitational constant and M is the mass of the sun. The numerical value of $2f/c^2$ is very small:

$$\frac{2f}{c^2} = 1.481 \times 10^{-28} \frac{\text{cm.}}{\text{gm.}} \quad (911.28)$$

If we multiply by the mass of the sun, we obtain a length, called the "gravitational radius" of the sun, of the numerical value

$$\alpha = 2.94 \times 10^5 \text{ cm.} \quad (911.29)$$

This length is very small compared with planetary distances, which shows that the expansion into powers of α/r is actually justified.

In the case of the nearest planet, Mercury, the cumulative effect of the advance of the perihelion becomes well pronounced. If a is the semi-major axis of the planetary orbit and ϵ its eccentricity, we get

$$\rho_0 = \frac{1}{a(1 - \epsilon^2)} \quad (911.30)$$

and thus the advance of the perihelion for every revolution becomes, according to (911.24),

$$\eta = \frac{6\pi f M}{(1 - \epsilon^2)a}. \tag{911.31}$$

This is Einstein's celebrated result, which was found to be in fullest agreement with the observed perihelion precession in the case of Mercury (advance per century: calculated 43.03 ± 0.03 seconds; observed 42.56 ± 0.94 seconds).[1]

It is of interest to observe that the precession effect checks the theoretical expression of g_{44} in the first and second approximation, and the expression of g_{11} in the first approximation.

12. The gravitational bending of light rays. The propagation of light rays can also be considered as a dynamical problem since the light rays have to be conceived as straight lines which belong to the null cone, that is geodesics whose length is zero. It is clear that in the formulation of the geodesic principle we could have multiplied the Lagrangian (910.3) by a constant m. In that case the identity (910.4) becomes m^2 instead of 1 on the right side. As m becomes smaller and smaller, the particle will approach more and more the light cone and in the limit $m = 0$ it will actually move on the light cone. In that case the constant of the Hamiltonian function becomes zero.

We can thus include the theory of the null geodesics in our previous considerations by changing the constant $H = 1$ to $H = 0$. The only difference that occurs in (91.19) is that the constant $-A^{-2}$ in the denominator disappears. We now obtain

$$\theta = -\int \frac{d\rho}{\sqrt{1 - \alpha\rho}\sqrt{B^2/(1 - \alpha\rho) - \rho^2}}. \tag{912.1}$$

Let us first neglect α. Then integration gives

$$\theta = \text{arc cos}\frac{\rho}{B}$$

or

[1]Cf. G. M. Clemence, Revs. Mod. Phys. *19*, 361 (1947).

$$\frac{1}{r} = B \cos \theta,$$

which means

$$r \cos \theta = \frac{1}{B}. \tag{912.2}$$

This is the polar equation of a *straight line*, and B is established as the reciprocal of the smallest distance r_0 at which the light ray passes the sun:

$$B = \frac{1}{r_0}. \tag{912.3}$$

If we now expand in powers of $\alpha\rho$, the second-order terms become negligibly small, and it suffices to integrate the simpler expression

$$\theta = - \int \frac{1 + \frac{1}{2}\alpha\rho}{\sqrt{B^2 + \alpha B^2 \rho - \rho^2}} \, d\rho. \tag{912.4}$$

We now put

$$\rho - \frac{1}{2}\alpha B^2 = \xi. \tag{912.5}$$

In the new variable we have, neglecting second-order terms in α:

$$\theta = - \int \frac{1 + \frac{1}{2}\alpha\xi}{\sqrt{B^2 - \xi^2}} \, d\xi, \tag{912.6}$$

which gives

$$\theta = \arccos \frac{\xi}{B} + \frac{1}{2}\alpha\sqrt{B^2 - \rho^2}. \tag{912.7}$$

Let us now go to the limit $r = \infty$, that is $\rho = 0$ and

$$\frac{\xi}{B} = - \frac{1}{2}\alpha B = - \frac{\alpha}{2r_0}. \tag{912.8}$$

We have to find an angle whose cosine has turned slightly negative and which is thus slightly greater than $\frac{1}{2}\pi$. We obtain in sufficient approximation

$$\theta = \frac{1}{2}\pi + \frac{\alpha}{2r_0}. \tag{912.9}$$

The deflection is towards the sun and corresponds to the motion on a very steep hyperbola. The angle of deflection becomes

$$\eta = \frac{\alpha}{2r_0} = \frac{fM}{c^2 r_0}, \qquad (912.10)$$

which corresponds to the deflection of a light ray subjected to the force of gravity, if we calculate in the elementary Newtonian fashion. This was the deflection of light predicted by Einstein on the basis of the "equivalence hypothesis," which was the guiding principle of his early speculations and which demonstrated to him that the Minkowskian line-element cannot be maintained in the presence of gravity. As our deduction shows, this deflection is caused by the factor dx_4^2 in the line-element, i.e. the component g_{44}.

The second term on the right side of (912.7) shows that an added deflection comes from the component g_{11}. The amount of this deflection is once more $\alpha/2B$ and thus the resulting deflection is *twice* the previous amount:

$$\eta = \frac{\alpha}{r_0} = \frac{2fM}{c^2 r_0}. \qquad (912.11)$$

Hence the non-Euclidean nature of the space part and the time part of the line-element participate in *equal amounts* in the deflection of a light ray near the limb of the sun. The total deflection in the position of a star is twice the previous amount, since it exists symmetrically on both sides. The total light deflection is thus

$$\delta = 2\eta = \frac{4fM}{c^2 r_0}. \qquad (912.12)$$

For a light ray passing the limb of the sun we obtain

$$\delta = 1.75 \text{ seconds.} \qquad (912.13)$$

The astronomical observations, obtained under more or less favourable conditions on the rare occasions of total eclipses, show clearly the existence of the light deflection in approximately the correct amount, the agreement being as good as one can expect under the given circumstances.

13. The gravitational red-shift of the spectral lines. In section 2 we have discussed the theory of the Doppler effect. This effect comes about because two observers in relative motion to each other who measure with their own clocks the time which passes between two light signals obtain different results. If the time measured by the two observers is t and t', then the Doppler effect, expressed in frequencies, becomes

$$\frac{\nu'}{\nu} = \frac{t'}{t}. \tag{913.1}$$

Let us now consider two light signals emitted on the sun and received on the earth. In view of the static nature of the sun's gravitational field (i.e. the g_{ik} do not depend on x_4), a certain Δx_4 recorded on the sun will come down to the earth unchanged. But the *proper time* measured by a clock placed on the sun and on the earth will record the times

$$t = \sqrt{g_{44}}\,\Delta x_4, \qquad t' = \sqrt{g_{44}'}\,\Delta x_4, \tag{913.2}$$

and thus

$$\frac{\nu'}{\nu} = \frac{\sqrt{g_{44}'}}{\sqrt{g_{44}}}. \tag{913.3}$$

In view of the smallness of α we have sufficiently closely

$$\sqrt{\frac{g_{44}'}{g_{44}}} = \frac{1 - \alpha/2r'}{1 - \alpha/2r} = \frac{\alpha}{2}\left(\frac{1}{r} - \frac{1}{r'}\right). \tag{913.4}$$

If we put

$$\nu' = \nu + \Delta\nu \tag{913.5}$$

and consider the smallness of the effect, we obtain

$$\frac{\Delta\nu}{\nu} = -\frac{\alpha}{2r} = -\frac{fM}{r} = -2.12 \times 10^{-6} \tag{913.6}$$

(since $1/r'$ is negligible in comparison with $1/r$). Hence the spectral lines emitted by the sun appear slightly shifted to smaller frequencies, i.e. longer wave-lengths, in consequence of the gravitational action of matter. This fundamental result of Einstein was first deduced from his "equivalence principle." The final theory established the same effect as a geometrical consequence of the four-dimensional Riemannian structure of the space-time world.

Under ordinary circumstances it is difficult to separate the gravitational Doppler shift from the kinematic Doppler shift. The gravitational shift for the sun, if expressed in kinematic terms, is not more than 0.64 km./sec., which is difficult to separate from other perturbations. But a spectacular verification of the effect was given by the spectral observation of the "dark companion" of the star Sirius. Here we know the masses of the two components from the orbital elements. We also know from the luminosity of the dark companion its radius and thus the gravitational potential on its surface. This star belongs to the category of the "white dwarfs." Although of the size of the earth, it has the mass of the sun, which means the fantastic density of 50,000 (compared with water). The gravitational red shift is thus 40 times larger than that from the sun and well measurable. The careful observations of W. S. Adams of the Mount Wilson Observatory[1] gave an excellent verification of the theoretical prediction (theoretical value, expressed in Doppler terms, 26 km./sec.; observed 23 km./sec.).

Bibliography

The following list of textbooks is not more than a cross-section of a very extensive literature.

Aharoni, J., *The Special Theory of Relativity* (Clarendon Press, 1959).

Bergmann, P. G., *Introduction to the Theory of Relativity* (Prentice-Hall, 1947).

Møller, C., *The Theory of Relativity* (Clarendon Press, 1952).

Pauli, W., *Theory of Relativity* (Pergamon Press, 1958).

Rainich, G. Y., *Mathematics of Relativity* (Wiley, 1950).

Rindler, W., *Special Relativity* (Oliver and Boyd, 1960).

Stephenson, G. and Kilmister, C. W., *Special Relativity for Physicists* (Longmans, Green, 1958).

Synge, J. L., *Relativity: The Special Theory* (North-Holland, 1956), *The General Theory* (North-Holland, 1960).

---

[1]Cf. Astrophysical Journal *67*, 195 (1928).

CHAPTER X

HISTORICAL SURVEY

A truly satisfactory and adequate history of the analytical principles of mechanics has never been written. The book by Dühring,[1] which claims to deal with this subject, contains little pertinent history. Mach's classical book on the development of mechanics[2] is primarily devoted to the physical principles of mechanics and is inadequate in its analytical aspects. Mach had so little sympathy with anything that smacked of *a priori* and rationalistic thinking that he could never reach a proper evaluation of the analytical method in its relation to the physical sciences. The predominant influence of the positivistic type of philosophy on the scientific thinking of the past fifty years has much to do with the fact that the development of the variational principles—this magnificent chapter in the evolution of human thought—never aroused the enthusiasm of scientific circles, but was merely tolerated as an efficient tool in describing mechanical phenomena. Hence it is understandable that no systematic historical account of this branch of mathematical physics follows the development up to our own day.* The short essay of A. Mayer[3] covers only the Maupertuis-Euler-Lagrange episode and

[1] E. E. Dühring, *Kritische Geschichte der allgemeinen Principien der Mechanik* (Leipzig, 1877).

[2] E. Mach, *The Science of Mechanics* (Chicago: Open Court Publishing Company, 1919).

[3] A. Mayer, *Geschichte des Princips der kleinsten Aktion* (Leipzig, 1877).

*This can no longer be maintained since the appearance of the outstanding and monumental historical study of René Dugas, *Histoire de la Mécanique* (Editions de Griffon, Neuchatel, 1950), which traces the evolution of mechanical ideas and principles from the time of the early Greeks to the present day. A sequel to this book, by the same author (and publisher), *La Mécanique au XVII Siècle* (1954), deals specifically with the great period of Kepler, Galileo, Descartes, Huygens, Newton, and Leibniz.

omits the Hamilton-Jacobi development. For this phase of the theory, which is essentially a creation of the nineteenth century, the occasional footnotes in the *Encyklopädie der mathematischen Wissenschaften*[1] were consulted by the author. The excellent articles in the *Handbuch der Physik*[2] of Geiger-Scheel were likewise of help. The present historical survey gives merely an outline of the principal developments of thought. For this reason it seems advisable to group the material around the leading individuals who played a critical part in these developments.

Aristotle (384-322 B.C.). The first veiled formulation of the principle of virtual work is contained in the *Physics* of Aristotle. He derives the law of the lever from the principle that "forces balance each other if they are inversely proportional to the velocities." Since it is the equilibrium of the lever which is in question, and yet the argument is based on velocities, the idea of a virtual displacement, brought about by a small disturbing force, is already clearly present. The name "virtual velocities" instead of "virtual displacements," widely used up to the 19th century, originates from Aristotle's formulation of the principle. The same principle—but in the new formulation: "What is gained in force is lost in velocity"—was used by Stevinus (1598-1620) for deducing the equilibrium of pulleys.

Galileo (1564-1642). The fundamental contributions of Galileo to mechanics are universally known. However, in paying exaggerated tribute to the empiricist Galileo as the father of the experimental method, one is frequently prone to overlook his great achievements as a theoretical physicist. He contributed essentially to theoretical mechanics by an improved formulation of Aristotle's principle. Galileo recognized the important fact that in formulating the principle of virtual velocities it is not the velocity, but only the velocity *in the direction of the force* which counts. His method amounts to the recognition of the "work"

[1]*Encyklopädie der mathematischen Wissenschaften*, IV, *Mechanik* (Leipzig, 1904-1935), articles 12 and 13 by G. Prange.

[2]*Handbuch der Physik* (Berlin, 1927), V, *Mechanics of Points and Rigid Bodies:* Chapter 1, "The Axioms of Mechanics," by G. Hamel, and Chapter 2 "The Principles of Mechanics," by L. Nordheim.

as the "product of the force and the displacement in the direction of the force." He applies the principal of virtual work to the equilibrium of a body on an inclined plane, and shows how his principle gives the same result that Stevinus found on the basis of the energy principle. He shows the validity of the same principle in hydrostatics where he deduces with its help the laws of hydrostatic pressure, established before by Archimedes through considerations based on the concept of the centre of mass.

John Bernoulli (1667-1748). In previous applications of the principle two forces were always engaged, the "moving force" and the "load." In these instances the law could be expressed by means of a proportion. John Bernoulli was the first to recognize the principle of virtual work as a general principle of statics with which all problems of equilibrium could be solved. He omits the use of proportions, and introduces the *product of the force and the virtual velocity in the direction of the force*, taken with a positive or negative sign according to the acute or obtuse angle between force and velocity. In a letter written in 1717 to Varignon, Bernoulli announced the general principle that for all possible infinitesimal displacements the sum of all these products must vanish if the forces balance each other. The principle now holds for any forces and under any mechanical conditions.

Remarkably ingenious also is another observation of Bernoulli. He compares the motion of a particle in a given field of force with the propagation of light in an optically heterogenous medium, and tries to give a mechanical theory of the refractive index on this basis. Bernoulli is thus the forerunner of that great theory of Hamilton which has shown that the principle of least action in mechanics and Fermat's principle of shortest time are strikingly analogous in their consequences, permitting the interpretation of optical phenomena in mechanical terms, and vice versa.

Newton (1642-1727). The fundamental equations of motion were formulated by Newton, based on previous researches of Leonardo and Galileo. The fundamental quantities of motion were established as the *momentum* and *the acting force*. The Newtonian law of motion solved the problem of motion of an isolated particle. It could be considered as the general solution of the problem of motion if one agreed to break up every assembly of masses into isolated particles. The difficulty arose, however, that the acting forces were not always known. The difficulty was partly overcome by the Third Law of Motion which announced the principle of action and reaction. This eliminated the unknown forces in the case of the motion of a rigid body, but the motion of mechanical systems with more complicated kinematical conditions was not always susceptible to the Newtonian analysis. The followers of Newton envisaged the Newtonian laws as absolute and universal laws of nature, interpreting them with a dogmatism to which their originator would never have subscribed. This dogmatic reverence of Newtonian particle mechanics prevented the physicists from an unprejudiced appreciation of the analytical principles which came into use during the 18th century, developed by the leading French mathematicians of that period. Even Hamilton's great contributions to mechanics were not recognized by his contemporaries on account of the prevalence of the Newtonian form of mechanics.

Leibniz (1646-1716). While Newton proposed to measure motion by the time rate of change of momentum, Leibniz argued for another quantity, the "vis viva," which—except for the factor 1/2—is identical with our "kinetic energy." Leibniz replaced the Newtonian equation by the equation that "the change of the kinetic energy is equal to the work done by the force." The ideas of Leibniz were in harmony with the later analytical developments. Both the kinetic energy and the work of the acting forces could easily be generalized from one single particle to an arbitrary system of particles, without the necessity of isolating the individual particles. The work of the forces could be replaced by another more fundamental quantity, the negative of the "potential energy"—a term coined by J. Rankine about the close of the 19th century. Both kinetic and potential energies were quantities which could characterize a system *as a whole*. This was one of their great advantages for an organic treatment of the dynamics of an interlocked system of particles. The other was that in spite of their scalar nature they were sufficient to include *all* the forces acting on the system, no matter how elaborate and numerous those forces were. This became possible by applying the principle of variation to the kinetic and the potential energies. The controversy between Newton and Leibniz concerning force is thus essentially a question of method. While the Newtonian definition may have been better suited to the vectorial treatment of mechanics, the concepts of Leibniz became the cornerstones of the analytical treatment.

D'Alembert (1717-1785). An important advance in the analytical treatment of mechanical systems was accomplished by d'Alembert, who reduced any problem of motion to a problem of equilibrium by the device of adding to the given acting forces a new force, namely, that created by the motion. This new force, the force of inertia, together with the other forces, produces equilibrium, which means that the principle of virtual work is applicable to any system in motion. All the equations of motion of an arbitrary mechanical system are thus comprised in one single variational principle.

Maupertuis (1698-1759). Maupertuis conceived the universal hypothesis that in all events of nature there is a certain quantity, called "action," which is a minimum. The bold universality of this assumption is admirable and well in line with the cosmic spirit of the 18th century. However, the mathematical powers of Maupertuis were far behind the high standards of his period. Maupertuis could not establish satisfactorily the quantity to be minimized. He applied the principle to the derivation of the laws of elastic collision. This phenomenon is very intricate if treated as a minimum problem and requires great skill in handling. Maupertuis obtained the correct result by an entirely incorrect method. More satisfactory is his treatment of the law of refraction, in which he shows how Fermat's principle of least time can be replaced by the principle of least action. This result was earlier recognized by John Bernoulli.

Of great interest is the Maupertuis-Euler episode, told by Mayer in his afore-mentioned paper. The priority of Maupertuis' discovery was assailed by Koenig, who claimed that Leibniz expressed the same idea in a private letter. This letter was never produced. In the ensuing controversy, Euler defended most emphatically the priority rights of Maupertuis. The peculiar thing in this defense is that Euler himself had discovered the principle at least one year before Maupertuis, and in an entirely correct form. In particular, Euler knew that both the actual and the varied motion have to satisfy the law of the conservation of energy. Without this auxiliary condition the action quantity of Maupertuis—even if corrected from a sum, the form in which Maupertuis used it, to an integral—loses all significance. Although Euler must have seen the weakness of Maupertuis' argument, he refrained from any criticism, and refrained from so much as mentioning his own achievements in this field, putting all his authority in favour of proclaiming Maupertuis as the inventor of the principle of least action. Even knowing Euler's extraordinarily generous and appreciative character, this self-effacing and self-denying modesty has no parallel in the entire history of science, which abounds in examples to the contrary.

Euler (1707-1783). Euler contributed very essentially to the development of theoretical mechanics. His treatment of the rotating rigid body makes the first use of kinematical variables by using the three components of angular velocity as auxiliary variables. In the variational treatment of mechanics Euler's pioneering work is of the highest quality. Euler started the systematic investigation of variational problems, sometimes called "isoperimetric problems." These maximum-minimum problems attracted the interest of the best minds—such as Newton, Leibniz, James and John Bernoulli—from the very start of the infinitesimal calculus. Euler found a differential equation which gave the implicit solution of an extended class of such problems. Although Euler did not express the principle of least action in the exact form in which it was first given by Lagrange, his application of the principle to mechanical phenomena is equivalent to Lagrange's explicit formulation.

Lagrange (1736-1813). The achievements of Lagrange, the greatest mathematician of the 18th century, parallel Euler's work in many respects. Lagrange gave the solution of isoperimetric problems quite independently of Euler, and with entirely new methods. He developed for this purpose a new calculus, the calculus of variations. He also recognized the superiority of variational principles on account of the freedom we have in characterizing the position of a mechanical system by any set of parameters we may choose (the "generalized coordinates"). If the chief advantage of the principle of virtual work and d'Alembert's principle has been that the mechanical system could be handled as an organic unity, without breaking it into isolated particles, the Lagrangian equations added a new feature of supreme importance: the invariance with respect to arbitrary coordinate transformations. This meant the freedom to choose any system of coordinates well suited to the nature of the problem. In his *Mécanique Analytique* (1788) Lagrange created a new and immensely powerful weapon which could solve any mechanical problem on the basis of pure calculation, without any reference to physical or geometrical considerations, provided that the kinetic energy and the potential energy of the system were given in abstract analytical form. He refers to this extraordinary achievement with his usual modesty when he says in the preface of his book: "The reader will find no figures in the work. The methods which I set forth do not require either constructions or geometrical or mechanical reasonings: but only algebraic operations, subject to a regular and uniform rule of procedure." Lagrange thus gave the program and foundation of analytical mechanics.

The method of treating auxiliary conditions with the help of the undetermined multiplier is another of Lagrange's immortal discoveries which plays a vital part in problems of theoretical mechanics. Hamilton, himself one of the most outstanding figures of analytical research, called Lagrange the "Shakespeare of mathematics," on account of the extraordinary beauty, elegance, and depth of the Lagrangian methods.

Hamilton (1805-1865). An entirely new world beyond the discoveries of Lagrange was opened by the investigations of Sir William Rowan Hamilton. The equations of Lagrange were rather complicated differential equations of the second order. Hamilton succeeded in transforming them into a set of differential equations of the first order with twice as many variables, considering the position coordinates and the momenta as independent variables. The differential equations of Hamilton are linear and separated in the derivatives. They represent the simplest and most desirable form into which the differential equations of a variational problem can be brought. Hence the name "canonical equations" by which Jacobi designated them.

It was likewise Hamilton who, by a transformation of d'Alembert's principle, gave the first exact formulation of the principle of least action. The form in which Euler and Lagrange employed the principle holds only for the conservative (scleronomic) case.

One of the most significant discoveries of Hamilton is his realization that problems of mechanics and of geometrical optics can be handled from a unified viewpoint. He operates with a "characteristic" or "principal" function in both optics and mechanics. This function has the property that by mere differentiation the path of a moving point, as well as the path of an optical ray, can be determined. Moreover, in both optics and mechanics, the characteristic function satisfies the same partial differential equation. The solution of this partial differential equation under the proper boundary conditions is equivalent to the solution of the equations of motion.

Jacobi (1804-1851). Jacobi was one of the few mathematicians who at once recognized the extraordinary importance and brilliance of the Hamiltonian methods. Jacobi developed the transformation theory of the canonical equations, the theory of "canonical transformations." He interpreted the characteristic function on the basis of the transformation theory, and showed that the function used by Hamilton is only one special case of a function which generates a suitable canonical transformation. He thus greatly extended the usefulness of Hamilton's partial differential equation by proving that *any* complete solution of that differential equation, without the specific boundary conditions required by Hamilton, is sufficient for the complete integration of the problem of motion.

Jacobi gave also a new formulation of the principle of least action for the time-independent case with which Euler and Lagrange were dealing. He criticized the formulation of Euler and Lagrange because the range of their integral did not satisfy the condition of varying between fixed limits. Although Euler and Lagrange applied their principle quite correctly, yet Jacobi's elimination of the time from the variational integral led to a new principle which determined the path of the moving point without saying anything about how the motion occurs in time. The analogy of this principle to Fermat's principle of least time, by which the path of an optical ray could be determined, established directly the analogy between optical and mechanical phenomena.

Gauss (1777-1855). Somewhat aside from the main path lies the "principle of least constraint," established by the eminent mathematician Gauss. This principle does not use a time-integral as the function to be minimized, but sets up, at any time, a certain positive quantity defined as the "constraint," and, considering position and velocity as given at that time, determines the accelerations by minimizing this quantity. The principle of Gauss has the virtue of being a genuine minimum—and not merely stationary value—principle. However, it has not the analytical advantages of the other principles, since his "constraint" involves the accelerations in addition to the position coordinates and velocities. *Hertz* gave a geometrical interpretation of the "constraint" of Gauss, conceiving it as the geodesic curvature of a space of $3N$ dimensions, the "configuration space." In this space the principle of Gauss can be interpreted as the "principle of the straightest path." This interpretation brings the Gaussian principle into close relationship with Jacobi's principle.

The later development of analytical mechanics. We have sketched the historical development of the variational principles of mechanics and thus our task, as far as the main topics of this book are concerned, is finished. However, many more scientists added their share in erecting the edifice, by furnishing the analytical tools and adding valuable details to the main theory, besides applying it to special cases. The following list enumerates the highlights of these contributions.

Poisson, together with Lagrange, introduced the bracket-symbols in the theory of perturbations, coming near to the general theory of canonical transformations.

Delaunay developed the analytical theory of multiply-periodic separable systems—a method which became later of basic importance in Bohr's atomic theory.

E. J. Routh, and independently *H. Helmholtz*, discovered the particular importance of the "speed variables" or "cyclic variables"—later called "kinosthenic" or "ignorable variables"—and the general process for their elimination.

J. J. Thomson thought of the possibility of a mechanics without forces, assuming the presence of kinosthenic coordinates which are "ignorable" because they can be eliminated.

H. Hertz developed this idea more systematically, conceiving the potential energy as the kinetic energy of hidden motions. It was likewise Hertz who conceived the geometrical aspects of mechanics by picturing the motion of a arbitrary mechanical system as the motion of a single particle in a space of many dimensions, the "configuration space."

S. Lie introduced the group viewpoint into the theory of canonical transformations, paying particular attention to the group of infinitesimal transformations.

H. Poincaré studied the "integral invariants" of the canonical equations. He added valuable contributions to the pertubation theory of astronomy, especially the three-body problem. Problems of the geometry of the configuration space "in the large" aroused Poincaré's interest and led him to fundamental topological investigations.

P. Appell studied the analytical implications of non-holonomic systems.

The contributions of many contemporary scientists to various advanced chapters of analytical mechanics are outside the scope of the present treatise, which is devoted to the fundamental principles only.

Analytical mechanics in its relation to contemporary physics. The two great achievements of contemporary physics, the theory of relativity and the quantum theory, are both closely allied to analytical mechanics. Einstein's theory of relativity revolutionized all branches of physics. It has shown that Newtonian mechanics has only an approximate validity for velocities which are small compared with the velocity of light. The analytical treatment on the basis of the principle of least action, however, remained intact. It was only the Lagrangian function which had to be modified, but the fact that the differential equations of motion can be derived from a minimum principle continues to hold. In fact, the complete independence of a variational principle from any special reference system made it particularly valuable for the formulation of equations which satisfy the principle of general relativity. This principle demands that the basic field equations of nature must remain invariant with respect to arbitrary coordinate transformations.

The other revolution in modern theoretical physics, the quantum theory, is likewise closely related to analytical mechanics, particularly in its Hamiltonian form. Bohr's theory of electronic orbits made excellent use of the Hamiltonian methods when the importance of separable systems for the formulation of quantum conditions was discovered. While in earlier days the study of the Hamiltonian methods was left to the astronomer, the formulation of the quantum conditions by Sommerfeld and Wilson in 1916, and the calculation of the Stark-effect in the same year by Epstein, emphatically demonstrated the importance of the Hamiltonian concepts for the structure of the atom.

The reinterpretation of the laws of nature in terms of the wave-mechanical theories of Schroedinger, Heisenberg and Dirac also grew out of Hamiltonian methods. Conjugate variables and canonical transformations are part of the foundation of the new theory. As a new feature, the matrix character of the q- and p-variables has been added to the theory of Heisenberg, Born and Jordan; while Dirac's theory considers conjugate variables as non-commutative quantities. Schroedinger, on the other hand, introduced the operational viewpoint and reinterpreted the partial differential equation of Hamilton-Jacobi as a wave equation. His starting point is the optico-mechanical analogy of Hamilton.

In spite of the radical departure of the new concepts from those of the older physics, the basic feature of the differential equations of wave-mechanics is their *self-adjoint* character, which means that they are derivable from a variational principle. Hence, in spite of all differences in the interpretation, the variational principles of mechanics continue to hold their ground in the description of all the phenomena of nature.

CHAPTER XI

MECHANICS OF THE CONTINUA

Introduction: In accordance with the general character of this book, the present chapter deals with the variational formulation of the mechanics of continua solely in its fundamental aspects. The very elaborate investigation concerning the solution of the basic equations, and their modification due to the presence of non-variational forces, are outside the scope of our discussion and have to be referred to the specialized literature (cf. Bibliography). For our aims the exposition of ideas is more important than manipulative details. This chapter can thus be considered as a condensed summary of the basic ideas of hydrodynamics (restricted to ideal fluids), elasticity (within the linear range), electromagnetism (Maxwellian theory), together with a discussion of the conservation laws derivable from the application of Noether's principle—a subject which has gained great importance in contemporary physics, although its true significance is often obscured or misunderstood.

1. The variation of volume integrals. The basic problem of finding the minimum (or at least stationary value) of a definite integral remains once more the central issue. The difference is that whereas previously the integral to be minimized involved a Lagrangian L (in the Hamiltonian formulation $L = T - V$) depending on quantities q_i, $\dot{q}_i$ which were functions of the *single variable t,* and the integration was extended over t alone, now, when we consider matter continuously distributed over a certain domain, the basic integral must be extended over a certain *volume* as well as the time t. Our new problem can thus be characterized in its general mathematical manifestation as follows. We have a certain action integral

$$A = \int L \, d\omega, \qquad (1.1)$$

where the volume element $d\omega$ is the element of a space of m dimensions. The Lagrangian function

$$L = L\left(\phi_i, \frac{\partial \phi_i}{\partial x_j}; x_k\right) \tag{1.2}$$

differs from the earlier Lagrangian of point mechanics (cf. p. 60) in that the $q_i = \phi_i$ are no longer functions of a single variable t, but of the m variables $x_1, x_2, \ldots, x_m$, frequently defined as the Cartesian coordinates of a space of m dimensions. Thus

$$d\omega = dx_1 dx_2 \ldots dx_m, \tag{1.3}$$

whereas in the case of arbitrary curvilinear coordinates we must put

$$d\omega = \left\|\frac{\partial \xi_i}{\partial x_k}\right\| dx_1 dx_2 \ldots dx_m. \tag{1.4}$$

The first factor in (1.4) denotes the functional determinant of the ξ_i (the Cartesian coordinates) expressed as functions of the curvilinear coordinates x_k:

$$\xi_i = \xi_i(x_1, x_2, \ldots, x_m), \tag{1.5}$$

i.e. the determinant of a matrix whose elements a_{ik} are defined by

$$a_{ik} = \frac{\partial \xi_i}{\partial x_k}. \tag{1.6}$$

The $\dot{q}_i$ of the previous treatment must now be replaced by the m partial derivatives

$$\frac{\partial \phi_i}{\partial x_1}, \frac{\partial \phi_i}{\partial x_2}, \ldots, \frac{\partial \phi_i}{\partial x_m}. \tag{1.7}$$

In physical applications the cases $m = 3$ (volume integrals) and $m = 4$ (integrals over volume and time) are of greatest interest.

Certain fundamental features of the calculus of variations of line integrals remain valid:

1. Variation and differentiation are interchangeable operations (cf. p. 57, equation (29.3)):

$$\delta \frac{\partial \phi_i}{\partial x_j} = \frac{\partial(\delta \phi_i)}{\partial x_j}. \tag{1.8}$$

2. Variation and integration are interchangeable operations (cf. p. 57, equation (29.5)):

$$\delta \int_{\Omega} = \int_{\Omega} \delta. \qquad (1.9)$$

3. Again it will be necessary to reduce the variation of a derivative to the variation of the function itself (cf. p. 58). This can be done, in analogy with the process (210.4), on the basis of a fundamental integral transformation (usually attributed to Gauss), which permits us to change a volume integral into a surface integral:

$$\int_{\Omega} \frac{\partial F}{\partial x_i} d\omega = \int_{S} F(S) \nu_i \, dS. \qquad (1.10)$$

Here the $\nu_1, \nu_2, \ldots, \nu_m$ denote the components of the normal ν to the boundary surface S (defined as a vector of length 1, perpendicular to the surface and pointing away from the region enclosed), and dS is the surface element. In three dimensions we shall denote the components of the normal by n_i and the volume element by $d\tau$.

4. As before, for the minimum of an integral, with reversible variations of the ϕ_i, the first variation (subject to certain boundary conditions) must vanish:

$$\delta \int_{\Omega} L \left(\phi_i, \frac{\partial \phi_i}{\partial x_j} ; x_k \right) = 0. \qquad (1.11)$$

2. Vector-analytic tools. The investigation of the mechanics of continua is practically inseparable from the general theory of vector and tensor analysis. This theory is based on the invariance of differential forms and has the great advantage that it permits the formulation of the basic equations in arbitrary coordinates. However, for many purposes the simpler tools of three-dimensional vector analysis suffice, which extol the value of certain often recurring basic operations (associated with rectangular coordinates), in which the vector is treated *as a whole*, without resolving it into components. They are derivable from Hamilton's "quaternion calculus" (cf. p. 304), made more flexible for physical applications by O. Heaviside (1894) and J. Gibbs (1901).

The operational rules of Hamilton's quaternions have the remarkable property that a quaternion remains unchanged in space if the orthogonal unit vectors $\mathbf{i}$, $\mathbf{j}$, $\mathbf{k}$ are replaced by any rotated set of vectors $\mathbf{i}'$, $\mathbf{j}'$, $\mathbf{k}'$ (provided that the determinant of the transformation is $+1$).

What we usually call a "vector $\mathbf{A}$" is a quaternion whose time part is zero. Gibbs replaced the quaternion product of two vectors $\mathbf{A}$ and $\mathbf{B}$ by the two quantities

$$\mathbf{A} \times \mathbf{B} = \tfrac{1}{2}(AB - BA) = -\mathbf{B} \times \mathbf{A} \tag{2.1}$$

(called the "vector product" or "cross-product" of $\mathbf{A}$ and $\mathbf{B}$), and

$$\mathbf{A} \cdot \mathbf{B} = -\tfrac{1}{2}(AB + BA) = \mathbf{B} \cdot \mathbf{A} \tag{2.2}$$

(called the "scalar product" or "dot product" of $\mathbf{A}$ and $\mathbf{B}$).

Hamilton introduced the fundamental "nabla operator" ∇ by putting

$$\nabla = \frac{\partial}{\partial x_1}\mathbf{i} + \frac{\partial}{\partial x_2}\mathbf{j} + \frac{\partial}{\partial x_3}\mathbf{k}; \tag{2.3}$$

this is now called the "gradient operator" and is denoted by "grad." Thus the gradient of a scalar function, i.e. the nabla operator applied to a mere field function ϕ, is defined by the field vector

$$\nabla\phi = \operatorname{grad}\,\phi = \frac{\partial\phi}{\partial x_1}\mathbf{i} + \frac{\partial\phi}{\partial x_2}\mathbf{j} + \frac{\partial\phi}{\partial x_3}\mathbf{k}, \tag{2.4}$$

whereas the dot product of ∇ applied to a vector $\mathbf{A}$ gives

$$\nabla \cdot \mathbf{A} = \operatorname{div}\,\mathbf{A} = \frac{\partial a_1}{\partial x_1} + \frac{\partial a_2}{\partial x_2} + \frac{\partial a_3}{\partial x_3}, \tag{2.5}$$

i.e. a field scalar, called the "divergence" of $\mathbf{A}$.

The cross-product of ∇ with a vector $\mathbf{A}$ is again a vector, called the "curl" of $\mathbf{A}$:

$$\nabla \times \mathbf{A} = \operatorname{curl}\,\mathbf{A} = \left(\frac{\partial a_3}{\partial x_2} - \frac{\partial a_2}{\partial x_3}\right)\mathbf{i}$$
$$+ \left(\frac{\partial a_1}{\partial x_3} - \frac{\partial a_3}{\partial x_1}\right)\mathbf{j} + \left(\frac{\partial a_2}{\partial x_1} - \frac{\partial a_1}{\partial x_2}\right)\mathbf{k}. \tag{2.6}$$

Combinations of the basic operations grad, curl, and div give rise to many important identities, easily provable by the basic definitions; e.g.

$$\text{div curl } \mathbf{A} = \mathbf{\nabla} \cdot (\mathbf{\nabla} \times \mathbf{A}) = 0, \tag{2.7}$$

$$\text{div grad } \phi = \mathbf{\nabla} \cdot \mathbf{\nabla}\phi = \mathbf{\nabla}^2\phi = \left(\frac{\partial^2}{\partial x_1{}^2} + \frac{\partial^2}{\partial x_2{}^2} + \frac{\partial^2}{\partial x_3{}^2} \right) \phi \tag{2.8}$$

(called the "Laplacian operator"),

$$\text{curl grad } \phi = \mathbf{\nabla} \times (\mathbf{\nabla}\phi) = 0, \tag{2.9}$$

$$\text{curl curl } \mathbf{A} = \mathbf{\nabla} \times (\mathbf{\nabla} \times \mathbf{A}) = - \mathbf{\nabla}^2 \cdot \mathbf{A} + \text{grad div } \mathbf{A}. \tag{2.10}$$

An important formula arises when the gradient is applied to the scalar product of two vectors. This formula can be derived by rotating the reference system into such a position that the vector $\mathbf{B}$ has the single component b_1. Then

$$
\begin{aligned}
\text{grad } \mathbf{A} \cdot \mathbf{B} &= \frac{\partial}{\partial x_1}(a_1 b_1)\mathbf{i} + \frac{\partial}{\partial x_2}(a_1 b_1)\mathbf{j} + \frac{\partial}{\partial x_3}(a_1 b_1)\mathbf{k} \\
&= \left(a_1 \frac{\partial b_1}{\partial x_1} + b_1 \frac{\partial a_1}{\partial x_1} \right)\mathbf{i} \\
&+ \left(a_1 \frac{\partial b_1}{\partial x_2} + b_1 \frac{\partial a_1}{\partial x_2} \right)\mathbf{j} \\
&+ \left(a_1 \frac{\partial b_1}{\partial x_3} + b_1 \frac{\partial a_1}{\partial x_3} \right)\mathbf{k}.
\end{aligned}
$$

The second column may be written as follows:

$$
\begin{aligned}
&\left(b_1 \frac{\partial}{\partial x_1} \right) a_1 \mathbf{i} \\
&+ \left(b_1 \frac{\partial}{\partial x_1} \right) a_2 \mathbf{j} + b_1 \left(\frac{\partial a_1}{\partial x_2} - \frac{\partial a_2}{\partial x_1} \right) \mathbf{j} \\
&+ \left(b_1 \frac{\partial}{\partial x_1} \right) a_3 \mathbf{k} + b_1 \left(\frac{\partial a_1}{\partial x_3} - \frac{\partial a_3}{\partial x_1} \right) \mathbf{k},
\end{aligned}
$$

which is the manifestation (applied to our special reference system) of the following vector operation:

$$(\mathbf{B} \cdot \text{grad})\mathbf{A} + \mathbf{B} \times \text{curl } \mathbf{A}.$$

The first column must be the same, exchanging $\mathbf{A}$ and $\mathbf{B}$, and thus we obtain

$$\text{grad}(\mathbf{A} \cdot \mathbf{B}) = (\mathbf{A} \cdot \text{grad})\mathbf{B} + \mathbf{A} \times \text{curl } \mathbf{B}$$
$$+ (\mathbf{B} \cdot \text{grad})\mathbf{A} + \mathbf{B} \times \text{curl } \mathbf{A}. \qquad (2.11)$$

In particular, for the case $\mathbf{A} = \mathbf{B}$,

$$\tfrac{1}{2} \text{grad } \mathbf{A}^2 = (\mathbf{A} \cdot \text{grad})\mathbf{A} + \mathbf{A} \times \text{curl } \mathbf{A}. \qquad (2.12)$$

(The vector product of two vectors has the peculiarity that it changes to its negative value under the influence of a reflection of the reference system. The same is true of the curl of a vector. But (2.11) and (2.12) are nevertheless genuine vector equations, because $(-1) \cdot (-1) = +1$.)

3. Integral theorems. Two integral transformations are of fundamental importance in theories of continuum mechanics. One is the theorem of Gauss, which permits the transformation of a volume integral into a surface integral, extended over the boundary S of the domain (cf. (1.10)):

$$\int_\tau \text{div } \mathbf{A} \, d\tau = \int_S \mathbf{A} \cdot \mathbf{n} \, dS. \qquad (3.1)$$

The other is the theorem of Stokes (cf. p. 211), which permits the transformation of a surface integral into a line integral, extended over the boundary curve of the (open) surface:

$$\int_S (\text{curl } \mathbf{A}) \cdot \mathbf{n} \, dS = \int_s \mathbf{A} \cdot \mathbf{t} \, ds, \qquad (3.2)$$

$\mathbf{t}$ being the tangent of the boundary curve s, taken as a vector of length 1 and oriented in such a way that when moving in the direction of $\mathbf{t}$, facing the normal $\mathbf{n}$, the surface S lies to the left.

The following theorem of Helmholtz (1858) is of great interest: An arbitrary vector field $\mathbf{u}$ (assumed to be continuous and differentiable) can be resolved into the sum of a gradient of a scalar field ("solenoidal field") and the curl of a vector field "vortex field"). This resolution is unique if $\mathbf{u}$ is given in the entire space:

$$\mathbf{u} = \text{grad } \psi + \text{curl } \mathbf{A} \qquad (3.3)$$

where the vortex field can be restricted by the condition

$$\text{div } \mathbf{A} = 0. \tag{3.4}$$

Taking the divergence of (3.3), we obtain (cf. (2.7–8))

$$\mathbf{\nabla}^2 \psi = \text{div } \mathbf{u}, \tag{3.5}$$

and taking the curl (cf. (2.9–10))

$$-\mathbf{\nabla}^2 \mathbf{A} = \text{curl } \mathbf{u}. \tag{3.6}$$

The potential equation with a given right side—called "Poisson's equation"—has a unique solution provided that u vanishes at infinity faster than r^{-1}.

In consequence of this resolution we can say that a divergence-free field is automatically a vortex field, whereas a curl-free field is automatically solenoidal.

4. The conservation of mass. In particle mechanics variation of the mechanical path led to the Euler-Lagrange equations (cf. pp. 60–62). In this process we varied the path of the particles but their *masses* were considered as given constants, remaining unvaried. If the masses are now continuously distributed, it is again imperative that we pay attention to the non-variation of the mass. Here the "density" ρ of the fluid is the fundamental quantity, defined by the fact that the mass contained in the volume element $d\tau$ is given by

$$dm = \rho \, d\tau. \tag{4.1}$$

Accordingly the total mass contained in the volume τ is given by the definite integral

$$m = \int_\tau \rho \, d\tau. \tag{4.2}$$

We might expect the constancy of the mass m to demand that this integral shall remain independent of the time t. This, however, would only be true if the volume τ were to *move with the fluid*, letting every particle follow its own stream line. If τ is a *definite* volume, *stationary* in space, then the quantity m will not remain constant because a certain amount of mass will flow out (or in) through the boundary of the volume τ. The amount of the fluid flowing out in unit time through the surface element dS is given by the scalar product $\rho \mathbf{v} \cdot \mathbf{n} \, dS$, where $\mathbf{v}$ is the velocity of the motion. Hence the conservation of mass demands that

$$\frac{dm}{dt} + \int_s \rho \mathbf{v} \cdot \mathbf{n} \, dS = 0. \tag{4.3}$$

The second term on the left can be transformed by the Gaussian integral transformation (3.7) into

$$\int_\tau \text{div}(\rho \mathbf{v}) d\tau, \tag{4.4}$$

and the law of conservation of mass then assumes the form

$$\int_\tau \left[\frac{\partial \rho}{\partial t} + \text{div}(\rho \mathbf{v}) \right] d\tau = 0. \tag{4.5}$$

This equation must hold for *arbitrarily chosen* volumes τ, which is only possible if the integrand vanishes. Hence the conservation of mass imposes the following condition on the density ρ of a fluid (continuity equation):

$$\frac{\partial \rho}{\partial t} + \text{div}(\rho \mathbf{v}) = 0. \tag{4.6}$$

5. Hydrodynamics of ideal fluids. We shall not consider the microscopic structure of fluids and the atomistic aspects of fluid particles, but treat the fluid purely in its macroscopic aspects. Frictional forces (viscosity) which originate from a transfer of macroscopic into microscopic motions demand an increase in the number of degrees of freedom and the application of statistical principles. They are thus automatically beyond the macroscopic variational treatment. We talk of an "ideal fluid" (including gases), in which viscosity effects can be considered as of negligible size.

The advantage of the variational treatment is that we do not have to know the specific nature of the forces which lead to a certain kinematic condition. In the case of fluids it suffices to know that the forces exerted against a change in shape of a certain fluid volume are negligible as long as the total volume is preserved, whereas strong forces resist any change in the volume τ. We are thus free to vary the position of a fluid particle, provided that we introduce the auxiliary condition that the volume

$$\tau = \int dx_1\, dx_2\, dx_3 \tag{5.1}$$

remains unchanged.

Now the change in position coordinates x_i to $x_i + \delta x_i$ can be considered as an infinitesimal transformation from rectangular to curvilinear coordinates. The functional determinant (2.6) has in its diagonal terms the elements

$$a_{ii} = 1 + \frac{\partial(\delta x_i)}{\partial x_i} \tag{5.2}$$

and the value of the determinant becomes (neglecting quantities of higher than first order)

$$||a_{ik}|| = 1 + \sum_{i=1}^{3} \frac{\partial \delta x_i}{\partial x_i} = 1 + \operatorname{div} \delta \mathbf{x}. \tag{5.3}$$

Hence the variation of the volume τ becomes

$$\delta\tau = \int \operatorname{div} \delta \mathbf{x}\, d\tau \tag{5.4}$$

and the condition that the variation of any volume vanishes entails the auxiliary condition

$$\operatorname{div} \delta \mathbf{x} = 0. \tag{5.5}$$

6. The hydrodynamic equations in Lagrangian formulation. Lagrange treated the problem of fluids in full analogy with the motion of particles. He considered the fluid contained in the infinitesimal volume $d\tau$, with the mass

$$dm = \rho\, d\tau \tag{6.1}$$

and the kinetic energy

$$T = \tfrac{1}{2}\rho v^2 d\tau. \tag{6.2}$$

The potential energy of this fluid element, caused by the weight of the fluid, becomes

$$V = \rho \phi\, d\tau \tag{6.3}$$

if we denote by ϕ the Newtonian potential of gravity. Hence the variational integral becomes

$$\int_{t_1}^{t_2} L\,dt = \int_{t_1}^{t_2} \rho(\tfrac{1}{2}v^2 - \phi)d\tau\,dt. \tag{6.4}$$

Combining all the fluid particles in a mechanical system we obtain the final variational integral in the form of a fourfold integral, extended over space and time:

$$A = \int_{t_1}^{t_2} \int_{\tau} \rho(\tfrac{1}{2}v^2 - \phi)d\tau\,dt. \tag{6.5}$$

To this has to be added the auxiliary condition (5.5), which gives rise to a Lagrangian multiplier p and the added integral

$$\int_{t_1}^{t_2} \int_{\tau} p \operatorname{div} \delta\mathbf{x}\,d\tau\,dt. \tag{6.6}$$

But

$$p \operatorname{div} \delta\mathbf{x} = \operatorname{div}(p\delta\mathbf{x}) - \operatorname{grad} p \cdot \delta\mathbf{x}. \tag{6.7}$$

The first term allows the transformation into a surface integral by the theorem of Gauss (3.1) and thus can have no influence on the resulting equations. Our action integral can therefore be written in the form

$$A' = \int_{t_1}^{t_2} \int_{\tau} [\rho(\tfrac{1}{2}v^2 - \phi) - \operatorname{grad} p \cdot \delta\mathbf{x}]d\tau\,dt. \tag{6.8}$$

The variation is restricted by the condition that the mass should not be varied, which means that

$$\delta(\rho d\tau) = 0. \tag{6.9}$$

Performing the variation in the usual Lagrangian fashion, we obtain the resulting equations of motion in the form

$$-\rho \frac{d\mathbf{v}}{dt} - \rho \operatorname{grad} \phi - \operatorname{grad} p = 0 \tag{6.10}$$

or

$$\frac{d\mathbf{v}}{dt} = -\operatorname{grad} \phi - \frac{\operatorname{grad} p}{\rho}. \tag{6.11}$$

The remarkable feature of these equations is that, although deduced from the assumption of incompressibility, they are valid even in the *compressible* case. We have here an excellent demonstration of the true significance of the Lagrangian multiplier (cf. p. 144) which comes about due to a "microscopic violation" of the given constraint. In our case the violation can assume macroscopic magnitude, yet the potential wall created by the forces opposing the compression is still so high that the higher-order terms become negligibly small.

7. Hydrostatics. If the fluid is in equilibrium, $\mathbf{v} = 0$ and we obtain

$$\operatorname{grad} \phi + \frac{\operatorname{grad} p}{\rho} = 0. \tag{7.1}$$

In the case of an incompressible fluid, ρ is a constant $= \rho_0$ and (7.1) becomes

$$\phi + \frac{p}{\rho_0} = \text{const.} \tag{7.2}$$

The Newtonian potential ϕ below a water surface at the depth z is

$$\phi = - gz \tag{7.3}$$

and thus p, the "hydrostatic pressure," becomes

$$p = \rho_0 gz \tag{7.4}$$

(Archimedes, 200 B.C.). In the case of air at a height z we have

$$\phi = gz \tag{7.5}$$

and

$$p = p_0 - \rho_0 gz, \tag{7.6}$$

where p_0 is the pressure at sea level, which can be measured by the compensating height of a 76 cm mercury column. The density of dry air at sea level (at 0°C) is $\rho_0 = 0.001224$ g/cm³, and the density of mercury 13.60 g/cm³; hence the atmospheric pressure at the height z measured in millimetres of mercury is

$$p = 760 - 0.900z \tag{7.7}$$

if z is measured in metres (this shows that the barometric pressure drops by 1 millimetre for every 11 metres of elevation above sea level; Pascal, 1648).

Air, however, is compressible and there exists a very definite relation between density and pressure. Under isothermic conditions

$$\frac{p}{\rho} = \frac{p_0}{\rho_0}. \tag{7.8}$$

If ρ is a function of p ("barotropic fluids"), then

$$\frac{\operatorname{grad} p}{\rho} = \frac{\operatorname{grad} p}{f(p)} = \operatorname{grad} P, \tag{7.9}$$

where

$$P = \int \frac{dp}{f(p)}. \tag{7.10}$$

This P is called the "barotropic pressure." For an isothermic atmosphere

$$P = \frac{p_0}{\rho_0} \int \frac{dp}{p} = \frac{p_0}{\rho_0} \log \frac{p}{p_0} = \frac{p_0}{\rho_0} \log \frac{\rho}{\rho_0}. \tag{7.11}$$

The condition of hydrostatic equilibrium is now

$$\phi + P = \text{const.}, \tag{7.12}$$

which for the atmosphere gives

$$gz + \frac{p_0}{\rho_0} \log \frac{p}{p_0} = 0 \tag{7.13}$$

or

$$p = p_0 e^{-\alpha z} \tag{7.14}$$

where

$$\alpha = \frac{\rho_0 g}{p_0} = 0.1184 \times 10^{-5} \, \text{cm}^{-1}.$$

Hence under isothermic conditions (at 0°C) the atmospheric pressure at the height z, measured in kilometres, is

$$p = p_0 e^{-0.1184 z} \tag{7.15}$$

(Laplace, 1817).

8. The circulation theorem. Let us lay a closed line $\mathscr{L}$ through certain fluid particles which carry this line along during their motion. We then speak of a "material line." We characterize this line in parametric form by putting

$$x_i = x_i(s), \tag{8.1}$$

which shall hold at the time t. Then, at the time $t + dt$, the same material line becomes

$$\bar{x}_i = x_i(s) + v_i(s)dt. \tag{8.2}$$

We now consider the "circulation" through this closed line, defined by

$$\int_s \mathbf{v} \cdot \mathbf{t}\, ds = \int_s \sum_{i=1}^{3} v_i\, dx_i; \tag{8.3}$$

at the time $t + dt$ we obtain

$$\int_s \sum (v_i + \dot{v}_i dt)\left(dx_i + \frac{dv_i}{ds}\, ds\, dt\right). \tag{8.4}$$

The infinitesimal difference between the two circulations is

$$\int_s \sum \left(\dot{v}_i dx_i + v_i \frac{dv_i}{ds}\, ds\right) dt. \tag{8.5}$$

The second term involves the quantity

$$\int_s d(\tfrac{1}{2}\mathbf{v}^2) = \tfrac{1}{2}(\mathbf{v}^2)_s - \tfrac{1}{2}(\mathbf{v}^2)_0 = 0 \tag{8.6}$$

in view of the fact that we have integrated over a *closed* curve, whose beginning and end coincide. What remains then is

$$\int_s \sum \dot{v}_i dx_i dt = \int_s \mathbf{v} \cdot d\mathbf{x}\, dt. \tag{8.7}$$

But the Lagrangian equations of motion (6.11) show that this quantity can be written

$$-\int_s \text{grad}(\phi + P) \cdot d\mathbf{x}\, dt$$

$$= [(\phi + P)_0 - (\phi + P)_s]dt = 0, \tag{8.8}$$

which means that *the circulation around the material line $\mathscr{L}$ remains unchanged in time*; this is true for *any* material line (Helmholtz, 1858). It demonstrates the conservation of circulation for an ideal fluid, thus explaining the destructive force of tornados.

We now lay an arbitrary material surface S through the fluid, whose boundary is the material line $\mathscr{L}$. Then by Stokes' theorem (3.18) the line integral (8.3) can be changed to the surface integral

$$\int_s (\operatorname{curl} \mathbf{v}) \cdot \mathbf{n} \, dS. \tag{8.9}$$

This quantity, carried along by the material surface S, is an *invariant* of the motion.

If it so happens that at the beginning of the motion the circulation around any closed line is zero, then also

$$\operatorname{curl} \mathbf{v} = 0. \tag{8.10}$$

We then speak of an *irrotational motion*, and this condition is *permanently maintained*.

9. Euler's form of the hydrodynamic equations. In the Lagrangian form (6.11) of the hydrodynamic equations we considered a moving fluid particle and its acceleration. The operation d/dt refers to two consecutive positions of the same particle. Euler considers a rectangular reference system which is *at rest*. The velocity field

$$\mathbf{v} = \mathbf{v}(x_1, x_2, x_3, t) \tag{9.1}$$

is now considered at a *fixed point* of space. The same is true of all other field quantities, such as pressure, density, etc. Now the operation d/dt bears the following relation to the operation $\partial/\partial t$ for any arbitrary (scalar or vector) quantity:

$$\frac{d}{dt} = \frac{\partial}{\partial t} + v_1 \frac{\partial}{\partial x_1} + v_2 \frac{\partial}{\partial x_2} + v_3 \frac{\partial}{\partial x_3} = \frac{\partial}{\partial t} + \mathbf{v} \cdot \operatorname{grad}. \tag{9.2}$$

Hence the Lagrangian equation (6.11) becomes, in the Eulerian formulation,

$$\frac{\partial \mathbf{v}}{\partial t} + (\mathbf{v} \cdot \operatorname{grad})\mathbf{v} = -\operatorname{grad}(\phi + P). \tag{9.3}$$

In the second term on the left we can make use of the vector identity (2.12) and put

$$(\mathbf{v} \cdot \text{grad})\mathbf{v} = \tfrac{1}{2} \text{grad } \mathbf{v}^2 - \mathbf{v} \times \text{curl } \mathbf{v}. \tag{9.4}$$

Euler's form of the hydrodynamic equations of a perfect fluid thus becomes

$$\frac{\partial \mathbf{v}}{\partial t} = \mathbf{v} \times \text{curl } \mathbf{v} - \text{grad}(\tfrac{1}{2}v^2 + \phi + P). \tag{9.5}$$

The non-linear terms on the right side clearly demonstrate the extraordinary difficulty of obtaining exact solutions of the hydrodynamics, even if we restrict ourselves to ideal fluids. For sufficiently small velocities the quadratic terms may be neglected and we can put

$$\frac{\partial \mathbf{v}}{\partial t} = -\text{grad}(\phi + P). \tag{9.6}$$

Moreover, the continuity equation (4.6) is in itself non-linear, but for small changes of ρ we can put it in the form

$$\frac{\partial \rho}{\partial t} + \rho_0 \text{ div } \mathbf{v} = 0, \tag{9.7}$$

in which case

$$\frac{\partial^2 \rho}{\partial t^2} + \rho_0 \frac{\partial}{\partial t} \text{ div } \mathbf{v} = 0. \tag{9.8}$$

Then, taking the divergence of (9.6), we obtain

$$\frac{\partial^2 \rho}{\partial t^2} = \rho_0 \nabla^2 P \tag{9.9}$$

since the ∇^2 of the Newtonian potential vanishes.

For ideal gases we can equate P to (7.11) if isothermic conditions prevail. Furthermore, since the change in ρ is usually small compared with ρ_0, we can put

$$P = \frac{p_0}{\rho_0} \log\left(1 + \frac{\Delta\rho}{\rho_0}\right) = \frac{p_0}{\rho_0^2} \Delta\rho, \tag{9.10}$$

in which case the relation (9.9) becomes

$$\frac{p_0}{\rho_0} \nabla^2 \Delta\rho = \frac{\partial^2 \Delta\rho}{\partial t^2}. \tag{9.11}$$

This is the customary *wave equation* for $\Delta\rho = \rho_1$:

$$\nabla^2\rho_1 = \frac{1}{c^2}\frac{\partial^2\rho_1}{\partial t^2} \tag{9.12}$$

if we put

$$c = \sqrt{(p_0/\rho_0)}. \tag{9.13}$$

This shows that, in the atmosphere, waves of density (and pressure) changes are generated (of particular interest to acoustics) which propagate (in dry air at 0°C) with the velocity

$$c = \sqrt{\frac{76 \times 13.6 \times 981}{0.001224}} = \sqrt{8.0918} \times 10^4$$

$$= 2.845 \times 10^4\,\frac{\text{cm}}{\text{sec}} = 284.5\,\frac{\text{metres}}{\text{sec}}.$$

This is the formula obtained by Newton which, however, gave too low a value for the propagation of sound waves. The correction was made by Laplace (1817), who observed that the oscillatory compression of a gas is not isothermic but an *adiabatic* phenomenon, since no time is left for a heat exchange and the gas heats up under compression (and correspondingly cools down in expansion). The proper relation between pressure and density under adiabatic conditions becomes

$$p = p_0\left(\frac{\rho}{\rho_0}\right)^\gamma, \qquad \rho = \rho_0\left(\frac{p}{p_0}\right)^{1/\gamma}, \tag{9.14}$$

where γ denotes the ratio c_p/c_v of the two types of specific heats: under constant pressure (c_p) and under constant volume (c_v). In this case

$$P = \int\frac{dp}{\rho} = \frac{p_0^{1/\gamma}}{\rho_0}\frac{p^{1-1/\gamma}}{1-1/\gamma} = \frac{p_0}{\rho_0}\left(\frac{\rho}{\rho_0}\right)^{\gamma-1}\frac{\gamma}{\gamma-1}. \tag{9.15}$$

Again, for small changes $\Delta\rho$ in the density, we can put

$$P = \frac{p_0}{\rho_0}\left(1+\frac{\Delta\rho}{\rho_0}\right)^{\gamma-1}\frac{\gamma}{\gamma-1} = \frac{p_0}{\rho_0}\frac{\gamma}{\gamma-1} + \frac{p_0\Delta\rho}{\rho_0^2}\gamma. \tag{9.16}$$

The relation (9.9) then changes to

$$\frac{\partial^2\rho_1}{\partial t^2} = \gamma\frac{p_0}{\rho_0}\nabla^2\rho_1 \tag{9.17}$$

and the propagation velocity of the oscillations becomes

$$c = \sqrt{\gamma \frac{p_0}{\rho_0}}. \tag{9.18}$$

The numerical value of γ for the atmosphere is $\gamma = 1.41$ and now we obtain

$$c = \sqrt{11.41} \times 10^4 = 3.377 \times 10^4 \frac{cm}{sec} = 337.7 \frac{metres}{sec},$$

which indeed agrees with the experimentally observed value.

10. The conservation of energy. Since in all mechanical phenomena the sum of the kinetic and potential energy is a constant, we would expect it to be easy to establish the corresponding relation for ideal fluids. In fact a complication arises in consequence of the *mutual potential energy*, which exists between the fluid particles. If in Lagrange's equation (6.11) we multiply on both sides by $\mathbf{v}dt = d\mathbf{x}$ we seem to obtain

$$d(\tfrac{1}{2}v^2) + d\phi + dP = 0, \tag{10.1}$$

which means that

$$\tfrac{1}{2}v^2 + \phi + P = \text{const.} \tag{10.2}$$

(D. Bernoulli, 1738). However, in actual fact P is generally a function of the time t, which is disregarded in the derivation of the relation (10.2). Only in the special case of a *steady motion* can the equation (10.2) be justified. A fluid motion is said to be *steady* if

$$\frac{\partial \mathbf{v}}{\partial t} = 0. \tag{10.3}$$

Similarly all other partial derivatives with respect to t vanish, i.e. we find at a certain point of space the same hydrodynamic quantities because, although the fluid particles move, their positions are immediately taken up by other particles. In this case P becomes independent of t and Bernoulli's equation (10.2) holds. We have to add, however, that the constant on the right side of (10.2) refers only to the motion along a definite stream line; it changes from stream line to stream line.

More can be said in the case of an *irrotational motion* (cf. 8.10). Here the velocity must become a pure gradient (cf. 3.3–6):

$$\mathbf{v} = \text{grad } \Phi. \tag{10.4}$$

Φ is called the "velocity potential." Euler's equation (9.5) now gives

$$\text{grad}\left(\frac{\partial \Phi}{\partial t} + \tfrac{1}{2}v^2 + \phi + P\right) = 0, \tag{10.5}$$

which means that

$$\frac{\partial \Phi}{\partial t} + \tfrac{1}{2}v^2 + \phi + P = \gamma(t). \tag{10.6}$$

The "constant" on the right side, although a function of time, is universal for *all stream lines*. If in addition our motion is *steady*, then even more can be said. Now $\partial \Phi / \partial t = 0$ and we obtain

$$\tfrac{1}{2}v^2 + \phi + P = \text{const.}, \tag{10.7}$$

where the constant on the right side is the same for all stream lines and for all times. This is the strongest form of Bernoulli's equation.

Bibliography

Lamb, H., *Hydrodynamics*, 6th ed. (Cambridge University Press, 1932).

McLeod, E. B., *Introduction to Fluid Dynamics* (Pergamon Press, 1963).

Milne-Thomson, L. M., *Theoretical Hydrodynamics* (Macmillan, London, 1949).

Serrin, J., *Mathematical Principles of Classical Fluid Mechanics*, Encyclopedia of Physics, VIII/1 (Springer, Berlin, 1959).

11. Elasticity. Mathematical tools. Whereas the theory of fluids can be based completely on vectors, the theory of elasticity demands working with *tensors* as a basic mathematical tool. Vectors can be visualized in space, and operating with a vector as a whole rather than with its components can be justified. This is no longer so in the domain of tensors. Little can be gained and a great deal lost in clarity if we try to operate with the tensor as a whole rather than its components. The classical tool of tensor analysis is the operation with *subscripts*. To this has to be added the "sum convention," which was introduced and consistently employed by Einstein (1915). This convention, now generally accepted, agrees that if in a formula of tensor analysis a certain subscript appears twice, one should automatically *sum* over that subscript (in general curvilinear coordinates the summation occurs over indices in opposite positions, one subscript and one superscript, but we shall restrict ourselves to Cartesian coordinates, in which case only subscripts will occur). This convention does away with the cumbersome summation signs which made the formulae of tensor calculus so formidable in earlier days. The summation includes the three space coordinates in pre-relativistic and the four space-time coordinates in relativistic formulae. We shall consistently adhere to this convention in all the following sections. Tensor analysis operates with arbitrary curvilinear coordinates (in any number of dimensions), involving arbitrary point transformations, but we shall restrict ourselves to the Cartesian group of *orthogonal* transformations (rectangular coordinates).

A vector **a** can be defined analytically by the linear form

$$\sum_{i=1}^{3} a_i x_i = a_i x_i,$$

demanding that any orthogonal transformation of the x_i to a new rectangular reference system x_i' shall leave the given linear form invariant:

$$a_i x_i \equiv a_j' x_j'. \tag{11.1}$$

Instead of a linear form we can now consider a *quadratic* form

$$a_{ik} x_i x_k \equiv a_{ik}' x_i' x_k', \tag{11.2}$$

thus defining $a_{ik} = a_{ki}$ as the components of a *tensor*, in this case a symmetric tensor of the "second rank." (For the definition of a non-symmetric tensor $a_{ik} \neq a_{ki}$ we can similarly use the invariance of the "bilinear form" $a_{ik} x_i y_k$, and generally for tensors of rank r the invariance of forms which are linear in r sets of variables, their coefficients having r subscripts.) There is no difficulty in showing that the transformation of the a_{ik} to a_{ik}' follows the same law as the transformation of the product $x_i x_k$ to $x_i' x_k'$.

The definition of a tensor shows directly that tensors of the same rank can be added, e.g.

$$a_{ik} + b_{ik} = c_{ik}. \tag{11.3}$$

Moreover, tensors of any rank can be multiplied, resulting in a new tensor whose rank is the sum of the ranks of the two factors, e.g.

$$a_i b_{km} = c_{ikm}. \tag{11.4}$$

Consider a tensor of any rank, for example a_{ikmn}. Now let us equate two indices and sum over them:

$$a_{iimn}. \tag{11.5}$$

The transformation law of these quantities coincides with the transformation law of

$$x_i y_i z_m u_n. \tag{11.6}$$

But $x_i y_i$ represents the *dot product* of the two vectors x and y which is an *invariant* of an orthogonal transformation. Hence the quantities (11.6) transform like $z_m u_n$, which means that the a_{iimn} form the components of a tensor of the *second rank*. The operation of equating two indices and forming the sum over them is called *contraction*. It lowers the rank of a tensor by two.

Kronecker's symbol δ_{ik}, defined by

$$\begin{aligned} \delta_{ik} &= 1, \quad \text{for } i = k, \\ &= 0, \quad \text{for } i \neq k, \end{aligned} \tag{11.7}$$

represents the components of a symmetric tensor of the second rank for all orthogonal transformations, because

$$\delta_{ik} x_i x_k = x_i^2 = x_i'^2 = \delta_{ik} x_i' x_k'. \tag{11.8}$$

Since the dx_i follow the same transformation law as the x_i, we can replace the algebraic form $a_{ik} y_i z_k$ by the differential form

$$a_{ik} dy_i dz_k \equiv a_{ik}' dy_i' dz_k'. \tag{11.9}$$

If the tensor components a_{ik} form a tensor *field* by changing from point to point, being given as continuous and differentiable functions of the rectangular coordinates x_j (and similarly a_{ik}' are functions of the transformed coordinates x_j'), we obtain, by writing down the equation (11.3) at two neighbouring points and taking the difference,

$$\frac{\partial a_{ik}}{\partial x_m} dx_m dy_i dz_k \equiv \frac{\partial a_{ik}'}{\partial x_m'} dx_m' dy_i' dz_k'. \qquad (11.10)$$

By definition we have obtained a tensor *of rank three*. The partial derivatives of a tensor field thus define a new tensor field, whose rank has been *increased by one*.

12. The strain tensor. In solid bodies strong forces are exerted against any change in shape of the body, not only a change in volume. Only a mere translation and rotation of a part of the body will not give rise to counteracting forces. The problem of elasticity is essentially a *geometrical* problem. We consider the *elastic displacement* $u_i(x_1, x_2, x_3)$, which changes the position of the point x_i to

$$\bar{x}_i = x_i + u_i, \qquad (12.1)$$

and the infinitesimal distance between two neighbouring points (called the "line element," cf. p. 18), before and after the elastic displacement. In rectangular coordinates

$$ds^2 = dx_1^2 + dx_2^2 + dx_3^2 = dx_i^2 = \delta_{ik} dx_i dx_k. \qquad (12.2)$$

The corresponding distance after the elastic displacement is

$$d\bar{s}^2 = (dx_i + du_i)^2 = dx_i^2 + 2dx_i du_i + du_i^2$$

$$= \left[\delta_{ik} + \left(\frac{\partial u_i}{\partial x_k} + \frac{\partial u_k}{\partial x_i} \right) + \frac{\partial u_m}{\partial x_i} \frac{\partial u_m}{\partial x_k} \right] dx_i dx_k. \qquad (12.3)$$

The difference between these two quadratic forms gives rise to a tensor

$$2\sigma_{ik} = \frac{\partial u_i}{\partial x_k} + \frac{\partial u_k}{\partial x_i} + \frac{\partial u_m}{\partial x_i} \frac{\partial u_m}{\partial x_k}, \qquad (12.4)$$

which is called the "strain tensor." The strain components

(which are dimensionless numbers) are usually small and the last non-linear term can be omitted. We shall thus put

$$\sigma_{ik} = \frac{1}{2}\left(\frac{\partial u_i}{\partial x_k} + \frac{\partial u_k}{\partial x_i}\right). \tag{12.5}$$

The potential energy density of the elastic forces is quadratic in the strain components, which we can write in the form

$$W = \tfrac{1}{2} c_{ikmn}\,\sigma_{ik}\,\sigma_{mn}. \tag{12.6}$$

Since W is a mere scalar and thus invariant with respect to rotations of the reference system, the coefficients c_{ikmn} must form the components of a tensor of rank four. The number of $\sigma_{ik} = \sigma_{ki}$ components is six so that we have a quadratic form in a six-dimensional space with

$$\tfrac{1}{2}(6.7) = 21$$

independent coefficients. Generally in the elastic behaviour of crystals all these coefficients occur, although the number can be reduced to 18 by the choice of a properly adapted reference system; and under higher symmetry conditions it can be even further reduced. For example, if axial symmetry prevails, with x_3 chosen as the axis of symmetry, then W must have the form

$$W = \tfrac{1}{2}[c_1(\sigma_{11}{}^2 + 2\sigma_{12}{}^2 + \sigma_{22}{}^2) + c_2(\sigma_{11} + \sigma_{22})^2 + c_3\sigma_{33}{}^2$$
$$+ c_4(\sigma_{11} + \sigma_{22})\sigma_{33} + c_5(\sigma_{13}{}^2 + \sigma_{23}{}^2)]. \tag{12.7}$$

Axial symmetry in elastic behaviour thus reduces the consideration to only five constants.[1]

For an isotropic body no preferential directions remain. In this case we must obtain, from the fourth-rank tensor $\sigma_{ik}\sigma_{mn}$, a scalar by the tools of tensor algebra. This is only possible by two contractions. Equating i and k and likewise m and n results in the scalar

$$\sigma_{ii}\sigma_{kk} = (\sigma_{ii})^2 = \sigma^2$$

when we denote the scalar σ_{ii} by σ (not to be confounded with "Poisson's ratio," see later in (16.2)). On the other hand, equating i and m (and likewise k and n) yields the scalar

[1] Cf. J. L. Synge, *Journal of Mathematics and Physics*, *35* (1957), 323.

$$\sigma_{ik}\sigma_{ik} = (\sigma_{ik})^2. \tag{12.8}$$

Hence under isotopic conditions the elastic energy assumes the form

$$W = \tfrac{1}{2}[\lambda\sigma^2 + 2\mu(\sigma_{ik})^2]. \tag{12.9}$$

λ and μ are called "Lamé's constants" (J. Lamé, 1852), or the two "moduli" of elasticity.

Many solids are actually composed of a large number of randomly oriented crystals. Under these conditions the solid is *macroscopically isotropic*, although it is microscopically crystalline.

13. The stress tensor. We shall consider the equilibrium of an elastically stressed body. Equilibrium demands that the potential energy be a minimum; we have therefore to find the first variation of the elastic energy:

$$\delta\int W d\tau = \int [\lambda\sigma\delta\sigma + 2\mu\sigma_{ik}\delta\sigma_{ik}]d\tau. \tag{13.1}$$

Now

$$\delta\sigma = \delta\sigma_{ii} = \frac{\partial\delta u_i}{\partial x_i}, \quad \delta\sigma_{ik} = \frac{1}{2}\left(\frac{\partial\delta u_i}{\partial x_k} + \frac{\partial\delta u_k}{\partial x_i}\right). \tag{13.2}$$

Making use of the Gaussian integral transformation (3.1) we obtain

$$\delta\int W d\tau = -\int_\tau \left[\lambda\frac{\partial\sigma}{\partial x_i}\delta u_i + 2\mu\frac{\partial\sigma_{ik}}{\partial x_k}\delta u_i\right]d\tau$$

$$+ \int_s [\lambda\sigma n_k\delta u_k + 2\mu\sigma_{ik} n_k\delta u_i]dS. \tag{13.3}$$

The integrand of the volume integral, if written in vector notation, becomes

$$-[\lambda\,\mathrm{grad\,div}\,\mathbf{u} + \mu(\nabla^2\mathbf{u} + \mathrm{grad\,div}\,\mathbf{u})]\cdot\delta\mathbf{u}. \tag{13.4}$$

To this has to be added the variation of the potential energy of gravity, which gives

$$\rho\,\mathrm{grad}\,\phi\cdot\delta\mathbf{u}. \tag{13.5}$$

The condition of elastic equilibrium thus demands the condition

$$\mu\nabla^2\mathbf{u} + (\lambda + \mu)\,\text{grad div }\mathbf{u} = \rho\,\text{grad }\phi. \qquad (13.6)$$

If in equation (13.4) we do not substitute for the strain tensor its explicit expression (12.5), we can write (13.6) in the form

$$\lambda\frac{\partial\sigma}{\partial x_i} + 2\mu\frac{\partial\sigma_{ik}}{\partial x_k} = \rho\frac{\partial\phi}{\partial x_i}, \qquad (13.7)$$

and the left side can be interpreted as the gradient operation $\partial/\partial x_k$:

$$\frac{\partial\tau_{ik}}{\partial x_k} = \rho\frac{\partial\phi}{\partial x_i}, \qquad (13.8)$$

applied to the tensor

$$\tau_{ik} = \lambda\sigma\delta_{ik} + 2\mu\sigma_{ik}; \qquad (13.9)$$

τ_{ik} is called the "stress tensor" (Cauchy, 1822).

The vectorial differential equation (13.6) has to be complemented by the proper boundary conditions. If the boundary is fixed, then the elastic displacement $\mathbf{u}$ is prescribed as zero on the boundary (or part of the boundary), which gives the boundary condition

$$\mathbf{u}(S) = 0. \qquad (13.10)$$

On a free part S' of the boundary the surface integral (13.3) gives the boundary condition

$$\tau_{ik}n_k = 0. \qquad (13.11)$$

If a given surface force f_i (force per cm²) acts on the boundary, then the virtual work of this force becomes

$$\int f_i\,\delta u_i\,dS, \qquad (13.12)$$

which gives the boundary condition

$$\tau_{ik}n_k = f_i. \qquad (13.13)$$

14. Small elastic vibrations. The addition of the kinetic energy to our variational problem (in the usual sense of $L = T - V$) contributes to the equation of equilibrium (13.6) an acceleration term so that for small displacements we obtain the equation of motion in the form

$$\rho \frac{\partial^2 \mathbf{u}}{\partial t^2} = \mu \nabla^2 \mathbf{u} + (\lambda + \mu) \text{ grad div } \mathbf{u} - \rho \text{ grad } \phi \qquad (14.1)$$

(the last term is usually negligibly small).

If we resolve the vector $\mathbf{u}$ into the sum of a gradient and a curl (cf. (3.3)):

$$\mathbf{u} = \text{curl } \mathbf{A} + \text{grad } \psi$$

we obtain for the vector $\mathbf{A}$ the wave equation

$$\rho \frac{\partial^2 \mathbf{A}}{\partial t^2} = \mu \nabla^2 \mathbf{A},$$

and for the scalar ψ

$$\rho \frac{\partial^2 \psi}{\partial t^2} = (\lambda + 2\mu) \nabla^2 \psi.$$

The scalar (solenoidal) wave propagates with the velocity

$$c_1 = \sqrt{[(\lambda + 2\mu)/\rho]},$$

and the vectorial (vortex) wave propagates with the velocity

$$c_2 = \sqrt{(\mu/\rho)}.$$

Both waves can be observed in seismological observatories on the occasion of earthquakes.

15. The Hamiltonization of variational problems. Hamilton's ingenious method of changing the second-order Lagrangian equations into a double number of first-order equations (cf. Appendix I) is again applicable, although the increased number of equations is not necessarily twice the original number. The basic idea carries over, however, to the realm of partial operators. We can write the elastic energy density W in the purely algebraic form

$$W = \tfrac{1}{2}(\lambda \sigma_{ii} \sigma_{kk} + 2\mu \sigma_{ik}^2) \qquad (15.1)$$

if we add the auxiliary condition

$$\frac{1}{2}\left(\frac{\partial u_i}{\partial x_k} + \frac{\partial u_k}{\partial x_i}\right) - \sigma_{ik} = 0. \tag{15.2}$$

This gives rise to the Lagrangian multiplier τ_{ik} and the added term

$$\tau_{ik}\left[\frac{1}{2}\left(\frac{\partial u_i}{\partial x_k} + \frac{\partial u_k}{\partial x_i}\right) - \sigma_{ik}\right], \tag{15.3}$$

which, in view of the symmetry condition $\tau_{ik} = \tau_{ki}$, is reducible to

$$\tau_{ik}\left(\frac{\partial u_i}{\partial x_k} - \sigma_{ik}\right). \tag{15.4}$$

We thus have the modified Lagrangian

$$W' = \tfrac{1}{2}[\lambda\sigma^2 + 2\mu\sigma_{ik}{}^2] + \tau_{ik}\left(\frac{\partial u_i}{\partial x_k} - \sigma_{ik}\right) \tag{15.5}$$

with the action variables u_i, σ_{ik}, τ_{ik}. Since, however, W' is purely algebraic in the σ_{ik}, they can be eliminated (cf. p. 128, Problem 3):

$$\frac{\partial W'}{\partial \sigma_{ik}} = \lambda\sigma\delta_{ik} + 2\mu\sigma_{ik} - \tau_{ik} = 0. \tag{15.6}$$

Our variational integrand now becomes

$$W' = \tau_{ik}\frac{\partial u_i}{\partial x_k} - H, \tag{15.7}$$

where H (the Hamiltonian function) is

$$H = \tau_{ik}\sigma_{ik} - \tfrac{1}{2}\tau_{ik}\sigma_{ik} = \tfrac{1}{2}\tau_{ik}\sigma_{ik}$$
$$= \frac{1}{4\mu}\left(\tau_{ik}{}^2 - \frac{\lambda}{3\lambda + 2\mu}\tau^2\right). \tag{15.8}$$

In this formulation the importance of Cauchy's stress tensor comes directly into evidence, because the multiplier τ_{ik} is in fact *identical with the stress tensor*. The variation of u_i gives rise to the boundary integral

$$\int \tau_{ik} n_k \delta u_i dS \tag{15.9}$$

and the operator

$$- \frac{\partial \tau_{ik}}{\partial x_k} . \tag{15.10}$$

Moreover, the relation of the stress tensor to the displacement is now a consequence of the variational principle, since variation with respect to τ_{ik} yields

$$\frac{1}{2} \left(\frac{\partial u_i}{\partial x_k} + \frac{\partial u_k}{\partial x_i} \right) = \frac{1}{2\mu} \left(\tau_{ik} - \frac{\lambda}{3\lambda + 2\mu} \tau \delta_{ik} \right), \tag{15.11}$$

which means, denoting the left side by σ_{ik}:

$$\sigma_{ik} = \frac{1}{2\mu} \left(\tau_{ik} - \frac{\lambda}{3\lambda + 2\mu} \tau \delta_{ik} \right), \tag{15.12}$$

$$\sigma = \frac{1}{2\mu} \left(\tau - \frac{3\lambda}{3\lambda + 2\mu} \tau \right) = \frac{\tau}{3\lambda + 2\mu}, \tag{15.13}$$

and thus

$$\tau_{ik} = 2\mu \sigma_{ik} + \lambda \sigma \delta_{ik}$$

in agreement with (13.8).

16. Young's modulus, Poisson's ratio. The theoretically convenient constants λ and μ are not adapted to easy experimental measurements. Th. Young (1807) measured the elastic elongation of a cylindrical rod, loaded by the weight Q. If l is the length of the rod and q its cross-section, then

$$\frac{Q}{q} = E \frac{\Delta l}{l}, \tag{16.1}$$

where E is called "Young's modulus." Poisson (1819) showed that associated with this elongation Δl is a small shrinking of the radius of the rod, in the sense of

$$- \frac{\Delta r}{r} \div \frac{\Delta l}{l} = \sigma. \tag{16.2}$$

This pure number (restricted to the range 0 to $\frac{1}{2}$) is called "Poisson's ratio." Lamé's constants λ and μ bear the following relation to E and σ:

$$\mu = \frac{1}{2} \frac{E}{1 + \sigma}, \quad \lambda = \frac{E\sigma}{(1 + \sigma)(1 - 2\sigma)}. \tag{16.3}$$

17. Elastic stability. Let us assume that we strike an elastic structure with a hammer. This will excite the characteristic vibrations of the system, which start with a lowest frequency and increase (theoretically) to infinity. Since the potential energy of an elastic structure is always positive definite, even its minimum is positive and all the eigenvalues can only be positive. If now we increase the external load of the structure and repeat the experiment, we shall find that all the frequencies decrease. At some particular load the point might come when the lowest eigenvalue becomes zero and then goes over to the negative range, which means that the trigonometric functions change to exponential functions, and the vibrations around the state of equilibrium change to a permanent departure from the original state of lowest equilibrium to a new state of still lower potential energy. This sudden change of an elastic structure into a completely new shape is called "buckling."

Buckling could never be explained on the basis of linear elasticity, i.e. a quadratic Lagrangian, because such a Lagrangian possesses inherent eigenvalues which cannot change with the external load. However, if the strain tensor increases, the quadratic Lagrangian ceases to hold and we enter the domain of non-linear elasticity. Even now the elastic energy remains positive definite, but if we again strike the structure with a hammer and thus apply a small *perturbation*, the perturbation energy is once more purely *quadratic* in the displacement components and the perturbation equations are once more *linear*. The difference is, however, that the perturbation Lagrangian and the original Lagrangian no longer coincide so that it is entirely possible that, although the original Lagrangian remains positive definite, the perturbation Lagrangian ceases to be so. If a load is reached at which the perturbation energy for some small added displacement can vanish, then the elastic structure loses its stability and "buckles" over into a new shape. This explains why an elastic structure can suddenly collapse; it may seem to do so because the limit of elasticity has been exceeded, but in actual fact it does so because of the highly non-linear nature of the equilibrium equations in the case of large displacements. The difficulty with the mathematical theory of buckling is thus primarily due to the fact that the theory of large elastic displacements (non-linear elasticity) still poses some unsolved problems.

Bibliography

Condon, E. U., and Odishaw, H., eds., *Handbook of Physics*, Part 3, Mechanics of Deformable Bodies (McGraw-Hill, 1958).

Sechler, E. E., *Elasticity in Engineering* (John Wiley & Sons, 1952).

Sokolnikoff, I. S., *Mathematical Theory of Elasticity* (McGraw-Hill, 1946).

Timoshenko, S., and Goodier, J. N., *Theory of Elasticity* (McGraw-Hill, 1951).

18. Electromagnetism. Mathematical tools. Einstein's discovery of the relative nature of time, and the formulation of special relativity by Minkowski as a four-dimensional Euclidean world (with the fourth coordinate $x_4 = ict$; cf. Chapter IX, section 3), make it imperative that the usual three-dimensional world of vectors and tensors be extended to four dimensions. Hence in our summation convention the subscripts should go consistently from 1 to 4, the components with subscript 4 being automatically associated with $x_4 = it$, because we want to measure the time t in "natural units" in which the velocity of light c is equal to 1 (we shall adhere to this convention in all the following discussions).

The Maxwellian theory of electromagnetism (1873) now automatically falls in line with the usual theory of tensors in a four-dimensional Euclidean world. In contradiction to elasticity, the basic field quantity F_{ik} is not a symmetric but an *antisymmetric* tensor of the second rank:

$$F_{ik} = \frac{\partial \phi_k}{\partial x_i} - \frac{\partial \phi_i}{\partial x_k}. \tag{18.1}$$

The space part of the four-vector ϕ_i can no longer be interpreted as a "displacement" (although this was still attempted in Maxwell's time), but as a field vector which we accept as a basic quantity of the electromagnetic field, without any specific kinematic significance.

19. The Maxwell equations. We consider the basic Lagrangian

$$L = \frac{1}{4}\left(\frac{\partial \phi_k}{\partial x_i} - \frac{\partial \phi_i}{\partial x_k}\right)^2 - J_i\phi_i, \tag{19.1}$$

where J_i, the "four-current," is assumed to be given as a function of the variables x_i. The variation of ϕ_i yields the field equations

$$\frac{\partial F_{ik}}{\partial x_k} = J_i. \tag{19.2}$$

Since F_{ik} is antisymmetric, the operation $\partial/\partial x_i$, applied to the left side, gives identically zero; this means that the four-current J_i must satisfy the condition

$$\frac{\partial J_i}{\partial x_i} = 0. \tag{19.3}$$

This equation expresses (in analogy with the conservation of mass; cf. (4.6)) the *conservation of the electric charge.*

Since ϕ_i is only defined up to the gradient of a scalar, we can subject it to the condition

$$\frac{\partial \phi_i}{\partial x_i} = 0. \tag{19.4}$$

(the "Lorentz condition," 1905). Here the equation (19.2) assumes the form

$$\Delta\phi_i = -J_i; \tag{19.5}$$

ϕ_i is called the "four-dimensional vector potential." The operation Δ is defined by

$$\Delta = \frac{\partial^2}{\partial x_k^2} \quad (k = 1, 2, 3, 4) \tag{19.6}$$

and is identical with the *wave operator* (propagation velocity $= 1$).

The Hamiltonian form of our variational principle is again obtained by writing it in the form

$$L = \tfrac{1}{4}F_{ik}^2 - J_i\phi_i \tag{19.7}$$

with the auxiliary condition

$$\frac{\partial \phi_k}{\partial x_i} - \frac{\partial \phi_i}{\partial x_k} - F_{ik} = 0. \tag{19.8}$$

This gives rise to the Lagrangian multiplier $\frac{1}{2}\tau_{ik}$ and the modified Lagrangian

$$L' = \tfrac{1}{4}F_{ik}^{\ 2} - J_i\phi_i + \tfrac{1}{2}\tau_{ik}\left(\frac{\partial \phi_k}{\partial x_i} - \frac{\partial \phi_i}{\partial x_k} - F_{ik}\right). \tag{19.9}$$

Eliminating the purely algebraic F_{ik}, we obtain

$$\tfrac{1}{2}F_{ik} - \tfrac{1}{2}\tau_{ik} = 0 \tag{19.10}$$

and

$$L' = \tfrac{1}{2}\tau_{ik}\left(\frac{\partial \phi_k}{\partial x_i} - \frac{\partial \phi_i}{\partial x_k}\right) - H, \tag{19.11}$$

where

$$H = \tfrac{1}{2}\tau_{ik}F_{ik} - \tfrac{1}{4}F_{ik}^{\ 2} + J_i\phi_i = \tfrac{1}{4}\tau_{ik}^{\ 2} + J_i\phi_i. \tag{19.12}$$

The variation of ϕ_i now yields

$$\frac{\partial \tau_{ik}}{\partial x_k} = J_i, \tag{19.13}$$

whereas the variation of τ_{ik} yields

$$\frac{\partial \phi_k}{\partial x_i} - \frac{\partial \phi_i}{\partial x_k} - \tau_{ik} = 0. \tag{19.14}$$

This double system of first-order equations is equivalent to the second-order system (19.5). The definition of the field strength τ_{ik} in terms of the vector potential is now a consequence of the variational principle.

Maxwell, of course, did not operate with relativistic concepts but with the vectors and scalars of ordinary three-dimensional space. The four-potential ϕ_i was broken up into the vector potential $\mathbf{A}$ and the scalar potential Φ, and likewise the four-current J_i into the electric current density $\mathbf{j}$ and charge density ρ:

$$\begin{aligned}
(\phi_1, \phi_2, \phi_3; \phi_4) &= (A_1, A_2, A_3; i\Phi), \\
(J_1, J_2, J_3; J_4) &= (j_1, j_2, j_3; i\rho).
\end{aligned} \tag{19.15}$$

Moreover, the antisymmetric tensor $F_{ik} = \tau_{ik}$ (with its six independent components) was replaced by the two vectors $\mathbf{E}$ and $\mathbf{H}$, according to the following scheme:

$$F_{14} = -iE_1, \quad F_{24} = -iE_2, \quad F_{34} = -iE_3,$$
$$F_{12} = H_3, \quad F_{23} = H_1, \quad F_{31} = H_2. \tag{19.16}$$

The equations (19.13) now become, if multiplied by -1:

$$\frac{\partial \mathbf{E}}{\partial t} - \operatorname{curl} \mathbf{H} = -\mathbf{j},$$
$$\operatorname{div} \mathbf{E} = -\rho \tag{19.17}$$

and the equations (19.14) become

$$\operatorname{curl} \mathbf{A} - \mathbf{H} = 0,$$
$$\operatorname{grad} \Phi + \frac{\partial \mathbf{A}}{\partial t} + \mathbf{E} = 0. \tag{19.18}$$

$\mathbf{A}$ and Φ can be eliminated from these equations, replacing them by the "second set of Maxwellian equations"

$$\frac{\partial \mathbf{H}}{\partial t} + \operatorname{curl} \mathbf{E} = 0,$$
$$\operatorname{div} \mathbf{H} = 0. \tag{19.19}$$

The eight equations (19.17) and (19.19) are Maxwell's equations, in the simplified form in which Lorentz presented them in his Electron Theory (1895).

The Maxwellian scheme operates solely with the electric and magnetic field strengths $\mathbf{E}$ and $\mathbf{H}$, and the question can be raised whether these equations in the given form might not be derivable from a variational principle, without reference to the vector four-potential. This, however, cannot be done in a natural way, since we cannot expect to obtain eight equations by varying six quantities. It is of interest, however, that, when the two divergence equations are omitted, the remaining two vector equations for the sourceless (right side equal to zero) case can in fact be derived from the following Lagrangian:

$$L = \mathbf{F}^* \left(\frac{\partial \mathbf{F}}{i \partial t} - \operatorname{curl} \mathbf{F} \right), \tag{19.20}$$

where $\mathbf{F}$ denotes the complex vector

$$\mathbf{F} = \mathbf{H} + i\mathbf{E}$$

and the asterisk refers to a change of i to $-i$. (This Lagrangian has no relativistic significance.) The resulting equation

$$\frac{\partial \mathbf{F}}{i \partial t} - \operatorname{curl} \mathbf{F} = 0$$

(if we add the omitted equation div $\mathbf{F} = 0$) combines the eight Maxwellian equations for the vacuum into four equations for the complex vector $\mathbf{F} = \mathbf{H} + i\mathbf{E}$.

Bibliography

Bleaney, B. I., and Bleaney, B. *Electricity and Magnetism* (Oxford University Press, 1957).

Jeans, J. H., *The Mathematical Theory of Electricity and Magnetism* (Cambridge University Press, 1925).

Lorentz, H. A., *The Theory of Electrons* (1905; Dover Reprint, New York, 1952).

Smythe, W. R., *Static and Dynamic Electricity* (McGraw-Hill, 1950).

Stratton, J. A., *Electromagnetic Theory* (McGraw-Hill, 1941).

20. Noether's principle. In the realm of point mechanics we have encountered (cf. Appendix II, p. 401) the ingenious idea of E. Noether (1918) that any infinitesimal transformation of either the action variables, or the independent variable, involving a constant parameter α, which leaves the Lagrangian unchanged, leads automatically to a certain conservation law. We deduced on this basis the conservation laws of momentum, energy, and angular momentum. In the realm of partial operators Noether's principle is equally valid and widely employed in contemporary physics. We shall demonstrate the application of this principle in a few characteristic and particularly important examples.

The Maxwellian equations. In the Lagrangian (19.1) the differential operator remains unchanged if ϕ_i is transformed to $\phi_i + \partial\alpha/\partial x_i$. We can consider α as an added field variable. Our Lagrangian (19.1) is now complemented by the term

$$-J_i \frac{\partial\alpha}{\partial x_i}$$

and the variation of α yields the condition

$$\frac{\partial J_i}{\partial x_i} = 0, \tag{20.1}$$

which expresses the conservation of electric charge. We have obtained this result before, by manipulating the field equations, but here it is obtained as a matter of principle, as a consequence of Noether's principle.

Let us consider the Lagrangian (19.20). The transformation of $\mathbf{F}$ to $\mathbf{F}e^{i\alpha}$ leaves the Lagrangian unchanged. It suffices, however, to consider α as infinitesimal, in which case our transformation becomes

$$\bar{\mathbf{F}} = \mathbf{F}(1 + i\alpha). \tag{20.2}$$

We now change α to an *added field variable* which is no longer a constant. Then

$$L' = L + \mathbf{F}^*\mathbf{F}\frac{\partial i\alpha}{i\partial t} + i\,\mathrm{grad}\,\alpha \cdot (\mathbf{F}^* \times \mathbf{F}). \tag{20.3}$$

Variation of α yields the equation

$$\frac{\partial(\mathbf{F}^* \cdot \mathbf{F})}{\partial t} + i\,\mathrm{div}(\mathbf{F}^* \times \mathbf{F}) = 0, \tag{20.4}$$

which in real terms becomes

$$\frac{1}{2}\frac{\partial(\mathbf{E}^2 + \mathbf{H}^2)}{\partial t} + \mathrm{div}(\mathbf{E} \times \mathbf{H}) = 0. \tag{20.5}$$

This fundamental relation expresses the *conservation of energy* of the (sourceless) electromagnetic field.

The Schrödinger equation. The Schrödinger equation (cf. p. 279), if we include the time t, is derivable from the following Lagrangian:

$$L = \psi^* \left(\frac{\partial \psi}{i \partial t} + V\psi \right) + \frac{1}{2m} \operatorname{grad} \psi^* \cdot \operatorname{grad} \psi \qquad (20.6)$$

($h/2\pi$ is normalized to 1). Here again the Lagrangian remains invariant with respect to a complex phase factor (assumed to be infinitesimal) applied to ψ:

$$\bar{\psi} = \psi(1 + i\alpha). \qquad (20.7)$$

Making α an added field variable, we obtain

$$L' = L + \psi^*\psi \frac{\partial \alpha}{\partial t}$$

$$+ \frac{i}{2m} \operatorname{grad} \alpha \, (\psi \operatorname{grad} \psi^* - \psi^* \operatorname{grad} \psi). \qquad (20.8)$$

Variation of α yields the relation

$$\frac{\partial}{\partial t} (\psi^*\psi) + \operatorname{div} \frac{i}{2m} (\psi \operatorname{grad} \psi^* - \psi^* \operatorname{grad} \psi). \qquad (20.9)$$

This represents the conservation of the probabilistic electric charge in Schrödinger's theory.

Generally, if in a transformation of the field variables an infinitesimal parameter α cancels out in the Lagrangian, then, making α an added field variable, the Euler-Lagrange equation associated with this variable can only take the form

$$\frac{\partial \rho}{\partial t} + \operatorname{div} \mathbf{A} = 0, \qquad (20.10)$$

which can always be interpreted as a *conservation law*, considering ρ as the density and $\mathbf{A}$ as the current of a certain physically significant quantity.

21. Transformation of the coordinates. Noether's principle is applicable also when we leave the action variables unchanged but transform the coordinates x_1, x_2, x_3, x_4. We want to assume that the Lagrangian does not contain the x_i explicitly in the form of given functions (such as, for example, the four-current J_i in the case of the Maxwell equations; cf. (19.1)). We now transform the rectangular coordinates x_i to curvilinear coordinates $\bar{x}_i$ by the infinitesimal transformation

$$\bar{x}_i = x_i + \epsilon \xi_i(x_1, \ldots, x_4) \tag{21.1}$$

and consider L as a function of the $\bar{x}_i$. Then the operation $\partial/\partial x_k$ has to be transformed by the rule of implicit differentiation as follows:

$$\frac{\partial}{\partial x_k} = \frac{\partial}{\partial \bar{x}_\alpha} \cdot \frac{\partial \bar{x}_\alpha}{\partial x_k} = \left(\delta_{\alpha k} + \epsilon \frac{\partial \xi_\alpha}{\partial x_k}\right) \frac{\partial}{\partial \bar{x}_\alpha}. \tag{21.2}$$

In addition, the volume element $d\tau_4$ has to be transformed according to (cf. (5.3))

$$d\tau_4 = d\tau_4' \left(1 - \epsilon \frac{\partial \xi_\alpha}{\partial x_\alpha}\right). \tag{21.3}$$

Hence the infinitesimal change of the action integral in consequence of the infinitesimal coordinate transformation (21.1) can be written as follows:

$$\int L'(\bar{x}_i) d\tau_4' = \int L \, d\tau_4$$
$$+ \epsilon \int \left(\frac{\partial L}{\partial \phi_{i,\alpha}} \phi_{i,k} \frac{\partial \xi_k}{\partial x_\alpha} - L \frac{\partial \xi_\alpha}{\partial x_\alpha}\right) d\tau_4 \tag{21.4}$$

with the notation

$$\phi_{i,\alpha} = \frac{\partial \phi_i}{\partial x_\alpha}.$$

We can consider the ξ_k as added field variables, which now participate in the variational problem. Variation with respect to ξ_k gives the conservation law

$$\frac{\partial}{\partial x_\alpha}\left(\frac{\partial L}{\partial \phi_{i,\alpha}} \phi_{i,k}\right) - \frac{\partial L}{\partial x_k} = 0, \tag{21.5}$$

which may also be written in the form

$$\frac{\partial t_{ik}}{\partial x_k} = 0 \tag{21.6}$$

with

$$t_{ik} = \frac{\partial L}{\partial \phi_{\alpha,k}} \phi_{\alpha,i} - L\delta_{ik}. \tag{21.7}$$

If the number of dimensions is reduced to one and the co-ordinates x_i to the single coordinate $x_4 = t$, then i and k are reduced to the single value $i = k = 4$ and the tensor (21.7) to the single component

$$t_{44} = \frac{\partial L}{\partial \phi_\alpha} \phi_\alpha - L \tag{21.8}$$

in which we recognize the total energy H of point mechanics (cf. p. 123). The equation (21.6) now takes the form

$$\frac{dt_{44}}{dt} = 0, \tag{21.9}$$

which expresses the energy conservation law. Since in particle mechanics the conservation of energy can be conceived as a consequence of Noether's principle due to the translation invariance of L with respect to t, while the conservation of momentum follows from the translation invariance with respect to the origin of the space coordinates (cf. pp. 402–404), it seems proper that we interpret the four equations (21.6) in field mechanics as the *conservation of momentum* ($i = 1, 2, 3$) *and energy* ($i = 4$).

As an example let us consider Newton's gravitational theory which is based on the scalar potential φ, subjected (in empty space) to the Laplacian equation

$$\nabla^2 \varphi = 0. \tag{21.10}$$

We bring this equation into harmony with the demands of (special) relativity (see p. 302) by complementing the Newtonian vacuum equation by a fourth coordinate $x_4 = it$:

$$\frac{\partial^2 \varphi}{\partial x_k^2} = 0 \quad (k = 1, 2, 3, 4) \tag{21.11}$$

expressing the fact (dictated by relativity) that gravitation (as all other forces of nature) propagates with the velocity of light and not with infinite velocity. The Lagrangian associated with this equation is

$$L = -\frac{1}{2}\left(\frac{\partial \varphi}{\partial x_i}\right)^2. \tag{21.12}$$

Now the expression (21.7) gives the following result for the t_{ik} (the sum index α being restricted to 1):

$$t_{ik} = -\frac{\partial \varphi}{\partial x_k}\frac{\partial \varphi}{\partial x_i} + \frac{1}{2}\left(\frac{\partial \varphi}{\partial x_m}\right)^2 \delta_{ik}. \tag{21.13}$$

We notice that in our case t_{ik} becomes a *symmetric* tensor

$$t_{ik} = t_{ki}, \tag{21.14}$$

which is generally not the case. The tensor t_{ik} defined by (21.7) is usually referred to as the "canonical energy-momentum tensor" (Pauli, 1920).

22. The symmetric energy-momentum tensor.

In the derivation of the tensor (21.7) we made use of the transformation of coordinates, without transforming the field variables ϕ_i. This means that we made an infinitesimal coordinate transformation, in which the field variables ϕ_i were treated as *scalars*. This, however, seldom fits the physical situation. In most field theories we encounter vectors or tensors (or possibly "spinors") which transform in a very definite manner when the coordinates are transformed. To treat them as scalars when in fact they are components of a tensor or spinor is obviously logically untenable. For this reason we must question the value of the "canonical energy-momentum tensor" obtained previously, even though this is occasionally overlooked (for example, Einstein in his "general relativity" defines the energy-momentum tensor of the gravitational field on the basis of the previous tensor (21.7), although the field variables are the ten components g_{ik} of the metrical tensor).

We shall assume that the basic Lagrangian satisfies the demand of special relativity that it be an invariant of an arbitrary Lorentz transformation (cf. p. 303), i.e. it remains unchanged if the four coordinates $x_1, x_2, x_3, x_4 = it$ are subjected to an arbitrary four-dimensional rotation. For our purposes it suffices to consider only *infinitesimal* rotations.

Now *two* changes occur. We have the transformation of the operation $\partial \phi_i/\partial x_k$ considered before (plus the infinitesimal change of the volume element), but we must also transform the field variables ϕ_i in the proper manner. This latter transformation will depend on the specific nature of the ϕ_i (whether they should be considered as scalars or tensors or spinors). At all events the general form of the transformation will be

$$\phi_i = \bar{\phi}_i + \epsilon \rho_{imjk} \frac{\partial \xi_j}{\partial x_k} \bar{\phi}_m, \tag{22.1}$$

where ρ_{ijkm} are given constants. Now the transformation of the operation $\partial \phi_i / \partial x_\alpha$ will give rise to terms in which also the *second derivatives* $\partial^2 \xi_j / \partial x_\alpha \, \partial x_\beta$ occur. Hence the resulting L' will take the form

$$L' = L + \epsilon \left(A_{ik} \frac{\partial \xi_i}{\partial x_k} + B_{ikm} \frac{\partial^2 \xi_m}{\partial x_i \partial x_k} - L \frac{\partial \xi_i}{\partial x_i} \right). \tag{22.2}$$

Here the A_{ik} and B_{ikm} are some given functions of the ϕ_i and their partial derivatives $\partial \phi_i / \partial x_k$.

The first term inside the parentheses can be split into a symmetric and an antisymmetric part:

$$\epsilon A_{ik} \frac{\partial \xi_i}{\partial x_k} = \frac{\epsilon}{4} (A_{ik} + A_{ki}) \left(\frac{\partial \xi_i}{\partial x_k} + \frac{\partial \xi_k}{\partial x_i} \right)$$

$$+ \frac{\epsilon}{4} (A_{ik} - A_{ki}) \left(\frac{\partial \xi_i}{\partial x_k} - \frac{\partial \xi_k}{\partial x_i} \right). \tag{22.3}$$

The quantity

$$\epsilon \left(\frac{\partial \xi_i}{\partial x_k} - \frac{\partial \xi_k}{\partial x_i} \right) \tag{22.4}$$

appearing in the second line has the meaning of an infinitesimal *rotation* of the reference system. The demand that L must not change under the influence of a rotation of the reference system means that the second line of (22.3) *must vanish*, with the result that in the expression (22.2) we can consider both A_{ik} and B_{ikm} as *symmetric* in i, k.

The variation of the ξ_i gives the conservation law

$$\frac{\partial T_{ik}}{\partial x_k} = 0 \tag{22.5}$$

with

$$-T_{ik} = A_{ik} - L\delta_{ik} - \frac{\partial B_{\alpha ki}}{\partial x_\alpha}. \tag{22.6}$$

The last term is not symmetric in i, k, but let us suppose that we modify it as follows:

$$-\frac{\partial}{\partial x_\alpha}(B_{\alpha ki} + B_{\alpha ik} - B_{kia}) \tag{22.7}$$

The symmetry in i, k is then restored. Moreover, if we differentiate with respect to x_k, the contribution from the added second and third terms *vanishes*, because they are antisymmetric in α, k and both of these indices are now summation indices.

We thus see that any Lorentz-invariant L gives rise to a *symmetric* tensor

$$-T_{ik} = A_{ik} - L\delta_{ik} - \frac{\partial}{\partial x_\alpha}(B_{\alpha ki} + B_{\alpha ik} - B_{kia}) \tag{22.8}$$

whose divergence is zero. We call this the *symmetric energy-momentum tensor* (and it is this tensor which has physical significance).

As an example we consider the sourceless Maxwell field, derivable from the Langrangian (cf. (19.1))

$$L = \frac{1}{4}\left(\frac{\partial\phi_i}{\partial x_k} - \frac{\partial\phi_k}{\partial x_i}\right)^2. \tag{22.9}$$

Under the influence of the infinitesimal transformation (21.1) the vector ϕ_i transforms as follows:

$$\phi_i = \bar{\phi}_i + \epsilon\bar{\phi}_\alpha\frac{\partial\xi_\alpha}{\partial x_i} \tag{22.10}$$

and

$$\frac{\partial\phi_i}{\partial x_k} = \frac{\partial\bar{\phi}_i}{\partial x_k} + \epsilon\left(\frac{\partial\bar{\phi}_i}{\partial x_\alpha}\frac{\partial\xi_\alpha}{\partial x_k} + \frac{\partial\bar{\phi}_\alpha}{\partial x_k}\frac{\partial\xi_\alpha}{\partial x_i}\right) + \epsilon\bar{\phi}_\alpha\frac{\partial^2\xi_\alpha}{\partial x_i\partial x_k}. \tag{22.11}$$

This gives for the transformation of the electromagnetic field strength (18.1):

$$F_{ki} = \bar{F}_{ki} + \epsilon\left(F_{\alpha i}\frac{\partial\xi_\alpha}{\partial x_k} - F_{\alpha k}\frac{\partial\xi_\alpha}{\partial x_i}\right) \tag{22.12}$$

and thus

$$L' = L + \frac{\epsilon}{2} F_{ki} \left(F_{\alpha i} \frac{\partial \xi_\alpha}{\partial x_k} - F_{\alpha k} \frac{\partial \xi_\alpha}{\partial x_i} \right) - L \frac{\partial \xi_\alpha}{\partial x_\alpha}$$

$$= L + \epsilon F_{i\alpha} F_{k\alpha} \frac{\partial \xi_i}{\partial x_k} - \tfrac{1}{4} F_{\alpha\beta}{}^2 \frac{\partial \xi_i}{\partial x_i} . \quad (22.13)$$

Hence in our case

$$A_{ik} = F_{i\alpha} F_{k\alpha},$$

whereas the tensor B_{ikm} is absent, in consequence of the *negative* sign in the definition of F_{ik}. (If F_{ik} were *symmetric* instead of antisymmetric, the energy-momentum tensor would contain also the first partial derivatives of the F_{ik}.)

The symmetric tensor

$$T_{ik} = -F_{i\alpha} F_{k\alpha} + \tfrac{1}{4} F_{\alpha\beta}{}^2 \delta_{ik} \quad (22.14)$$

was discovered (in non-relativistic interpretation) by Maxwell (1873) as expressing the energy-momentum tensor of the electromagnetic field.

Generally, if we actually evaluate the tensors A_{ik} and $B_{\alpha ik}$, which appear in formula (22.8), and collect terms, we obtain the following relation between the tensor t_{ik}, defined by (21.7), and the symmetric energy-momentum tensor (22.8):

$$-T_{ik} = t_{ik} + \frac{\partial}{\partial x_m} \left(\sigma_{mik} + \sigma_{imk} + \sigma_{kmi} \right)$$

$$= \tfrac{1}{2}(t_{ik} + t_{ki}) + \frac{\partial}{\partial x_m} \left(\sigma_{imk} + \sigma_{kmi} \right), \quad (22.15)$$

where we have introduced the tensor σ_{ikm} (antisymmetric in k, m) by putting

$$\sigma_{ikm} = \frac{1}{2} \frac{\partial L}{\partial \phi_{\alpha, i}} \phi_\beta (\rho_{\alpha\beta km} - \rho_{\alpha\beta mk}). \quad (22.16)$$

(The physical energy-momentum tensor is not necessarily *equal* to, but only *proportional* to this T_{ik}, the factor of proportionality depending on the units in which the energy is measured and the *sign* being determined by the condition that the energy density T_{44} must come out as *positive*.)

23. The ten conservation laws. Let us consider the conservation laws (22.5) by integrating over a certain three-dimensional volume τ and applying the Gaussian integral transformation (3.1):

$$\frac{1}{i}\frac{d}{dt}\int T_{k4}\,d\tau + \int T_{k\alpha}n_\alpha\,dS = 0. \tag{23.1}$$

In an "isolated system," i.e. a system which has the property that the boundary integrals (i.e. the flux of momentum and energy through the boundary) vanish,

$$P_k = i\int T_{k4}\,d\tau = \text{const.} \quad (k = 1, 2, 3), \tag{23.2}$$

which means that the total *momentum* of the system remains constant. Likewise

$$E = \int T_{44}\,d\tau = \text{const.}, \tag{23.3}$$

which means that the total *energy* of the system remains constant. If T_{ik} were not symmetric, this is all that we could say. However, the symmetry of T_{ik} gives rise to *six added conservation laws*:

$$\frac{\partial(x_i T_{jk} - x_j T_{ik})}{\partial x_k} = 0. \tag{23.4}$$

In consequence of these equations we obtain (again restricting ourselves to an isolated system) six added quantities which do not change in time

$$\int (x_k T_{j4} - x_j T_{k4})d\tau = \text{const.} \tag{23.5}$$

If k, j take the values 1, 2, 3, we obtain the *conservation of angular momentum*. If k takes the values 1, 2, 3 and $j = 4$, we obtain the last three conservation laws in the form

$$\int x_k T_{44}\,d\tau - it\int T_{k4}\,d\tau = 0. \tag{23.6}$$

If we define the *centre of energy* of the system by putting

$$\xi_k = \frac{\int x_k T_{44}\,d\tau}{\int T_{44}\,d\tau}, \tag{23.7}$$

then (23.6) becomes

$$E\xi_k - tP_k = 0 \qquad (23.8)$$

or

$$P_k = E\frac{d\xi_k}{dt}. \qquad (23.9)$$

Thus we have obtained a purely *kinematic* definition of the total momentum, in analogy with Newton's definition: "The total momentum of a body is equal to the total energy times the velocity of the centre of energy." Newton speaks of the *mass* of the body instead of its *energy*. The replacement of the word "mass" by "energy" is in complete harmony with Einstein's fundamental discovery—derived from the principle of relativity (1907; cf. p. 292)—that *mass and energy are identical* (in a time scale $c = 1$, which we have used consistently). It was Planck in 1909 who pointed out that the field theoretical interpretation of Einstein's principle can only be the *symmetry of the energy-momentum tensor*. If the T_{i4} $(i = 1, 2, 3)$ (i.e. the momentum density) and the T_{4i}, the energy current, did not agree, then the conservation of mass and energy would follow different laws and the principle $m = E$ could not be maintained.

Nor could a non-symmetric energy-momentum tensor guarantee the law of inertia, according to which the centre of mass of an isolated system moves in a straight line with constant velocity.

24. The dynamical law in field theoretical derivation. We now consider a system which is no longer isolated; i.e. the flux of momentum and energy through the boundary no longer vanishes. Our system is in interaction with the external world. The conservation of momentum and energy takes the form

$$\frac{dP_k}{dt} = -\int T_{k\alpha} n_\alpha \, dS \quad (k, \alpha = 1, 2, 3), \qquad (24.1)$$

$$\frac{dE}{idt} = -\int T_{4\alpha} n_\alpha \, dS. \qquad (24.2)$$

The first equation can be interpreted as Newton's law of motion: "The time rate of change of the momentum equals the moving force." The "moving force" appears in the form of a surface integral, extended over the boundary of the region:

$$F_k = -\int T_{k\alpha} n_\alpha \, dS \quad (k = 1, 2, 3). \tag{24.3}$$

If we operate with a reference system in which the centre of mass is momentarily at rest, then the equation of motion appears in the form: "Mass times acceleration equals moving force."

The great importance of this method of obtaining the dynamical law in field theoretical derivation can be appreciated from the following circumstance. Most of the field theories with which we are familiar are of a *linear* nature. This means that they permit the principle of superposition and thus exclude any dynamical interaction. At the same time they lead to singularities, i.e. to points in space at which the field quantities become infinite. The fact of dynamical interaction between fields demonstrates that nature cannot be purely linear, nor can it tolerate singular points. The basic Lagrangian can certainly not be purely quadratic, but must contain higher-order terms which become of decisive importance near the core of the material particles. In ignorance of these higher-order terms we put something on the right side of our linear field equations as a substitute for the non-linearity which is hidden from our knowledge ("asylum ignorantiae" in Einstein's terminology). Fortunately, these domains are very small and become of importance only if we inquire into the structure of elementary particles.

The advantage of the conservation laws in their integrated form is that the surface integral (24.3) can be pushed out to a distance at which the field is already practically replaceable by its linear approximation. Hence in principle we can obtain the "moving force" acting on a particle if we know only the field in a domain in which the linear approximation is still valid. For example, in the Maxwellian case we find that in a reference system of instantaneous rest the law of motion of an electron becomes

$$m_0 \frac{d^2 \xi_k}{dt^2} = e E_k,$$

where e is the charge of the electron and E_k is the external electric field strength. Transforming this equation to a moving reference system, the same law becomes

$$m_0 \frac{d}{dt} \frac{\mathbf{v}}{\sqrt{(1 - v^2)}} = e(\mathbf{E} + \mathbf{v} \times \mathbf{H}).$$

The force on the right side is called the "Lorentz force."

However, we cannot expect the law thus obtained to be of absolute validity The Lorentz force comes about by an interaction of the external field with the internal field of the electron, evaluated on the basis of the surface integral (24.3). An added force arises, however, by an interaction of the internal field *with itself*. In Newtonian mechanics the self-field cannot act on itself because the internal forces annihilate each other in view of the law "action equals reaction." This is no longer so if the finite propagation velocity of electromagnetic action is taken into account. But to evaluate the resulting reaction of the inner field on itself, this field would have to be known with microscopic accuracy, and this is impossible, considering the unknown structure of the (non-linear) core of the particle.

Bibliography

Jauch, J. M., and Rohrlich, R., *The Theory of Photons and Electrons* (Addison-Wesley, Reading, 1955).

Lanczos, C., "Variational Principles," in R. C. Clark and G. H. Derick, eds., *Mathematical Methods in Solid State Physics and Superfluid Theory* (Oliver & Boyd, Edinburgh, 1969).

Roman, P., *Theory of Elementary Particles* (North-Holland, Amsterdam, 1960).

Wentzel, G., *Quantum Theory of Fields* (Interscience, New York, 1949).

APPENDIX I*

THE important transition from the Lagrangian to the Hamiltonian form of dynamics can be accomplished in a more direct way, without the use of the Legendre transformation, basing the argument solely on the method of the Lagrangian multiplier. Consider the given Lagrangian function $L = L(q_1, \ldots, q_n; \dot{q}_1, \ldots, \dot{q}_n; t)$. Let us assume that we want to treat the variables $\dot{q}_i$ as a second set of independent variables w_i; i.e. we wish to write

$$L = L(q_1, \ldots, q_n; w_1, \ldots, w_n; t). \tag{1}$$

This is permissible, provided that we add the conditions

$$\dot{q}_i - w_i = 0 \tag{2}$$

as *auxiliary conditions* to our problem. The new formulation is clearly identical with the original problem, since the replacement of the w_i on the basis of the auxiliary conditions leads us back to the original problem.

But now we can handle the new problem with the help of the Lagrangian multiplier method, modifying the original Lagrangian to

$$L' = L + \sum_{i=1}^{n} p_i(\dot{q}_i - w_i), \tag{3}$$

where the Lagrangian multipliers have been denoted by p_i. Our new problem has the triple number of variables, namely the q_i, p_i, and w_i. We notice, however, that the w_i appear solely as *algebraic* variables, without derivatives. Such dynamical variables can be eliminated in advance (cf. p. 128, problem 3). The elimination of a variable means that a relation which holds for

*To chapter VI, section 4, p. 172.

the actual motion is assumed to hold also for the varied motion. This is generally *not* allowed because it might violate the restricting condition that the variation must vanish at the two end-points of the range. But in the case of an algebraic variable no integration by parts is involved and thus the restrictive clause "the variation must vanish at the two end-points" can be dropped. Hence the *a priori* elimination of algebraic variables is always permitted.

Now the elimination of the w_i on the basis of the equation

$$\frac{\partial L'}{\partial w_i} = 0 \tag{4}$$

means that we should solve the equations

$$\frac{\partial L}{\partial w_i} - p_i = 0 \tag{5}$$

for the w_i. But this is exactly the same as saying that we should solve the equations

$$\frac{\partial L}{\partial \dot{q}_i} = p_i \tag{6}$$

(with the original L) for the $\dot{q}_i$, expressing them in terms of the p_i and q_i.

Our L' has the form

$$L' = \sum_{i=1}^{n} p_i \dot{q}_i - H(q_i, p_i, t)$$

if we put

$$H = \sum_{i=1}^{n} p_i w_i - L(q_\alpha, w_\alpha; t). \tag{8}$$

Since the w_i will be eliminated anyway, we can just as well define

$$H = \sum_{i=1}^{n} p_i \dot{q}_i - L(q_\alpha, \dot{q}_\alpha; t) \tag{9}$$

with the added condition that the $\dot{q}_i$ have to be eliminated on the basis of the equations (6). *This is exactly the programme by which a Lagrangian problem is transformed into a Hamiltonian problem.*

That we can equally go in the opposite direction, starting with the Hamiltonian formulation and transforming it into the Lagrangian formulation, follows from the fact that the Hamiltonian integrand (7) is purely algebraic in the p_i, which can thus be eliminated. This means that we should solve the equations

$$\frac{\partial L'}{\partial p_i} = \dot{q}_i - \frac{\partial H}{\partial p_i} = 0, \tag{10}$$

expressing the p_i in terms of the q_i and $\dot{q}_i$. Substitution of these p_i into L' gives a function of the q_i and $\dot{q}_i$ which is nothing but the original Lagrangian L. The duality of the transformation is thus established, but our results have been obtained solely by making use of the Lagrangian multiplier method, together with the elimination of algebraic variables.

There is the further advantage in this procedure that we obtain at once the transformation of a variational problem with higher derivatives into the canonical form, without going through the successive elimination process of pp. 170–1. Let us assume, for example, that we have the Lagrangian

$$L = L(y, \dot{y}, \ddot{y}, \ldots, y^{(n)}; t). \tag{11}$$

Now we put

$$\begin{aligned} y &= q_1, \\ \dot{q}_1 &= q_2, \\ \dot{q}_2 &= q_3, \\ &\vdots \\ \dot{q}_n &= q_{n+1}, \end{aligned} \tag{12}$$

$$\begin{aligned} L' &= L(q_1, q_2, \ldots, q_{n+1}; t) \\ &+ p_1(\dot{q}_1 - q_2) + p_2(\dot{q}_2 - q_3) + \ldots + p_n(\dot{q}_n - q_{n+1}), \end{aligned} \tag{13}$$

which means

$$L' = \sum_{i=1}^{n} p_i \dot{q}_i - H, \tag{14}$$

where

$$H = p_1 q_2 + p_2 q_3 + \ldots + p_n q_{n+1} - L(q_1, \ldots, q_{n+1}; t). \tag{15}$$

The only algebraic variable we should eliminate here is q_{n+1}, by making use of the equation

$$\frac{\partial L}{\partial q_{n+1}} = p_n. \tag{16}$$

The result is a first-order canonical system in the n pairs of variables $q_1, q_2, \ldots, q_n; p_1, p_2, \ldots, p_n$.

APPENDIX II

Noether's invariant variational problems. The eminent algebraist Emmy Noether obtained important results in relation to a certain group of variational problems. Her paper, published in 1918,* had far-reaching repercussions in its application to problems of contemporary field physics, in both relativity and quantum theory. Noether's theory can be conceived as an application of the theory of ignorable variables (cf. p. 125), although it is not treated usually from this viewpoint.

Noether considers variational problems having the property that the action integral remains invariant with respect to a group of transformations, applied either to the dependent or the independent variables. She shows that every parameter associated with such transformations leads to a corresponding conservation law. In fact it suffices to consider a group of *infinitesimal* transformations.

In order to explain Noether's method, we shall first consider an example treated before, namely the derivation of the *conservation of energy* (cf. p. 133) for a system whose Lagrangian does not depend explicitly on the time t. For this purpose we transform the independent variable t by the transformation

$$t = t' + \alpha, \tag{1}$$

where α is an infinitesimal constant. Now the $q_i(t)$ of the variational problem change by this transformation, but we can equally well consider them as functions of the new variable t', abandoning the previous variable t. Then our Lagrangian takes the form

$$L = L(q_i, q_i', t' + \alpha). \tag{2}$$

Let us assume, however, that L does *not* depend explicitly on t.

*E. Noether, "Invariante Variationsprobleme," *Goett. Nachr.* 1918, pp. 235–257.

In that case our new variational integral becomes

$$A = \int_{t_1-a}^{t_2-a} L(q_i, q_i')dt' \tag{3}$$

and we see that the constant α does not appear in the variational integral, except in the limits. Hence we have an example of the type of variational problems that Noether considered.

We shall now depart from the assumption that α is a constant and consider α as a *function of* t', which satisfies the boundary conditions

$$\alpha(t_1) = \alpha(t_2) = 0. \tag{4}$$

In this case the action integral will become a function of α, but only $\alpha'(t')$ will appear explicitly in the integral. We now have to write

$$\frac{dq_i}{dt} = \frac{dq_i}{dt'}(1 - \alpha'), \tag{5}$$

and the Lagrangian of the transformed problem becomes (considering the infinitesimal nature of $\alpha(t')$):

$$L(q_i, \dot{q}_i) = L(q_i, q_i'(1 - \alpha'))$$
$$= L(q_i, q_i') - \left(\sum_{i=1}^{n} \frac{\partial L}{\partial \dot{q}_i} \dot{q}_i \right) \alpha'. \tag{6}$$

Moreover,

$$dt = (1 + \alpha')dt'. \tag{7}$$

Hence our new variational integral becomes, if we neglect quantities of higher than first order in $\alpha(t')$:

$$A = \int_{t_1}^{t_2} L(q_i, q_i')dt' - \int_{t_1}^{t_2} \left(\sum_{i=1}^{n} \frac{\partial L}{\partial \dot{q}_i} \dot{q}_i - L \right) \alpha'dt'. \tag{8}$$

The transformation (1) did not change our variational problem, and we cannot expect the new form of the action integral to give us more information than the previous one. But in the new form (8) we have an *added degree of freedom* at our disposal, viz. $\alpha(t')$, which has been added to the previous $q_i(t')$. The Lagrangian equation associated with $\alpha(t')$ yields

$$\frac{d}{dt'}\left(\sum_{i=1}^{n} \frac{\partial L}{\partial \dot{q}_i} \dot{q}_i - L \right) = 0, \qquad (9)$$

that is

$$\sum_{i=1}^{n} p_i \dot{q}_i - L = E, \qquad (10)$$

which is the well-known *energy theorem* of classical mechanics (cf. 53.12 and 56.3). It is not an added equation, which must hold in addition to the Lagrangian equations, but rather a *consequence* of the Lagrangian equations, now deduced from Noether's principle of a transformation that leaves the variational integral unchanged.

As a second example we consider the case of a Lagrangian that is "translation-invariant." Let us operate with rectangular coordinates x_i, y_i, z_i and assume that the potential energy of the given mechanical system depends only on the *difference* of the position coordinates of the particles:

$$V = V(x_i - x_k, y_i - y_k, z_i - z_k) \qquad (11)$$

(in accordance with Newton's third law of motion: "action equals reaction"). In this case the transformation

$$\begin{aligned} x_i &= x_i' + \alpha, \\ y_i &= y_i' + \beta, \\ z_i &= z_i' + \gamma, \end{aligned} \qquad (12)$$

where α, β, γ are constants, changes neither the potential nor the kinetic energy of the system, and again we have an example of a variational integral that remains invariant with respect to a certain transformation (in the present case the transformation is applied to the dynamical variables and the transformation (12) has the significance of a mere *translation* of the coordinate system).

Once more we shall change the constants α, β, γ of the transformation to functions of t, which adds three new degrees of freedom to our variational problem. These added degrees do not appear in the potential energy, but the kinetic energy of the system now becomes

$$T = \frac{1}{2} \sum_{i=1}^{N} m_i(\dot{x}_i{}^2 + \dot{y}_i{}^2 + \dot{z}_i{}^2)$$

$$= \frac{1}{2} \sum_{i=1}^{N} m_i[(\dot{x}_i{}' + \dot{\alpha})^2 + (\dot{y}_i{}' + \dot{\beta})^2 + (\dot{z}_i{}' + \dot{\gamma})^2], \qquad (13)$$

and assuming that α, β, γ are infinitesimal, we obtain (cf. 51.7 and 51.11):

$$A = \int_{t_1}^{t_2} (T - V)dt$$

$$= \int_{t_1}^{t_2} L dt + \int_{t_1}^{t_2} \sum_{i=1}^{N} m_i(\dot{x}_i \dot{\alpha} + \dot{y}_i \dot{\beta} + \dot{z}_i \dot{\gamma})dt. \qquad (14)$$

The Lagrangian equations with respect to the new action variables α, β, γ yield

$$\sum_{i=1}^{N} m_i \dot{x}_i = c_1,$$

$$\sum_{i=1}^{N} m_i \dot{y}_i = c_1, \qquad (15)$$

$$\sum_{i=1}^{N} m_i \dot{z}_i = c_3.$$

These equations express the well-known "conservation of momentum" of classical mechanics.

As a third example we consider a Lagrangian that is "rotation-invariant." If the potential energy V is the result of "central forces," then V depends only on the *distance* between two particles, i.e. on the quantity

$$r_{ik} = \sqrt{(x_i - x_k)^2 + (y_i - y_k)^2 + (z_i - z_k)^2}. \qquad (16)$$

In this case not only a constant translation but even a constant *rotation* of the reference system leaves both potential and kinetic energy invariant, and once more we have an example of a Noether type of problem. An infinitesimal rotation of the coordinates may be written as follows, if we make use of the symbolism of vector analysis (cf. 32.4):

$$\mathbf{r} = \mathbf{r}' + \mathbf{\Omega} \times \mathbf{r}', \tag{17}$$

where $\mathbf{\Omega}$ is an arbitrary infinitesimal vector. If one assumes again that $\mathbf{\Omega}$ becomes a function of t, the potential energy becomes independent of $\mathbf{\Omega}$, but the kinetic energy takes the form

$$T = \frac{1}{2} \sum_{i=1}^{N} m_i \dot{\mathbf{r}}_i{}^2 = \frac{1}{2} \sum_{i=1}^{N} m_i (\dot{\mathbf{r}}_i{}' + \dot{\mathbf{\Omega}} \times \mathbf{r}_i{}')^2$$

$$= \frac{1}{2} \sum_{i=1}^{N} m_i \dot{\mathbf{r}}_i{}'^2 + \dot{\mathbf{\Omega}} \sum_{i=1}^{N} m_i (\mathbf{r}_i \times \dot{\mathbf{r}}_i). \tag{18}$$

Again we have added three degrees of freedom by adding the free vector $\mathbf{\Omega}$ to our action variables. The Lagrangian equations associated with the vector $\mathbf{\Omega}$ now yield

$$\sum_{i=1}^{N} m_i (\mathbf{r}_i \times \dot{\mathbf{r}}_i) = \sum_{i=1}^{N} (\mathbf{r}_i \times m_i \mathbf{v}_i) = \overline{\mathbf{M}} = \text{const.}, \tag{19}$$

which in physical terms expresses the conservation of the *total angular momentum*, valid for any mechanical system acted upon by central forces (application to the solar system, for example, yields Kepler's area law).

BIBLIOGRAPHY

THERE are many text-books on Analytical Mechanics; the following selection, in the author's opinion, may be consulted beneficially for collateral reading.

Ames, J. S., and Murnaghan, F. D., *Theoretical Mechanics* (Ginn & Co., 1929).

Coe, C. J., *Theoretical Mechanics: A Vectorial Treatment* (Macmillan, 1938).

Courant, R. and Hilbert, D., *Methoden der mathematischen Physik* (Springer, vol. I, 1931; vol. II, 1937).

MacMillan, W. D., *Statics and Dynamics of a Particle* (McGraw-Hill, 1927).

Phillips, H. B., *Vector Analysis* (Wiley, 1933).

Synge, J. L., "Classical Dynamics," *Encyclopedia of Physics,* vol. III/1, pp. 1-225 (Springer, 1960).

Synge, J. L. and Griffith, B. A., *Principles of Mechanics* (McGraw-Hill, 1942).

Wintner, Aurel. *The Analytical Foundations of Celestial Mechanics* (Princeton, 1941).

For advanced reading:

Appell, P. E., *Traité de mécanique rationelle*, vols. I and II (Gauthier-Villars, 1909).

Birkhoff, G. D., *Dynamical Systems* (American Math. Soc., 1927).

Cartan, E., *Leçons sur les invariants intégraux* (Hermann, 1922).

Nordheim, L., "Die Prinzipe der Dynamik" (*Handbuch der Physik*, vol. V, pp. 43-90).

Nordheim, L. and Fues, E., "Die Hamilton-Jacobische Theorie der Dynamik" (*Handbuch der Physik*, vol. V, pp. 91-130).

Prange, G., "Die allgemeinen Integrationsmethoden der analytischen Mechanik" (*Encyklopädie der mathematischen Wissenschaften*, vol. IV, no. 2, issue 4).

Whittaker, E. T., *A Treatise on the Analytical Dynamics of Particles and Rigid Bodies* (Cambridge, 1917).

An excellent historical review of the older literature on theoretical dynamics up to 1860 can be found in:

Cayley, A., *Collected Mathematical Papers* (Cambridge, 1890). "Report on the Recent Progress of Theoretical Dynamics," vol. III, pp. 156-204, and a second report on special problems, vol IV, pp. 513-93.

INDEX

A CATALOG OF SELECTED
DOVER BOOKS
IN SCIENCE AND MATHEMATICS

Astronomy

BURNHAM'S CELESTIAL HANDBOOK, Robert Burnham, Jr. Thorough guide to the stars beyond our solar system. Exhaustive treatment. Alphabetical by constellation: Andromeda to Cetus in Vol. 1; Chamaeleon to Orion in Vol. 2; and Pavo to Vulpecula in Vol. 3. Hundreds of illustrations. Index in Vol. 3. 2,000pp. 6⅛ x 9¼.
Vol. I: 0-486-23567-X
Vol. II: 0-486-23568-8
Vol. III: 0-486-23673-0

EXPLORING THE MOON THROUGH BINOCULARS AND SMALL TELESCOPES, Ernest H. Cherrington, Jr. Informative, profusely illustrated guide to locating and identifying craters, rills, seas, mountains, other lunar features. Newly revised and updated with special section of new photos. Over 100 photos and diagrams. 240pp. 8¼ x 11. 0-486-24491-1

THE EXTRATERRESTRIAL LIFE DEBATE, 1750–1900, Michael J. Crowe. First detailed, scholarly study in English of the many ideas that developed from 1750 to 1900 regarding the existence of intelligent extraterrestrial life. Examines ideas of Kant, Herschel, Voltaire, Percival Lowell, many other scientists and thinkers. 16 illustrations. 704pp. 5⅜ x 8½. 0-486-40675-X

THEORIES OF THE WORLD FROM ANTIQUITY TO THE COPERNICAN REVOLUTION, Michael J. Crowe. Newly revised edition of an accessible, enlightening book recreates the change from an earth-centered to a sun-centered conception of the solar system. 242pp. 5⅜ x 8½. 0-486-41444-2

A HISTORY OF ASTRONOMY, A. Pannekoek. Well-balanced, carefully reasoned study covers such topics as Ptolemaic theory, work of Copernicus, Kepler, Newton, Eddington's work on stars, much more. Illustrated. References. 521pp. 5⅜ x 8½.
0-486-65994-1

A COMPLETE MANUAL OF AMATEUR ASTRONOMY: TOOLS AND TECHNIQUES FOR ASTRONOMICAL OBSERVATIONS, P. Clay Sherrod with Thomas L. Koed. Concise, highly readable book discusses: selecting, setting up and maintaining a telescope; amateur studies of the sun; lunar topography and occultations; observations of Mars, Jupiter, Saturn, the minor planets and the stars; an introduction to photoelectric photometry; more. 1981 ed. 124 figures. 25 halftones. 37 tables. 335pp. 6½ x 9¼. 0-486-40675-X

AMATEUR ASTRONOMER'S HANDBOOK, J. B. Sidgwick. Timeless, comprehensive coverage of telescopes, mirrors, lenses, mountings, telescope drives, micrometers, spectroscopes, more. 189 illustrations. 576pp. 5⅝ x 8¼. (Available in U.S. only.)
0-486-24034-7

STARS AND RELATIVITY, Ya. B. Zel'dovich and I. D. Novikov. Vol. 1 of *Relativistic Astrophysics* by famed Russian scientists. General relativity, properties of matter under astrophysical conditions, stars, and stellar systems. Deep physical insights, clear presentation. 1971 edition. References. 544pp. 5⅝ x 8¼. 0-486-69424-0

Chemistry

THE SCEPTICAL CHYMIST: THE CLASSIC 1661 TEXT, Robert Boyle. Boyle defines the term "element," asserting that all natural phenomena can be explained by the motion and organization of primary particles. 1911 ed. viii+232pp. 5⅜ x 8½.
0-486-42825-7

RADIOACTIVE SUBSTANCES, Marie Curie. Here is the celebrated scientist's doctoral thesis, the prelude to her receipt of the 1903 Nobel Prize. Curie discusses establishing atomic character of radioactivity found in compounds of uranium and thorium; extraction from pitchblende of polonium and radium; isolation of pure radium chloride; determination of atomic weight of radium; plus electric, photographic, luminous, heat, color effects of radioactivity. ii+94pp. 5⅜ x 8½.   0-486-42550-9

CHEMICAL MAGIC, Leonard A. Ford. Second Edition, Revised by E. Winston Grundmeier. Over 100 unusual stunts demonstrating cold fire, dust explosions, much more. Text explains scientific principles and stresses safety precautions. 128pp. 5⅜ x 8½.   0-486-67628-5

THE DEVELOPMENT OF MODERN CHEMISTRY, Aaron J. Ihde. Authoritative history of chemistry from ancient Greek theory to 20th-century innovation. Covers major chemists and their discoveries. 209 illustrations. 14 tables. Bibliographies. Indices. Appendices. 851pp. 5⅜ x 8½.   0-486-64235-6

CATALYSIS IN CHEMISTRY AND ENZYMOLOGY, William P. Jencks. Exceptionally clear coverage of mechanisms for catalysis, forces in aqueous solution, carbonyl- and acyl-group reactions, practical kinetics, more. 864pp. 5⅜ x 8½.
0-486-65460-5

ELEMENTS OF CHEMISTRY, Antoine Lavoisier. Monumental classic by founder of modern chemistry in remarkable reprint of rare 1790 Kerr translation. A must for every student of chemistry or the history of science. 539pp. 5⅜ x 8½. 0-486-64624-6

THE HISTORICAL BACKGROUND OF CHEMISTRY, Henry M. Leicester. Evolution of ideas, not individual biography. Concentrates on formulation of a coherent set of chemical laws. 260pp. 5⅜ x 8½.   0-486-61053-5

A SHORT HISTORY OF CHEMISTRY, J. R. Partington. Classic exposition explores origins of chemistry, alchemy, early medical chemistry, nature of atmosphere, theory of valency, laws and structure of atomic theory, much more. 428pp. 5⅜ x 8½. (Available in U.S. only.)   0-486-65977-1

GENERAL CHEMISTRY, Linus Pauling. Revised 3rd edition of classic first-year text by Nobel laureate. Atomic and molecular structure, quantum mechanics, statistical mechanics, thermodynamics correlated with descriptive chemistry. Problems. 992pp. 5⅜ x 8½.   0-486-65622-5

FROM ALCHEMY TO CHEMISTRY, John Read. Broad, humanistic treatment focuses on great figures of chemistry and ideas that revolutionized the science. 50 illustrations. 240pp. 5⅜ x 8½.   0-486-28690-8

Engineering

DE RE METALLICA, Georgius Agricola. The famous Hoover translation of greatest treatise on technological chemistry, engineering, geology, mining of early modern times (1556). All 289 original woodcuts. 638pp. 6¾ x 11.   0-486-60006-8

FUNDAMENTALS OF ASTRODYNAMICS, Roger Bate et al. Modern approach developed by U.S. Air Force Academy. Designed as a first course. Problems, exercises. Numerous illustrations. 455pp. 5⅜ x 8½.   0-486-60061-0

DYNAMICS OF FLUIDS IN POROUS MEDIA, Jacob Bear. For advanced students of ground water hydrology, soil mechanics and physics, drainage and irrigation engineering and more. 335 illustrations. Exercises, with answers. 784pp. 6⅛ x 9¼.
0-486-65675-6

THEORY OF VISCOELASTICITY (Second Edition), Richard M. Christensen. Complete consistent description of the linear theory of the viscoelastic behavior of materials. Problem-solving techniques discussed. 1982 edition. 29 figures. xiv+364pp. 6⅛ x 9¼.   0-486-42880-X

MECHANICS, J. P. Den Hartog. A classic introductory text or refresher. Hundreds of applications and design problems illuminate fundamentals of trusses, loaded beams and cables, etc. 334 answered problems. 462pp. 5⅜ x 8½.   0-486-60754-2

MECHANICAL VIBRATIONS, J. P. Den Hartog. Classic textbook offers lucid explanations and illustrative models, applying theories of vibrations to a variety of practical industrial engineering problems. Numerous figures. 233 problems, solutions. Appendix. Index. Preface. 436pp. 5⅜ x 8½.   0-486-64785-4

STRENGTH OF MATERIALS, J. P. Den Hartog. Full, clear treatment of basic material (tension, torsion, bending, etc.) plus advanced material on engineering methods, applications. 350 answered problems. 323pp. 5⅜ x 8½.   0-486-60755-0

A HISTORY OF MECHANICS, René Dugas. Monumental study of mechanical principles from antiquity to quantum mechanics. Contributions of ancient Greeks, Galileo, Leonardo, Kepler, Lagrange, many others. 671pp. 5⅜ x 8½. 0-486-65632-2

STABILITY THEORY AND ITS APPLICATIONS TO STRUCTURAL MECHANICS, Clive L. Dym. Self-contained text focuses on Koiter postbuckling analyses, with mathematical notions of stability of motion. Basing minimum energy principles for static stability upon dynamic concepts of stability of motion, it develops asymptotic buckling and postbuckling analyses from potential energy considerations, with applications to columns, plates, and arches. 1974 ed. 208pp. 5⅜ x 8½.
0-486-42541-X

METAL FATIGUE, N. E. Frost, K. J. Marsh, and L. P. Pook. Definitive, clearly written, and well-illustrated volume addresses all aspects of the subject, from the historical development of understanding metal fatigue to vital concepts of the cyclic stress that causes a crack to grow. Includes 7 appendixes. 544pp. 5⅜ x 8½. 0-486-40927-9

ROCKETS, Robert Goddard. Two of the most significant publications in the history of rocketry and jet propulsion: "A Method of Reaching Extreme Altitudes" (1919) and "Liquid Propellant Rocket Development" (1936). 128pp. 5⅜ x 8½.   0-486-42537-1

STATISTICAL MECHANICS: PRINCIPLES AND APPLICATIONS, Terrell L. Hill. Standard text covers fundamentals of statistical mechanics, applications to fluctuation theory, imperfect gases, distribution functions, more. 448pp. 5⅜ x 8½.
0-486-65390-0

ENGINEERING AND TECHNOLOGY 1650–1750: ILLUSTRATIONS AND TEXTS FROM ORIGINAL SOURCES, Martin Jensen. Highly readable text with more than 200 contemporary drawings and detailed engravings of engineering projects dealing with surveying, leveling, materials, hand tools, lifting equipment, transport and erection, piling, bailing, water supply, hydraulic engineering, and more. Among the specific projects outlined-transporting a 50-ton stone to the Louvre, erecting an obelisk, building timber locks, and dredging canals. 207pp. 8⅜ x 11¼.
0-486-42232-1

THE VARIATIONAL PRINCIPLES OF MECHANICS, Cornelius Lanczos. Graduate level coverage of calculus of variations, equations of motion, relativistic mechanics, more. First inexpensive paperbound edition of classic treatise. Index. Bibliography. 418pp. 5⅜ x 8½.   0-486-65067-7

PROTECTION OF ELECTRONIC CIRCUITS FROM OVERVOLTAGES, Ronald B. Standler. Five-part treatment presents practical rules and strategies for circuits designed to protect electronic systems from damage by transient overvoltages. 1989 ed. xxiv+434pp. 6⅛ x 9¼.   0-486-42552-5

ROTARY WING AERODYNAMICS, W. Z. Stepniewski. Clear, concise text covers aerodynamic phenomena of the rotor and offers guidelines for helicopter performance evaluation. Originally prepared for NASA. 537 figures. 640pp. 6⅛ x 9¼.
0-486-64647-5

INTRODUCTION TO SPACE DYNAMICS, William Tyrrell Thomson. Comprehensive, classic introduction to space-flight engineering for advanced undergraduate and graduate students. Includes vector algebra, kinematics, transformation of coordinates. Bibliography. Index. 352pp. 5⅜ x 8½.   0-486-65113-4

HISTORY OF STRENGTH OF MATERIALS, Stephen P. Timoshenko. Excellent historical survey of the strength of materials with many references to the theories of elasticity and structure. 245 figures. 452pp. 5⅜ x 8½.   0-486-61187-6

ANALYTICAL FRACTURE MECHANICS, David J. Unger. Self-contained text supplements standard fracture mechanics texts by focusing on analytical methods for determining crack-tip stress and strain fields. 336pp. 6⅛ x 9¼.   0-486-41737-9

STATISTICAL MECHANICS OF ELASTICITY, J. H. Weiner. Advanced, self-contained treatment illustrates general principles and elastic behavior of solids. Part 1, based on classical mechanics, studies thermoelastic behavior of crystalline and polymeric solids. Part 2, based on quantum mechanics, focuses on interatomic force laws, behavior of solids, and thermally activated processes. For students of physics and chemistry and for polymer physicists. 1983 ed. 96 figures. 496pp. 5⅜ x 8½.
0-486-42260-7

Mathematics

FUNCTIONAL ANALYSIS (Second Corrected Edition), George Bachman and Lawrence Narici. Excellent treatment of subject geared toward students with background in linear algebra, advanced calculus, physics and engineering. Text covers introduction to inner-product spaces, normed, metric spaces, and topological spaces; complete orthonormal sets, the Hahn-Banach Theorem and its consequences, and many other related subjects. 1966 ed. 544pp. 6⅛ x 9¼. 0-486-40251-7

ASYMPTOTIC EXPANSIONS OF INTEGRALS, Norman Bleistein & Richard A. Handelsman. Best introduction to important field with applications in a variety of scientific disciplines. New preface. Problems. Diagrams. Tables. Bibliography. Index. 448pp. 5⅜ x 8½. 0-486-65082-0

VECTOR AND TENSOR ANALYSIS WITH APPLICATIONS, A. I. Borisenko and I. E. Tarapov. Concise introduction. Worked-out problems, solutions, exercises. 257pp. 5⅝ x 8¼. 0-486-63833-2

AN INTRODUCTION TO ORDINARY DIFFERENTIAL EQUATIONS, Earl A. Coddington. A thorough and systematic first course in elementary differential equations for undergraduates in mathematics and science, with many exercises and problems (with answers). Index. 304pp. 5⅜ x 8½. 0-486-65942-9

FOURIER SERIES AND ORTHOGONAL FUNCTIONS, Harry F. Davis. An incisive text combining theory and practical example to introduce Fourier series, orthogonal functions and applications of the Fourier method to boundary-value problems. 570 exercises. Answers and notes. 416pp. 5⅜ x 8½. 0-486-65973-9

COMPUTABILITY AND UNSOLVABILITY, Martin Davis. Classic graduate-level introduction to theory of computability, usually referred to as theory of recurrent functions. New preface and appendix. 288pp. 5⅜ x 8½. 0-486-61471-9

ASYMPTOTIC METHODS IN ANALYSIS, N. G. de Bruijn. An inexpensive, comprehensive guide to asymptotic methods–the pioneering work that teaches by explaining worked examples in detail. Index. 224pp. 5⅜ x 8½ 0-486-64221-6

APPLIED COMPLEX VARIABLES, John W. Dettman. Step-by-step coverage of fundamentals of analytic function theory–plus lucid exposition of five important applications: Potential Theory; Ordinary Differential Equations; Fourier Transforms; Laplace Transforms; Asymptotic Expansions. 66 figures. Exercises at chapter ends. 512pp. 5⅜ x 8½. 0-486-64670-X

INTRODUCTION TO LINEAR ALGEBRA AND DIFFERENTIAL EQUATIONS, John W. Dettman. Excellent text covers complex numbers, determinants, orthonormal bases, Laplace transforms, much more. Exercises with solutions. Undergraduate level. 416pp. 5⅜ x 8½. 0-486-65191-6

RIEMANN'S ZETA FUNCTION, H. M. Edwards. Superb, high-level study of landmark 1859 publication entitled "On the Number of Primes Less Than a Given Magnitude" traces developments in mathematical theory that it inspired. xiv+315pp. 5⅜ x 8½. 0-486-41740-9

CALCULUS OF VARIATIONS WITH APPLICATIONS, George M. Ewing. Applications-oriented introduction to variational theory develops insight and promotes understanding of specialized books, research papers. Suitable for advanced undergraduate/graduate students as primary, supplementary text. 352pp. 5⅜ x 8½.
0-486-64856-7

COMPLEX VARIABLES, Francis J. Flanigan. Unusual approach, delaying complex algebra till harmonic functions have been analyzed from real variable viewpoint. Includes problems with answers. 364pp. 5⅜ x 8½.   0-486-61388-7

AN INTRODUCTION TO THE CALCULUS OF VARIATIONS, Charles Fox. Graduate-level text covers variations of an integral, isoperimetrical problems, least action, special relativity, approximations, more. References. 279pp. 5⅜ x 8½.
0-486-65499-0

COUNTEREXAMPLES IN ANALYSIS, Bernard R. Gelbaum and John M. H. Olmsted. These counterexamples deal mostly with the part of analysis known as "real variables." The first half covers the real number system, and the second half encompasses higher dimensions. 1962 edition. xxiv+198pp. 5⅜ x 8½. 0-486-42875-3

CATASTROPHE THEORY FOR SCIENTISTS AND ENGINEERS, Robert Gilmore. Advanced-level treatment describes mathematics of theory grounded in the work of Poincaré, R. Thom, other mathematicians. Also important applications to problems in mathematics, physics, chemistry and engineering. 1981 edition. References. 28 tables. 397 black-and-white illustrations. xvii + 666pp. 6⅛ x 9¼.
0-486-67539-4

INTRODUCTION TO DIFFERENCE EQUATIONS, Samuel Goldberg. Exceptionally clear exposition of important discipline with applications to sociology, psychology, economics. Many illustrative examples; over 250 problems. 260pp. 5⅜ x 8½.
0-486-65084-7

NUMERICAL METHODS FOR SCIENTISTS AND ENGINEERS, Richard Hamming. Classic text stresses frequency approach in coverage of algorithms, polynomial approximation, Fourier approximation, exponential approximation, other topics. Revised and enlarged 2nd edition. 721pp. 5⅜ x 8½.   0-486-65241-6

INTRODUCTION TO NUMERICAL ANALYSIS (2nd Edition), F. B. Hildebrand. Classic, fundamental treatment covers computation, approximation, interpolation, numerical differentiation and integration, other topics. 150 new problems. 669pp. 5⅜ x 8½.   0-486-65363-3

THREE PEARLS OF NUMBER THEORY, A. Y. Khinchin. Three compelling puzzles require proof of a basic law governing the world of numbers. Challenges concern van der Waerden's theorem, the Landau-Schnirelmann hypothesis and Mann's theorem, and a solution to Waring's problem. Solutions included. 64pp. 5⅜ x 8½.
0-486-40026-3

THE PHILOSOPHY OF MATHEMATICS: AN INTRODUCTORY ESSAY, Stephan Körner. Surveys the views of Plato, Aristotle, Leibniz & Kant concerning propositions and theories of applied and pure mathematics. Introduction. Two appendices. Index. 198pp. 5⅜ x 8½.   0-486-25048-2

INTRODUCTORY REAL ANALYSIS, A.N. Kolmogorov, S. V. Fomin. Translated by Richard A. Silverman. Self-contained, evenly paced introduction to real and functional analysis. Some 350 problems. 403pp. 5⅜ x 8½.   0-486-61226-0

APPLIED ANALYSIS, Cornelius Lanczos. Classic work on analysis and design of finite processes for approximating solution of analytical problems. Algebraic equations, matrices, harmonic analysis, quadrature methods, much more. 559pp. 5⅜ x 8½.
0-486-65656-X

AN INTRODUCTION TO ALGEBRAIC STRUCTURES, Joseph Landin. Superb self-contained text covers "abstract algebra": sets and numbers, theory of groups, theory of rings, much more. Numerous well-chosen examples, exercises. 247pp. 5⅜ x 8½.
0-486-65940-2

QUALITATIVE THEORY OF DIFFERENTIAL EQUATIONS, V. V. Nemytskii and V.V. Stepanov. Classic graduate-level text by two prominent Soviet mathematicians covers classical differential equations as well as topological dynamics and ergodic theory. Bibliographies. 523pp. 5⅜ x 8½.   0-486-65954-2

THEORY OF MATRICES, Sam Perlis. Outstanding text covering rank, nonsingularity and inverses in connection with the development of canonical matrices under the relation of equivalence, and without the intervention of determinants. Includes exercises. 237pp. 5⅜ x 8½.   0-486-66810-X

INTRODUCTION TO ANALYSIS, Maxwell Rosenlicht. Unusually clear, accessible coverage of set theory, real number system, metric spaces, continuous functions, Riemann integration, multiple integrals, more. Wide range of problems. Undergraduate level. Bibliography. 254pp. 5⅜ x 8½.   0-486-65038-3

MODERN NONLINEAR EQUATIONS, Thomas L. Saaty. Emphasizes practical solution of problems; covers seven types of equations. ". . . a welcome contribution to the existing literature...."–*Math Reviews.* 490pp. 5⅜ x 8½.   0-486-64232-1

MATRICES AND LINEAR ALGEBRA, Hans Schneider and George Phillip Barker. Basic textbook covers theory of matrices and its applications to systems of linear equations and related topics such as determinants, eigenvalues and differential equations. Numerous exercises. 432pp. 5⅜ x 8½.   0-486-66014-1

LINEAR ALGEBRA, Georgi E. Shilov. Determinants, linear spaces, matrix algebras, similar topics. For advanced undergraduates, graduates. Silverman translation. 387pp. 5⅜ x 8½.   0-486-63518-X

ELEMENTS OF REAL ANALYSIS, David A. Sprecher. Classic text covers fundamental concepts, real number system, point sets, functions of a real variable, Fourier series, much more. Over 500 exercises. 352pp. 5⅜ x 8½.   0-486-65385-4

SET THEORY AND LOGIC, Robert R. Stoll. Lucid introduction to unified theory of mathematical concepts. Set theory and logic seen as tools for conceptual understanding of real number system. 496pp. 5⅜ x 8¼.   0-486-63829-4

TENSOR CALCULUS, J.L. Synge and A. Schild. Widely used introductory text covers spaces and tensors, basic operations in Riemannian space, non-Riemannian spaces, etc. 324pp. 5⅜ x 8¼. 0-486-63612-7

ORDINARY DIFFERENTIAL EQUATIONS, Morris Tenenbaum and Harry Pollard. Exhaustive survey of ordinary differential equations for undergraduates in mathematics, engineering, science. Thorough analysis of theorems. Diagrams. Bibliography. Index. 818pp. 5⅜ x 8½. 0-486-64940-7

INTEGRAL EQUATIONS, F. G. Tricomi. Authoritative, well-written treatment of extremely useful mathematical tool with wide applications. Volterra Equations, Fredholm Equations, much more. Advanced undergraduate to graduate level. Exercises. Bibliography. 238pp. 5⅜ x 8½. 0-486-64828-1

FOURIER SERIES, Georgi P. Tolstov. Translated by Richard A. Silverman. A valuable addition to the literature on the subject, moving clearly from subject to subject and theorem to theorem. 107 problems, answers. 336pp. 5⅜ x 8½. 0-486-63317-9

INTRODUCTION TO MATHEMATICAL THINKING, Friedrich Waismann. Examinations of arithmetic, geometry, and theory of integers; rational and natural numbers; complete induction; limit and point of accumulation; remarkable curves; complex and hypercomplex numbers, more. 1959 ed. 27 figures. xii+260pp. 5⅜ x 8½. 0-486-63317-9

POPULAR LECTURES ON MATHEMATICAL LOGIC, Hao Wang. Noted logician's lucid treatment of historical developments, set theory, model theory, recursion theory and constructivism, proof theory, more. 3 appendixes. Bibliography. 1981 edition. ix + 283pp. 5⅜ x 8½. 0-486-67632-3

CALCULUS OF VARIATIONS, Robert Weinstock. Basic introduction covering isoperimetric problems, theory of elasticity, quantum mechanics, electrostatics, etc. Exercises throughout. 326pp. 5⅜ x 8½. 0-486-63069-2

THE CONTINUUM: A CRITICAL EXAMINATION OF THE FOUNDATION OF ANALYSIS, Hermann Weyl. Classic of 20th-century foundational research deals with the conceptual problem posed by the continuum. 156pp. 5⅜ x 8½.
0-486-67982-9

CHALLENGING MATHEMATICAL PROBLEMS WITH ELEMENTARY SOLUTIONS, A. M. Yaglom and I. M. Yaglom. Over 170 challenging problems on probability theory, combinatorial analysis, points and lines, topology, convex polygons, many other topics. Solutions. Total of 445pp. 5⅜ x 8½. Two-vol. set.
Vol. I: 0-486-65536-9 Vol. II: 0-486-65537-7

INTRODUCTION TO PARTIAL DIFFERENTIAL EQUATIONS WITH APPLICATIONS, E. C. Zachmanoglou and Dale W. Thoe. Essentials of partial differential equations applied to common problems in engineering and the physical sciences. Problems and answers. 416pp. 5⅜ x 8½. 0-486-65251-3

THE THEORY OF GROUPS, Hans J. Zassenhaus. Well-written graduate-level text acquaints reader with group-theoretic methods and demonstrates their usefulness in mathematics. Axioms, the calculus of complexes, homomorphic mapping, *p*-group theory, more. 276pp. 5⅜ x 8½. 0-486-40922-8

Math–Decision Theory, Statistics, Probability

ELEMENTARY DECISION THEORY, Herman Chernoff and Lincoln E. Moses. Clear introduction to statistics and statistical theory covers data processing, probability and random variables, testing hypotheses, much more. Exercises. 364pp. 5⅜ x 8½. 0-486-65218-1

STATISTICS MANUAL, Edwin L. Crow et al. Comprehensive, practical collection of classical and modern methods prepared by U.S. Naval Ordnance Test Station. Stress on use. Basics of statistics assumed. 288pp. 5⅜ x 8½. 0-486-60599-X

SOME THEORY OF SAMPLING, William Edwards Deming. Analysis of the problems, theory and design of sampling techniques for social scientists, industrial managers and others who find statistics important at work. 61 tables. 90 figures. xvii +602pp. 5⅜ x 8½. 0-486-64684-X

LINEAR PROGRAMMING AND ECONOMIC ANALYSIS, Robert Dorfman, Paul A. Samuelson and Robert M. Solow. First comprehensive treatment of linear programming in standard economic analysis. Game theory, modern welfare economics, Leontief input-output, more. 525pp. 5⅜ x 8½. 0-486-65491-5

PROBABILITY: AN INTRODUCTION, Samuel Goldberg. Excellent basic text covers set theory, probability theory for finite sample spaces, binomial theorem, much more. 360 problems. Bibliographies. 322pp. 5⅜ x 8½. 0-486-65252-1

GAMES AND DECISIONS: INTRODUCTION AND CRITICAL SURVEY, R. Duncan Luce and Howard Raiffa. Superb nontechnical introduction to game theory, primarily applied to social sciences. Utility theory, zero-sum games, n-person games, decision-making, much more. Bibliography. 509pp. 5⅜ x 8½. 0-486-65943-7

INTRODUCTION TO THE THEORY OF GAMES, J. C. C. McKinsey. This comprehensive overview of the mathematical theory of games illustrates applications to situations involving conflicts of interest, including economic, social, political, and military contexts. Appropriate for advanced undergraduate and graduate courses; advanced calculus a prerequisite. 1952 ed. x+372pp. 5⅜ x 8½. 0-486-42811-7

FIFTY CHALLENGING PROBLEMS IN PROBABILITY WITH SOLUTIONS, Frederick Mosteller. Remarkable puzzlers, graded in difficulty, illustrate elementary and advanced aspects of probability. Detailed solutions. 88pp. 5⅜ x 8½. 65355-2

PROBABILITY THEORY: A CONCISE COURSE, Y. A. Rozanov. Highly readable, self-contained introduction covers combination of events, dependent events, Bernoulli trials, etc. 148pp. 5⅜ x 8¼. 0-486-63544-9

STATISTICAL METHOD FROM THE VIEWPOINT OF QUALITY CONTROL, Walter A. Shewhart. Important text explains regulation of variables, uses of statistical control to achieve quality control in industry, agriculture, other areas. 192pp. 5⅜ x 8½. 0-486-65232-7

Math–Geometry and Topology

ELEMENTARY CONCEPTS OF TOPOLOGY, Paul Alexandroff. Elegant, intuitive approach to topology from set-theoretic topology to Betti groups; how concepts of topology are useful in math and physics. 25 figures. 57pp. 5⅜ x 8½. 0-486-60747-X

COMBINATORIAL TOPOLOGY, P. S. Alexandrov. Clearly written, well-organized, three-part text begins by dealing with certain classic problems without using the formal techniques of homology theory and advances to the central concept, the Betti groups. Numerous detailed examples. 654pp. 5⅜ x 8½. 0-486-40179-0

EXPERIMENTS IN TOPOLOGY, Stephen Barr. Classic, lively explanation of one of the byways of mathematics. Klein bottles, Moebius strips, projective planes, map coloring, problem of the Koenigsberg bridges, much more, described with clarity and wit. 43 figures. 210pp. 5⅜ x 8½. 0-486-25933-1

THE GEOMETRY OF RENÉ DESCARTES, René Descartes. The great work founded analytical geometry. Original French text, Descartes's own diagrams, together with definitive Smith-Latham translation. 244pp. 5⅜ x 8½. 0-486-60068-8

EUCLIDEAN GEOMETRY AND TRANSFORMATIONS, Clayton W. Dodge. This introduction to Euclidean geometry emphasizes transformations, particularly isometries and similarities. Suitable for undergraduate courses, it includes numerous examples, many with detailed answers. 1972 ed. viii+296pp. 6⅛ x 9¼. 0-486-43476-1

PRACTICAL CONIC SECTIONS: THE GEOMETRIC PROPERTIES OF ELLIPSES, PARABOLAS AND HYPERBOLAS, J. W. Downs. This text shows how to create ellipses, parabolas, and hyperbolas. It also presents historical background on their ancient origins and describes the reflective properties and roles of curves in design applications. 1993 ed. 98 figures. xii+100pp. 6½ x 9¼. 0-486-42876-1

THE THIRTEEN BOOKS OF EUCLID'S ELEMENTS, translated with introduction and commentary by Sir Thomas L. Heath. Definitive edition. Textual and linguistic notes, mathematical analysis. 2,500 years of critical commentary. Unabridged. 1,414pp. 5⅜ x 8½.. Three-vol. set.
Vol. I: 0-486-60088-2 Vol. II: 0-486-60089-0 Vol. III: 0-486-60090-4

SPACE AND GEOMETRY: IN THE LIGHT OF PHYSIOLOGICAL, PSYCHOLOGICAL AND PHYSICAL INQUIRY, Ernst Mach. Three essays by an eminent philosopher and scientist explore the nature, origin, and development of our concepts of space, with a distinctness and precision suitable for undergraduate students and other readers. 1906 ed. vi+148pp. 5⅜ x 8½. 0-486-43909-7

GEOMETRY OF COMPLEX NUMBERS, Hans Schwerdtfeger. Illuminating, widely praised book on analytic geometry of circles, the Moebius transformation, and two-dimensional non-Euclidean geometries. 200pp. 5⅝ x 8¼. 0-486-63830-8

DIFFERENTIAL GEOMETRY, Heinrich W. Guggenheimer. Local differential geometry as an application of advanced calculus and linear algebra. Curvature, transformation groups, surfaces, more. Exercises. 62 figures. 378pp. 5⅜ x 8½. 0-486-63433-7

History of Math

THE WORKS OF ARCHIMEDES, Archimedes (T. L. Heath, ed.). Topics include the famous problems of the ratio of the areas of a cylinder and an inscribed sphere; the measurement of a circle; the properties of conoids, spheroids, and spirals; and the quadrature of the parabola. Informative introduction. clxxxvi+326pp. 5⅜ x 8½.
0-486-42084-1

A SHORT ACCOUNT OF THE HISTORY OF MATHEMATICS, W. W. Rouse Ball. One of clearest, most authoritative surveys from the Egyptians and Phoenicians through 19th-century figures such as Grassman, Galois, Riemann. Fourth edition. 522pp. 5⅜ x 8½.
0-486-20630-0

THE HISTORY OF THE CALCULUS AND ITS CONCEPTUAL DEVELOP-MENT, Carl B. Boyer. Origins in antiquity, medieval contributions, work of Newton, Leibniz, rigorous formulation. Treatment is verbal. 346pp. 5⅜ x 8½. 0-486-60509-4

THE HISTORICAL ROOTS OF ELEMENTARY MATHEMATICS, Lucas N. H. Bunt, Phillip S. Jones, and Jack D. Bedient. Fundamental underpinnings of modern arithmetic, algebra, geometry and number systems derived from ancient civiliza-tions. 320pp. 5⅜ x 8½.
0-486-25563-8

A HISTORY OF MATHEMATICAL NOTATIONS, Florian Cajori. This classic study notes the first appearance of a mathematical symbol and its origin, the com-petition it encountered, its spread among writers in different countries, its rise to pop-ularity, its eventual decline or ultimate survival. Original 1929 two-volume edition presented here in one volume. xxviii+820pp. 5⅜ x 8½.
0-486-67766-4

GAMES, GODS & GAMBLING: A HISTORY OF PROBABILITY AND STATISTICAL IDEAS, F. N. David. Episodes from the lives of Galileo, Fermat, Pascal, and others illustrate this fascinating account of the roots of mathematics. Features thought-provoking references to classics, archaeology, biography, poetry. 1962 edition. 304pp. 5⅜ x 8½. (Available in U.S. only.)
0-486-40023-9

OF MEN AND NUMBERS: THE STORY OF THE GREAT MATHEMATICIANS, Jane Muir. Fascinating accounts of the lives and accom-plishments of history's greatest mathematical minds–Pythagoras, Descartes, Euler, Pascal, Cantor, many more. Anecdotal, illuminating. 30 diagrams. Bibliography. 256pp. 5⅜ x 8½.
0-486-28973-7

HISTORY OF MATHEMATICS, David E. Smith. Nontechnical survey from ancient Greece and Orient to late 19th century; evolution of arithmetic, geometry, trigonometry, calculating devices, algebra, the calculus. 362 illustrations. 1,355pp. 5⅜ x 8½. Two-vol. set.   Vol. I: 0-486-20429-4   Vol. II: 0-486-20430-8

A CONCISE HISTORY OF MATHEMATICS, Dirk J. Struik. The best brief his-tory of mathematics. Stresses origins and covers every major figure from ancient Near East to 19th century. 41 illustrations. 195pp. 5⅜ x 8½.
0-486-60255-9

Physics

OPTICAL RESONANCE AND TWO-LEVEL ATOMS, L. Allen and J. H. Eberly. Clear, comprehensive introduction to basic principles behind all quantum optical resonance phenomena. 53 illustrations. Preface. Index. 256pp. 5⅜ x 8½. 0-486-65533-4

QUANTUM THEORY, David Bohm. This advanced undergraduate-level text presents the quantum theory in terms of qualitative and imaginative concepts, followed by specific applications worked out in mathematical detail. Preface. Index. 655pp. 5⅜ x 8½. 0-486-65969-0

ATOMIC PHYSICS (8th EDITION), Max Born. Nobel laureate's lucid treatment of kinetic theory of gases, elementary particles, nuclear atom, wave-corpuscles, atomic structure and spectral lines, much more. Over 40 appendices, bibliography. 495pp. 5⅜ x 8½. 0-486-65984-4

A SOPHISTICATE'S PRIMER OF RELATIVITY, P. W. Bridgman. Geared toward readers already acquainted with special relativity, this book transcends the view of theory as a working tool to answer natural questions: What is a frame of reference? What is a "law of nature"? What is the role of the "observer"? Extensive treatment, written in terms accessible to those without a scientific background. 1983 ed. xlviii+172pp. 5⅜ x 8½. 0-486-42549-5

AN INTRODUCTION TO HAMILTONIAN OPTICS, H. A. Buchdahl. Detailed account of the Hamiltonian treatment of aberration theory in geometrical optics. Many classes of optical systems defined in terms of the symmetries they possess. Problems with detailed solutions. 1970 edition. xv + 360pp. 5⅜ x 8½. 0-486-67597-1

PRIMER OF QUANTUM MECHANICS, Marvin Chester. Introductory text examines the classical quantum bead on a track: its state and representations; operator eigenvalues; harmonic oscillator and bound bead in a symmetric force field; and bead in a spherical shell. Other topics include spin, matrices, and the structure of quantum mechanics; the simplest atom; indistinguishable particles; and stationary-state perturbation theory. 1992 ed. xiv+314pp. 6⅛ x 9¼. 0-486-42878-8

LECTURES ON QUANTUM MECHANICS, Paul A. M. Dirac. Four concise, brilliant lectures on mathematical methods in quantum mechanics from Nobel Prize-winning quantum pioneer build on idea of visualizing quantum theory through the use of classical mechanics. 96pp. 5⅜ x 8½. 0-486-41713-1

THIRTY YEARS THAT SHOOK PHYSICS: THE STORY OF QUANTUM THEORY, George Gamow. Lucid, accessible introduction to influential theory of energy and matter. Careful explanations of Dirac's anti-particles, Bohr's model of the atom, much more. 12 plates. Numerous drawings. 240pp. 5⅜ x 8½. 0-486-24895-X

ELECTRONIC STRUCTURE AND THE PROPERTIES OF SOLIDS: THE PHYSICS OF THE CHEMICAL BOND, Walter A. Harrison. Innovative text offers basic understanding of the electronic structure of covalent and ionic solids, simple metals, transition metals and their compounds. Problems. 1980 edition. 582pp. 6⅛ x 9¼. 0-486-66021-4

HYDRODYNAMIC AND HYDROMAGNETIC STABILITY, S. Chandrasekhar. Lucid examination of the Rayleigh-Benard problem; clear coverage of the theory of instabilities causing convection. 704pp. 5⅜ x 8¼. 0-486-64071-X

INVESTIGATIONS ON THE THEORY OF THE BROWNIAN MOVEMENT, Albert Einstein. Five papers (1905–8) investigating dynamics of Brownian motion and evolving elementary theory. Notes by R. Fürth. 122pp. 5⅜ x 8½. 0-486-60304-0

THE PHYSICS OF WAVES, William C. Elmore and Mark A. Heald. Unique overview of classical wave theory. Acoustics, optics, electromagnetic radiation, more. Ideal as classroom text or for self-study. Problems. 477pp. 5⅜ x 8½. 0-486-64926-1

GRAVITY, George Gamow. Distinguished physicist and teacher takes reader-friendly look at three scientists whose work unlocked many of the mysteries behind the laws of physics: Galileo, Newton, and Einstein. Most of the book focuses on Newton's ideas, with a concluding chapter on post-Einsteinian speculations concerning the relationship between gravity and other physical phenomena. 160pp. 5⅜ x 8½.
0-486-42563-0

PHYSICAL PRINCIPLES OF THE QUANTUM THEORY, Werner Heisenberg. Nobel Laureate discusses quantum theory, uncertainty, wave mechanics, work of Dirac, Schroedinger, Compton, Wilson, Einstein, etc. 184pp. 5⅜ x 8½. 0-486-60113-7

ATOMIC SPECTRA AND ATOMIC STRUCTURE, Gerhard Herzberg. One of best introductions; especially for specialist in other fields. Treatment is physical rather than mathematical. 80 illustrations. 257pp. 5⅜ x 8½. 0-486-60115-3

AN INTRODUCTION TO STATISTICAL THERMODYNAMICS, Terrell L. Hill. Excellent basic text offers wide-ranging coverage of quantum statistical mechanics, systems of interacting molecules, quantum statistics, more. 523pp. 5⅜ x 8½.
0-486-65242-4

THEORETICAL PHYSICS, Georg Joos, with Ira M. Freeman. Classic overview covers essential math, mechanics, electromagnetic theory, thermodynamics, quantum mechanics, nuclear physics, other topics. First paperback edition. xxiii + 885pp. 5⅜ x 8½. 0-486-65227-0

PROBLEMS AND SOLUTIONS IN QUANTUM CHEMISTRY AND PHYSICS, Charles S. Johnson, Jr. and Lee G. Pedersen. Unusually varied problems, detailed solutions in coverage of quantum mechanics, wave mechanics, angular momentum, molecular spectroscopy, more. 280 problems plus 139 supplementary exercises. 430pp. 6½ x 9¼. 0-486-65236-X

THEORETICAL SOLID STATE PHYSICS, Vol. 1: Perfect Lattices in Equilibrium; Vol. II: Non-Equilibrium and Disorder, William Jones and Norman H. March. Monumental reference work covers fundamental theory of equilibrium properties of perfect crystalline solids, non-equilibrium properties, defects and disordered systems. Appendices. Problems. Preface. Diagrams. Index. Bibliography. Total of 1,301pp. 5⅜ x 8½. Two volumes. Vol. I: 0-486-65015-4 Vol. II: 0-486-65016-2

WHAT IS RELATIVITY? L. D. Landau and G. B. Rumer. Written by a Nobel Prize physicist and his distinguished colleague, this compelling book explains the special theory of relativity to readers with no scientific background, using such familiar objects as trains, rulers, and clocks. 1960 ed. vi+72pp. 5⅜ x 8½. 0-486-42806-0

CATALOG OF DOVER BOOKS

A TREATISE ON ELECTRICITY AND MAGNETISM, James Clerk Maxwell. Important foundation work of modern physics. Brings to final form Maxwell's theory of electromagnetism and rigorously derives his general equations of field theory. 1,084pp. 5⅜ x 8½. Two-vol. set. Vol. I: 0-486-60636-8 Vol. II: 0-486-60637-6

QUANTUM MECHANICS: PRINCIPLES AND FORMALISM, Roy McWeeny. Graduate student-oriented volume develops subject as fundamental discipline, opening with review of origins of Schrödinger's equations and vector spaces. Focusing on main principles of quantum mechanics and their immediate consequences, it concludes with final generalizations covering alternative "languages" or representations. 1972 ed. 15 figures. xi+155pp. 5⅜ x 8½. 0-486-42829-X

INTRODUCTION TO QUANTUM MECHANICS With Applications to Chemistry, Linus Pauling & E. Bright Wilson, Jr. Classic undergraduate text by Nobel Prize winner applies quantum mechanics to chemical and physical problems. Numerous tables and figures enhance the text. Chapter bibliographies. Appendices. Index. 468pp. 5⅜ x 8½. 0-486-64871-0

METHODS OF THERMODYNAMICS, Howard Reiss. Outstanding text focuses on physical technique of thermodynamics, typical problem areas of understanding, and significance and use of thermodynamic potential. 1965 edition. 238pp. 5⅜ x 8½. 0-486-69445-3

THE ELECTROMAGNETIC FIELD, Albert Shadowitz. Comprehensive undergraduate text covers basics of electric and magnetic fields, builds up to electromagnetic theory. Also related topics, including relativity. Over 900 problems. 768pp. 5⅝ x 8¼. 0-486-65660-8

GREAT EXPERIMENTS IN PHYSICS: FIRSTHAND ACCOUNTS FROM GALILEO TO EINSTEIN, Morris H. Shamos (ed.). 25 crucial discoveries: Newton's laws of motion, Chadwick's study of the neutron, Hertz on electromagnetic waves, more. Original accounts clearly annotated. 370pp. 5⅜ x 8½. 0-486-25346-5

EINSTEIN'S LEGACY, Julian Schwinger. A Nobel Laureate relates fascinating story of Einstein and development of relativity theory in well-illustrated, nontechnical volume. Subjects include meaning of time, paradoxes of space travel, gravity and its effect on light, non-Euclidean geometry and curving of space-time, impact of radio astronomy and space-age discoveries, and more. 189 b/w illustrations. xiv+250pp. 8⅜ x 9¼. 0-486-41974-6

STATISTICAL PHYSICS, Gregory H. Wannier. Classic text combines thermodynamics, statistical mechanics and kinetic theory in one unified presentation of thermal physics. Problems with solutions. Bibliography. 532pp. 5⅜ x 8½. 0-486-65401-X

Paperbound unless otherwise indicated. Available at your book dealer, online at **www.doverpublications.com**, or by writing to Dept. GI, Dover Publications, Inc., 31 East 2nd Street, Mineola, NY 11501. For current price information or for free catalogues (please indicate field of interest), write to Dover Publications or log on to **www.doverpublications.com** and see every Dover book in print. Dover publishes more than 500 books each year on science, elementary and advanced mathematics, biology, music, art, literary history, social sciences, and other areas.